WHERE
SONG
BEGAN

ALSO BY TIM LOW

Wild Food Plants of Australia
Feral Future
The New Nature

WHERE SONG BEGAN

Australia's birds and how they changed the world

TIM LOW

Yale

UNIVERSITY PRESS

New Haven and London

Yale University Press books may be purchased in quantity for educational, business, or
promotional use. For information, please e-mail sales.press@yale.edu (U.S. office) or
sales@yaleup.co.uk (U.K. office).

Text design by Adam Laszczuk.
Printed in the United States of America.

Library of Congress Control Number: 2016939292
ISBN: 978-0-300-22166-4 (hardcover: alk. paper)

A catalogue record for this book is available from the British Library.

This paper meets the requirements of ANSI / NISO Z39.48-1992 (Permanence of Paper).

10 9 8 7 6 5 4 3 2 1

For Carol, with thanks

Contents

Preface to the Yale Edition

Europeans who settled in Australia from 1788 onwards soon decided that the local animals were inferior. The egg-laying platypus and echidna and the quirky marsupials were taken to be primitive mammals that had survived only because Australia was isolated from other continents. This thinking did not stop Australians from feeling pride in their wildlife, but it was tempered by a belief that the mammals, and to some extent all the animals, were backward. The birds fitted this picture by seeming to lack any aptitude for song. Australians were willing to think this way because it fitted the nation's status as a relatively small (by population) and parochial outpost of European culture to the south of Asia. Australia looked to Britain, and later the United States, for guidance about culture and foreign policy. There was a view that the best of everything came from the north. Not until the 1970s did Australians become culturally self-confident, deciding that their film directors, actors, novelists and scientists were as good as any in the world.

I am mentioning this in a preface for Northern Hemisphere readers to emphasise why the finding that Australia gave the world its songbirds is so significant in Australia. It proved a surprise all over the world, of course, but in Australia it suits nationalistic narratives about discarded notions of cultural inferiority and northern assumptions of superiority. I make the most of this in the book when I talk about 'northern orthodoxy' stunting thinking about birds.

We live in exciting times for bird research, which means that interesting new work has appeared since this book came out in Australia in the middle of 2014. Those who follow bird classification should know about two papers that attempt to show how all the world's bird orders fit together. The first, published in *Science* in 2014 by Erich Jarvis and colleagues, offers a new classificatory tree for the world's thirty-eight bird orders. In chapter seven I say we don't know where pigeons fit on the avian tree of life, but this paper places them on a third branch, just above waterfowl and landfowl, in the company of flamingos and another three orders.

The paper is based on forty-eight whole genomes of species from thirty-five bird orders, rather than on selected gene sequences, so we might hope it provides an accurate picture of relationships. I was excited when it came out, but my enthusiasm wavered early in 2015 when another paper appeared, in Nature, using different but also high-quality genetic sequencing, featuring an avian tree that is not the same, with, for example, pigeons in a different location. The tree produced by Jarvis looks a little more plausible, but the differences between these papers ensure that uncertainty lives on. What is important to my book is that both papers agree about a close relationship between perching birds and parrots. I described this relationship as looking very likely, but now it is beyond doubt.

Another paper in 2015, by Gerald Mayr, rejects earlier claims that fossils found in Europe and North America, dating back as much as 55 million years, represent early parrots. These fossils weren't easy to reconcile with all the evidence that parrots originated in the Southern Hemisphere, so it is pleasing to know the claims can be discounted.

A December 2015 paper saw Joel Cracraft revamp his bold theory that all of today's birds have Gondwanan origins because, he said, the giant Chicxulub meteorite, which smashed violently into Mexico 65 million years ago, obliterated birds everywhere else. With co-author Santiago Claramunt, Cracraft is now proposing that the only avian meteorite survivors were in South America. The paper doesn't provide new fossil or genetic evidence, and instead relies on establishing that the deeper branches of clades tend to be from South America. This is true of pigeons and falcons, for example, but to make the theory work Cracraft and Claramunt propose that New Zealand wrens started out in South America, and that ostrich ancestors did so as well, then spread north and west through Alaska (as part of Beringia). According to this hypothesis, birds had two pathways for spread around the world, one through North America and the other through Australia via Antarctica. Their paper is interesting but speculative, with a discussion containing statements prefaced by 'we hypothesize' and the like.

It will be interesting to see what future research brings.

References

Claramunt S and Cracraft J (2015) A new time tree reveals Earth history's imprint on the evolution of modern birds, *Science Advances* 1: e1501005.

Jarvis ED, Mirarab S, Aberer AJ, et al. (2014) Whole-genome analyses resolve early branches in the tree of life of modern birds, *Science* 346: 1320–31.

Mayr G (2015) A reassessment of Eocene parrotlike fossils indicates a previously undetected radiation of zygodactyl stem group representatives of passerines (Passeriformes), *Zoologica Scripta* 44: 587–602.

Prum RO, Berv JS, Dornburg A, et al. (2015) A comprehensive phylogeny of birds (Aves) using targeted next-generation DNA sequencing, *Nature* 526: 569–73.

Introduction

Australia has such unusual mammals that they have long cast a shadow over something of equal distinction, the birds. These stand out in so many different ways – in ecology, behaviour, evolution and biogeography – that we can learn more about Australia from its birds than its mammals.

The idea of the continent having exceptional birds is an old one, going back hundreds of years to Dutch sailors spying swans dark as coal and naturalists finding decorative bowers in the cedar brushes of New South Wales. Australian exceptionalism, if that's what we call it, peaked in a 1911 bird guide that was reprinted many times: 'Competent authorities have proposed to divide the world, biologically, into two parts – Australia and the rest of the world, and they have considered Australia the more interesting part.'

The expert behind this fanciful claim, John Leach, went on to describe Australia as the place where the bird world 'seems to reach its culminating point'. He was exaggerating in order to excite pride in the fauna, but all the same, Australia does have exceptional birds.[1]

Little progress was made in explaining why this might be so until the 1980s, when an ambitious American scientist developed the first molecular methods to decipher animal relationships. Charles Sibley was astounded by what his Australian samples showed. His work led straight to the discovery that Australia was the first home of the world's songbirds, the birds that can sing.[2]

Another step forward was a bold journal article published in 2007 which interpreted certain oddities about the birds – especially the domineering honeyeaters – by invoking the continent's infertile soils and distinctive vegetation. One of its authors, biologist Antoni Milewski, has said to me that conclusions in ecology can hide in plain sight, and it seems to me that Australians give too little thought to the unusual birds they live among, that they accept them at face value rather than wondering about the forces that brought them to be.[3]

For example, attacks by Australian magpies are accepted as a normal part of suburban life, when by global standards they are exceptional. In his book about magpie aggression, Griffith University's Darryl Jones told of one terrorised school in Brisbane where, over two weeks, a bird cut the faces of more than a hundred children. On one particular morning, 'throngs of screaming parents at the gates were trying to get their terrified children to run quickly across the open area to the main building where the school medical officer was waiting with the first aid kit'. By his estimate Brisbane has at least 500 people-attacking magpies, and maybe as many as 1700. Local newspapers publish an online map of conflict zones.[4]

The magpie is one of Australia's two or three most popular birds, despite statistics showing that magpies stab the eyes of one or two people each year, often causing permanent damage. A magpie in Toowoomba that blinded a boy in one eye was relocated rather than killed. Canadian biologist David Bird has nominated the Australian magpie as likely to be, for suburbanites, the most serious avian menace in the world.[5]

A single species can't tell us much about Australia's birds, but aggression is a force that occurs more widely. Attacks by one species of bird on another (interspecific aggression) reach levels unmatched anywhere else. In one morning in Australia I can see more clashes than I ever do in a month in the field on another continent. A 2012 article in *Ecology*, using language journals seldom allow, described 'despotic aggressiveness over sub-continental scales'. If one feature of Australia's birds stands out especially, it is this.[6]

Most of the fighting is initiated by a group whose aggression has helped it become Australia's largest family of birds, the honeyeaters. With parrots, they contribute greatly to Australia having a bird fauna that stands out on many levels.

Honeyeaters and parrots benefit from the vast supplies of sugar and starch available from eucalypts and other bird-adapted plants. Australia's infertile soils, its frequent fires and unreliable climate contribute to the carbohydrate abundance. The environment has shaped the birds so strongly that a causal link can even be drawn between harsh soils and harsh honeyeater calls, operating through the influence of the vegetation. Mammals and sucking insects, shaped themselves by the landscape, have in turn influenced the birds. Marsupials, by having done poorly in some ecological roles, have let birds such as parrots prosper in their place.[7]

When biologists compare Australia to the rest of the world they sometimes emphasise the similarities, by talking about Australia having its own versions of Northern Hemisphere robins, flycatchers and creepers, as if evolution walks the same walk everywhere. There is value in this, but the differences are more revealing. A reading of recent research shows that Australian birds are more likely than most to eat sweet foods, live in complex societies, lead long lives, attack other birds, and be intelligent and loud. The services they render to plants are also sometimes exceptional. Honeyeaters and parrots are globally outstanding as pollinators, and pigeons and cassowaries are unsurpassed at moving seeds; Australia is a land where bird–plant interactions are especially strong. Ornithologist Richard Schodde was right to declare that the bird life of Australia and its sister island New Guinea is 'the most distinct and "different" in the world'.[8]

The make-up of the bird fauna is a key part of what makes it distinctive. Before Sibley developed the first molecular methods, bird relationships had often been difficult to decipher because bird bones, to accommodate flight, are so flimsy they seldom survive as fossils, and because flight so limits the forms birds can take that experts often can't tell if similar-looking birds are closely related or not. Molecular comparisons, using DNA, have revolutionised our understanding of origins and relationships, with any number of genetic studies confirming that Australia has served as a major evolutionary centre. The world's scientific community would not accept this until 2004, although two Australians, following Sibley's lead, produced telling evidence in the 1980s. Their findings bear on key questions in biology, such as why, when Darwinian theory demands they compete, do some animals cooperate?[9]

The origins of Australia's birds can't be separated from the origin of Australia itself, a land that changed greatly over time by detaching from Gondwana, the southern supercontinent, and drifting north into the dry middle latitudes. Twenty million years ago Australia was wetter and largely clad in rainforest. For much of its past it was united with Antarctica by a peninsula of land that gradually narrowed and was finally severed some 45–38 million years ago. Eucalypt and platypus fossils found in South America show how important that land connection was.[10]

Three periods of the past in particular are crucial to understanding birds. The Cretaceous, ending 65 million years ago, saw the first

Era	Period	Epoch	Age
Cenozoic	Quaternary	Holocene	0.01
		Pleistocene	2.6
	Neogene	Pliocene	5
		Miocene	24
	Paleogene	Oligocene	34
		Eocene	56
		Paleocene	65
Mesozoic	Cretaceoous	Late	99
		Early	144

Numbers represent millions of years before present. Note that the epochs are of very unequal lengths.

'modern' birds appearing among dinosaurs. The early Eocene, 56–48 million years ago, provided the last favourable opportunities for overland travel between South America and Australia, owing to a hot wet climate and an Antarctic landbridge. The Pleistocene, which went from 2.59 million to 11 700 years ago, had the birds we see today, responding to ice age climate change. Glacial maxima are very dry, turning so much water into ice that sea levels dropped 100 metres or more, uniting most of Australia's islands with the mainland, with dire consequences for seabirds. (The term 'ice age' is one I use to refer to periods in time that were colder than today, when technically, the past 2.6 million years have been one continuous ice age, marked by some relatively warm periods, such as the interglacial we are living in today.)[11]

A few background comments are necessary before I delve properly into all of this. Biology is a difficult subject to write about because accuracy and readability so easily drag each other down. To represent the natural world with verisimilitude you need any number of qualifiers – *probably, usually, often, sometimes, possibly, presumably* – which can crowd the story. The temptation for the writer is to leave some out, but rather than compromise accuracy I have tried to remain ethical about the uncertainties, assumptions and exceptions I travelled with. Most of the biology I report on is consensus science, but some is not, which means making choices about which science to adopt, something I've endeavoured to balance by mentioning the theories I did not accept. Biology is in any event such a young and evolving science that some of what counts as dogma today will be jettisoned tomorrow. Because ecosystems are so complicated, we should beware of ecological stories that seem too neat. Future thinking will be served best if today's uncertainties are disclosed rather than concealed.

The language I've adopted is informal and conversational, but chosen with enough care that I feel safe inviting scientists to trust it. The vast majority of statements are referenced, and some are expanded on in the source notes. I encourage all readers to travel far beyond the pages of this book, because there is so much of interest that I could not include. Many journal articles are available free online.

From chapter four on, when I say 'Australia' I have New Guinea (with Papua) in mind as well, except in the last chapter, which is about 'Australians' and their birds. Including New Guinea as part of Australia, although it is the only way to make sense of Australia's birds, is not entirely satisfactory, but nor is any approach I've come across. The Solomon and Maluku islands are arguably also part of Australia, biologically speaking, but I do not include them when I provide numbers of Australian birds. Following common practice I use 'North America' to mean the United States and Canada, when the actual landmass continues to Panama. 'Southern Africa' is defined as Namibia, Botswana, Mozambique and South Africa.

Australian bird names and taxonomy follow Christidis and Boles (2008), with occasional departures to reflect new research. A recent trend is to divide the pied imperial-pigeon (as per Christidis and Boles) into two very similar species (Torresian and pied imperial-pigeons), a change that, if adopted here, would require some rewording in chapter eight, but without altering the conclusions it reaches, as the source notes explain. I omit adjectives where the context removes the need, often saying 'kookaburra' for 'laughing kookaburra', 'currawong' for 'pied currawong' and 'lyrebird' for 'superb lyrebird'. Birds are considered to be living dinosaurs but I use the word 'dinosaur' only for non-avian dinosaurs. I define 'endangered' and 'vulnerable' in accordance with federal legislation, which is based on criteria set by the International Union for Conservation of Nature (IUCN).

Australia's main bird organisation has changed its name multiple times, and the incarnations are mentioned here in the hope of pre-empting confusion. At its birth it was the Ornithologists' Union of Australia; in 1910 it became the Royal Ornithologists' Union of Australia; in 1966, Birds Australia; in 2012 it merged with the Bird Observer's Club of Australia to become BirdLife Australia. I have used whichever name was current for the events I describe.

One

Food Worth Defending

English pioneers settling in Australia found much to complain about. In place of proud British trees glowing golden in the fall were 'shadeless, grey, sombre-looking gum trees, stretching away in indefinite monotony', wrote artist Ellis Rowan in 1898. The birds bettered the trees by offering more colours than in England, but that counted for less than their failure to sing. Charges about harsh cries and poverty of song were often made. 'There are several chirpers,' remarked visiting naturalist Thomas Harvey in 1854, in a complaint that was typical for the time, 'a few Whistlers many screamers, Screechers, & yelpers, but no songsters among the birds here.'[1]

Even the great John Gould, who in lavish books did more than anyone to promote Australia's birds, could not warm to their calls. He noted that 'feebly indeed . . . are represented the melodious notes' that rendered spring in England so joyous. The parrots screamed and the honeyeaters were monotonous.[2]

Larks, nightingales and song thrushes were brought out to repair the situation. Melbourne University's professor of biology Frederick McCoy promised in 1862 that the savage silence would soon be ended by 'touching, joyous strains of Heaven-taught melody'. He declared that Europe's songsters could from evil turn 'even the veriest brute', as if, by hearing their own birds, Australians risked becoming barbarians. The larks thrived and the thrushes survived, but the four nightingales freed in Melbourne's botanic gardens did not last a full year, although they were heard singing in spring.[3]

Australians today like to lampoon the narrow-mindedness of their forebears, but those early jibes about the birds had a biological basis. Australia does sound harsher than your average field or forest overseas. Like a person vomiting was how Gould described the call of the little wattlebird, a large honeyeater. The great bowerbird's hissing is described in one 2001 guide as a cross between tearing paper and violent vomiting.[4]

Lorikeets roar like sports crowds when they swing into trees in screeching hordes, their every whisper a shout. Gould, whom I will turn to often as the first ornithologist to experience Australia's birds, obtained no respite from their screaming when he discharged a gun beneath their tree. Birds are not usually this loud and harsh. The flowers provide one reason why.

Sugar Fuels Noise

Far more nectar is available to birds in Australia than on other continents — enough to fight over, and harsh cries assert possession. Australia's eucalypts and paperbarks (*Melaleuca*) are the only bird-pollinated trees on earth to form vast forests, and wattlebirds and lorikeets are important pollinators of their flowers.[5]

In the Northern Hemisphere, north of the tropics, all the main trees — oaks, beeches, elms, conifers, and the like — are pollinated by wind. Insects pollinate the shrubs and smaller trees. One book on the subject expressed the general view that no bird-pollinated flowers exist in Europe, nor in Asia north of the Himalayas. That was proved wrong by a 2005 article, 'First Confirmation of a Native Bird-pollinated Plant in Europe', when warblers were seen on rare pea flowers in Spain, but the pool of examples remains ridiculously small.[6]

The contrast with Australia could hardly be greater. Many of its best known plants are bird-pollinated: banksias, grevilleas, bottlebrushes, grasstrees, paperbarks, hakeas and hundreds of eucalypts. Birds are not the only visitors of these flowers but often serve as their best pollinators. Australia's national floral emblem, the golden wattle, feeds birds, as do most of the official state flowers: the waratah (New South Wales), pink heath (Victoria), blue gum (Tasmania), Sturt's desert pea (South Australia), kangaroo paw (Western Australia). One pollinator, the brown honeyeater, is known to visit more than 300 different flowers in more than twenty-five families.[7]

Nectar birds are plentiful in Africa, the Americas and tropical Asia, but the vast majority are tiny hummingbirds and sunbirds. Cuba's bee hummingbird, at 1.8 grams, is the tiniest bird on earth. Hummingbirds attend a wide range of plants, but seldom trees. Australia's biggest honeyeater – the Tasmanian yellow wattlebird – is five times the weight of the largest nectar bird on another continent, the spectacled spiderhunter of South-East Asia, a large sunbird. So large are wattlebirds they were hunted in the past for the table. Seeing them in the Hobart markets, Gould compared them in size to European magpies. A closed season was imposed in 1903 from fear the species would not last. Hunting them remained legal until the 1960s, when the bag limit was set at fifteen birds a day. Illegal hunting went on until at least 1991, when a poacher told me about the delights of wattlebird stew: the yellow fat congealing on top was skimmed off as butter. Just think, nectar in flowers ending up as butter on the homestead table. Red wattlebirds sold in Sydney's poultry shops in large numbers. They were pronounced the best eating bird in the bush. Sugar produces a sweet meat that pleased many colonial bellies.[8]

Outside Australia nothing like this was possible. No hummingbird was ever hung in a poultry shop. Only in Australia are whole forests dominated by nectar birds. Honeyeaters can account for 80 per cent of the birds in Victoria's box-ironbark woodlands when the flowers peak. Twenty-four honeyeater and five parrot species use this sugar-rich domain. The travel routines of each species represent pollen delivery in many different directions, making it possible that on rare occasions pollen arrives from as far away as Tasmania and Queensland. Red wattlebirds sometimes roam in flocks of hundreds, and one aggregation of thousands has been seen, implying vast amounts of pollen moved by this species alone.[9]

Nectar birds are loud because their food repays defence. A fruit, seed or insect is gone after it is eaten, but flowers supply nectar hour after hour, often day after day. More than other foods, flowers reward aggression.[10]

Outside Australia I have witnessed any number of attacks by one bird species on another – involving such diverse combatants as broadbills, jaegers, vireos and woodhoopoes – but always at levels that place Australia, with all its nectar, far ahead. Small birds everywhere harass hawks, ravens, and other predators that steal from their nests. Birds of all sizes may defend a zone around their eggs and young, as lapwings do with zeal. But fights between species over food are much rarer, except around flowers, and sometimes near bunches of fruit.[11]

Birds all over the world fight over flowers. The aggression of hummingbirds and honeycreepers influenced theories about resource territoriality that I studied at university in the 1970s. But so small are hummingbirds that their calls are thin and their ire often aimed at butterflies and bees. The flowers they attend are often too scattered for effective defence. Australia also has small, widely spaced flowers that feed small honeyeaters, but as well as these it has big inflorescences that justify defence by larger birds. Eucalypts have unusually open canopies that suit aggression by exposing intruders to view. Relentless feuds at flowers ensure that pollen-dusted birds are ever coming and going as skirmishes are won and lost. If there was no combat the plants would benefit far less.[12]

Aggression begins when buds open, lasting until the nectar fails. The same flower may be visited ten or twenty times a day. A banksia

flowerhead may flow for twenty days, a eucalypt blossom for a week.
The territory of a New Holland honeyeater may come down to fourteen banksia heads, or 4 cubic metres in the crown of a tree. Twenty
honeyeaters may divide a eucalypt into as many territories, which
expand in size – and reduce in number – if insects or larger birds
(which resist eviction) take too much nectar. When she breeds, the
New Holland female assumes control of the blossoms near the nest,
forcing her partner to the periphery of their domain. But all too often
there are no territories as such, just aggression during feeding bouts.[13]

Sugar feeds Australian mammals as well, fur at night replacing
feathers by day at many of the same flowers. Australia's community of nectar-feeding mammals is the world's largest. It takes in
the tiny honey possum, the tellingly named sugar glider, the squirrel glider, feathertail glider, the rainforest-dwelling striped possum,
antechinuses, the rare dibbler, bush rats, and bats in several sizes,
all consuming nectar in large or small amounts. When the northern
paperbarks bloom, little red flying foxes form camps more than a
million strong.[14]

Aboriginal people tapped into this resource, sucking flowers and
preparing nectar drinks, including alcoholic brews. Nowhere outside
Australia did nectar feed so many people. In the south-western heathlands of Western Australia pioneer James Drummond claimed that
the Aborigines lived for five or six weeks on banksia honey. Nectar
was sometimes taken from birds rather than flowers, as James Dawson
recounted in 1881:

> sweet liquid is obtained by mischievous boys from young para-
> keets after they are fed by the old birds with honey dew, gathered
> from the blossom of the trees. When the nest is discovered in the
> hole of a gum tree, it is constantly visited, and the young birds
> pulled out, and held by their feet till they disgorge their food into
> the mouth of their unwelcome visitant.[15]

The importance of sugar grows when you factor in the sources other than flowers. Manna, honeydew and lerp – sweet substances found on leaves and bark – sustained millions more birds, as well as marsupials and people. When lerp coated leaves, Aboriginal people would 'thrive on it most amazingly', a situation that is scarcely known about today. To make sense of Australian ecology, all the sugar needs explaining.[16]

An attempt was not made until 2007, when an American ecologist teamed with a South African ex-patriot to produce a far-reaching article, 'Ecology of Australia: The Effects of Nutrient-poor Soils and Intense Fires'. One anomaly it addressed was why an unusually high proportion of plant species is pollinated by vertebrates of larger average size than on other continents. From the vantages afforded by different continents, Gordon Orians and Antoni Milewski offered a hypothesis about Australia, part of which goes like this:[17]

Australia's ample sunshine and depleted soils encourage plants to produce more carbohydrates than they can use. All the sugars produced by photosynthesis cannot be converted into tissues or seeds because the soil nutrients needed as additional ingredients are scarce. The surplus sugar is fed to birds as nectar in return for pollination. Birds can thrive on this sugar because they eat insects as well to provide missing nutrients. Honeyeaters will pursue flies so tiny that more energy is lost chasing them than is gained eating them, but they have bountiful sugar to fund their pursuit of rare phosphorus, zinc, iodine and cobalt. These minerals have become precious because Australia is so flat and geologically stable that there is little new soil created to replace the nutrients leached away by tens of millions of years of rain.[18]

If Orians and Milewski are correct, big nectar birds are a consequence of strong sunshine and poor soil. On fertile lands supporting rainforest, most flowers are small and attended only by insects, as their theory would imply, although this may be for different reasons. But where soils are poor, banksias and grevilleas sometimes overflow.

I have camped under a coast banksia that dripped noisily onto my tent all night.

But there is more to consider than this. Nectar birds provide special services to plants. The clearest evidence for this has emerged from research funded by the timber industry.

Successful Trees

The Hobart art gallery has a painting that shows the view in 1889 from one corner of Franklin Square in the city centre. Dominating the scene is a Tasmanian blue gum of profound proportions, with a trunk much wider than a passer-by in a top hat is tall. Davey Street in that scene is just a dirt track wending past a church and a distant wedge of forest. One tree from that wedge of forest (as well as the church) remains today. She's a craggy old blue gum, listed on Hobart's register of significant trees, whose shaggy mane hangs over the traffic that prowls Davey Street.[19]

Upon leaving the gallery and locating that tree, I was soon looking up at two sleek green birds with red-daubed faces lapping at her flowers. They were swift parrots, an endangered species which is down to a thousand or so pairs today. Endangered species do not usually feed above city traffic, but swift parrots are so regular in Hobart that collisions with windows pose a serious threat.[20]

A government brochure titled 'Prevent Window Collisions' offers helpful suggestions. 'Let your windows get dirty', it says. Grime on windows evidently saves endangered lives. 'Wherever possible draw drapes and blinds.' Wind chimes and pot plants should be placed outside windows. There are even parrot-safe building guidelines, which recommend, for example, windows angled to reflect the ground instead of open sky.

When Gould visited Hobart town in 1838 he watched these birds feeding in blue gums, their staple breeding-season food, a few feet above passing residents. From his biography I learned that he had his first lodgings in Davey Street; we may have seen swift parrots in the same tree, more than a century and a half apart.[21]

Were Gould alive today he would find the trees doing much better than the parrots. Tasmanian blue gums (*Eucalyptus globulus*) have become, you might say, silvicultural superstars. Stronger smelling than most trees, blue gums were planted by the millions across Africa, India, the Mediterranean region and Latin America when doctors decided in the nineteenth century that aromatic leaves dispelled malarial miasmas.[22]

A modern role for them materialised in the 1920s when a method was found to separate fibres in wood, allowing paper to be made cheaply from trees in place of the rags, cotton and shrubs previously used. Because they are hardy, grow fast, and yield strong flexible fibres that bleach white, blue gums have become the main source of paper in the temperate zones of the world. Perfect for photocopy and printer paper and for strong but pliable toilet paper, their fibre is a product for our times.[23]

Other eucalypts have their uses, and because they thrive in poor soil they are now the most widely grown trees on earth after pines (*Pinus*). India and Brazil have more than 4 million hectares of eucalypts each, China almost 3 million. Having evolved on the world's most infertile continent, these trees are pre-adapted to grow on degraded land. In Africa they tower above the local vegetation, birds of prey preferring them for nesting and waterbirds for roosting. Africa's tallest tree is claimed to be a eucalypt in South Africa. Europe's tallest tree is thought to be a eucalypt in Portugal.[24]

They helped one country reinvent itself. The Ethiopians had abandoned several capital cities after running out of wood, until the early twentieth century when Emperor Menelik II stabilised his nation by having masses of blue gums grown.[25]

They have transformed bird lives as well. Like Coca-Cola, eucalypt nectar is a global beverage. Consumers I have seen around the world include tanagers, bulbuls, sunbirds, American warblers, and any number of hummingbirds. Blue gum flowers are so large the nectar can be swilled around inside them, and this drink now sustains a nearly extinct hummingbird, the Juan Fernandez firecrown, which once shared its island home west of Chile with castaway Alexander Selkirk, an inspiration for the Robinson Crusoe story. Goats set free on the island and hunted by Selkirk helped ruin the vegetation, and most firecrowns in winter now rely on blue gum nectar or on garden flowers. Four hummingbird males will divide a eucalypt into four strongly defended territories. They are another endangered bird to depend on blue gums – on the opposite side of the Pacific Ocean.[26]

So important are these trees today that scientists are funded to study their genetics. From his office at the University of Tasmania, Brad Potts told me that eucalypts have exchanged genes on a scale unknown in other trees. The evidence suggests that Tasmanian blue gums evolved in Victoria then spread south several million years ago when the sea level fell. Upon reaching the sheltered coast where Hobart now stands they encountered other eucalypts from which they received genes to suit their new location. Twelve eucalypt species in this region have shared genetic combinations (haplotypes), implying a wide exchange of pollen effected by birds and insects.[27]

Blue gums have far larger flowers than Tasmania's other eucalypts, ensuring ample visits from swift parrots, wattlebirds and smaller honeyeaters, lorikeets, silvereyes, and occasional cockatoos and rosellas, not to mention insects that come to suck and lap. Many birds will take eucalypt nectar when it is bountiful and undefended – even magpies, finches and crows.[28]

Climate Survivors

Climate change is encouraging much talk about the need for trees to migrate. In the Northern Hemisphere during ice ages, trees shifted south and north as glaciers came and went. Australia is different. Some years ago at a climate change workshop in Canberra I heard geneticist

Margaret Byrne talk about various plants and lizards coming through the last ice age in Australia not by migrating large distances, but by surviving in small protected sites (refugia) within their current bounds. She found high genetic variation across the range of a species, rather than the homogeneity that tells of rapid spread. Fossil pollen tells the same story. The CSIRO's Mike Dunlop and Peter Brown noted in a major report 'no evidence for distribution shifts being widespread among Australian species in response to warming after the last glacial maximum'. In a climate change report for the Queensland government I concluded that large movements by trees are unlikely in future as they did not happen in the past.[29]

Different seas and soils explain why. The large landmasses in the north accumulated so much ice that plants had to relocate to survive. In the Southern Hemisphere, oceans moderated temperatures by carrying warm water south and persistence *in situ* was often possible. Australia is so infertile that many plants succeed by adapting to certain soils, reducing the value of migration because the soils vary so much from region to region. A strong flowering effort reduces the need to move by assisting genetic turnover in times of change. More pollen exchanged means more varied offspring, improving the chances that some of them will have the features that suit the future.[30]

Trees in northern lands were able to advance quickly after the glaciers melted because they had birds and mammals shifting their seeds, as well as vacant soil bared by glaciers in which to grow. Eucalypts stand out by lacking anything on their seeds to invite animals or wind to move them (apart from the small wings on *Corymbia* species). As Brad Potts and others have noted, eucalypts have mobile pollen rather than mobile seeds. Western Australia's heathland plants have elicited the same comment. Lorikeets winging 100 kilometres south or east can bring pollen from trees adapted to hotter and drier climates, like those expected in future. Eucalypt pollen is thought to remain viable for about a week, suiting long journeys. Flowering is typically erratic, obliging many birds to wander the landscape, resulting in pollen from one stand travelling in many different directions. Pollen mobility can explain the paradox of Australia today: a land in the recent grip of an ice age dominated by trees with no talent for travel.[31]

To protect the resource they depend on, forest managers should be conserving birds and bats, especially the long-range travellers, but the opposite happens. The flower bird that roams furthest, from Tasmania to Queensland in some years, is the aptly named swift parrot, yet Forestry Tasmania was controversially logging one of its few breeding sites, Wielangta Forest, until a moratorium was declared in 2008. These birds hit windows because they are built for rapid transit. No other pollinator crosses the sea to reach Tasmania, on flights thought to take as little as six hours. No other Tasmanian tree is so important to commerce. Some blue gums in eastern Tasmania have genes otherwise confined to Victorian blue gums, and swift parrots have been invoked to explain this. They return to blue gums in Tasmania at the very time blue gums in Victoria are flowering, suiting pollen transfer.[32]

The sheer number of eucalypt species – some 800-plus – is something to wonder about. *Eucalyptus* is the world's second-largest genus of large trees, after the figs, yet figs are spread over six continents while eucalypts are native to one. The Sydney region alone has more than twice as many eucalypt species (100-plus) as Britain has total tree species. Six or eight may grow together without clear ecological separation. (I define 'eucalypts' the modern way, as including the genera *Eucalyptus*, *Corymbia* and *Angophora*, but my statement about genus *Eucalyptus* holds true.) Wind-pollinated trees cannot mingle like this because too much pollen reaches the wrong stigmas. The diversity is a measure of high rates of species evolution and high rates of survival, and birds contribute to both, although I should point out that some eucalypts, mainly in Australia's south-east, have insects as their only pollinators.[33]

Eucalypts hybridise freely, and while hybrids often do poorly, in times of climate change, when birds may have to roam more widely to feed, new hybrid combinations may become the key to the future. As for the past, Brad Potts has written about eucalypt species following ice age climate change that 'may have merged and been resurrected to varying degrees'. The twelve species sharing pollen in the Hobart area survived big swings without leaving their refuge.[34]

A large pool of trees can sustain a large community of birds if flowering across a region is staggered to provide continuity. Bird pollination benefits from diverse forests in a way wind pollination does not. Australia's northern savannas have eucalypts flowering through the dry season and paperbarks in the wet, although yields vary. Honeyeaters eat insects, lerp, manna and fruits to survive gaps in the nectar supply. A delightful journal article, 'Thinking Honeyeater', with the inviting subtitle 'Nectar Maps for the Northern Territory', draws on vegetation maps to show where in each month the sugar flows. With big orange flowers and dominating 100000 square kilometres, Darwin woollybutt (*Eucalyptus miniata*) is the top producer, attracting thirteen species of bird that usually ignore nectar, including butcherbirds, which fumble at the flowers because their bills are the wrong shape.[35]

The banksias in south-western Australia provide an unusually reliable supply of nectar. Near Perth their flowering seasons are sequential and only slightly overlap. The presence of the honey possum, the world's only non-flying flower mammal, one that depends entirely on pollen for protein, attests to nonstop nectar for millions of years past. I watched one, stripes down its back, scurry past me at dusk like a sleek mouse, led by its long snout towards a prostrate banksia.[36]

The south-west has more than a thousand flower species visited by birds, the highest density on earth. This fits with its status as one of the world's oldest landscapes, with some of the most depleted soils, 290 million years having passed since the last mountain-building. Controlling the best sugar flows is the western wattlebird, described by Gould as 'very pugnacious, attacking every bird, both large and small, that approaches its domicile'. The poor soils it lords over suit floral variety because they prevent any one plant dominating, and the nectar flows generously because plants compete for visitors.[37]

Ancient Pollination

The south-west is a centre for one of the world's oldest plant families, one that adapted from wet rainforest to drought-prone land, and which is famed today for its iconic members: banksias, waratahs, grevilleas, hakeas, and the only significant food for humans to come out of Australia, the macadamia, a tree whose creamy racemes attract silvereyes and honeyeaters as well as native bees. I refer to the family Proteaceae.[38]

Several days after I saw those swift parrots in Hobart I went out with pollination ecologist Andrew Hingston to Mt Wellington, the peak behind the city. Tasmanian waratahs (*Telopea truncata*) hid their leathery leaves and rosy buds in the soft mist around us. Andrew knelt to show me jackjumpers, wasp-like ants whose nests mark openings among the stringybarks reached by the morning sun, a setting favoured by the waratahs as well.

Tasmanian waratah flower heads are much smaller – about the size of a toddler's fist – than those of the famed Sydney waratah, but their claim to antiquity is much greater. Thirty-million-year-old fossils found near Launceston have been identified as the same species. Identification was possible because the fossil imprints are exquisitely detailed, showing not just leaf veins, but the butts of hairs and stomata through which Oligocene air once flowed.[39]

Information can come in the tiniest packages. A Cretaceous pollen grain in Victoria caught the eyes of botanists Mary Dettman and David Jarzen, by providing a perfect match with Tasmanian waratah pollen – in size, sculpture, aperture anatomy and probably also in shape, although the 70-million-year-old fossil grain extracted from Victorian sediments is a little squashed. If the comparison is valid, birds were visiting Tasmanian waratahs in the time of dinosaurs.[40]

Comparing pollen old and new is no exact science, so I emailed Lynne Milne, an expert who is critical of some identifications. She told me to take this one seriously. No one is saying the Victorian grain came from *Telopea truncata*, but the plant that left it may have been its direct ancestor – with flowers that looked the same or nearly the same. Tasmania, Lynne reminded me, hosts many early lineages, the geology aiding long-term survival, with plants shifting up or down slopes in response to climate shifts.

Is the waratah nectar I licked a Cretaceous drink? The red flowers imply visits by birds rather than bees because the latter don't register this colour well, although some marsupials do. The rift valley that formed as eastern Australia and Antarctica parted was the place of proliferation of the banksia family (Proteaceae), judging by all the pollen found in the sediments it left. It could have been the place on earth where beaks and flowers first met, since it provides the oldest evidence of this. Botanists have described Cretaceous forests in southern Australia which, based on fossil pollen identifications, 'probably' had grevilleas in the canopy, waratahs in the understorey, and woollybushes (*Adenanthos*) on forest fringes or in the poorer soils. The picture is one of birds flitting past dinosaurs to lap at scarlet and orange sprays.[41]

Birds are the main pollinators of these plants today and probably always were. They offer advantages over bees by feeding on cold wet days when insects stay dormant, and by moving pollen further and in larger amounts. In a grove of banksias in Western Australia scientists found evidence of pollen moving through the landscape on a scale never recorded before. Banksias usually produce two seeds in a follicle, and genetic analysis showed that 96 per cent of the time the pollen to produce each seed came from a different plant. In other words, almost every follicle had two seeds with different 'fathers'. This 'exceptionally high multiplicity of paternity' encouraged Siegy Krauss and colleagues to propose that Australia has something unique, a 'honey-eater pollination syndrome' – a relationship between birds and plants marked by unmatched pollen mobility.[42]

At a road cutting in Victoria, fossilised wing feathers were unearthed proving that Australia had small birds back in the Cretaceous. Honeyeaters and lorikeets aren't that old, so there must have been earlier birds probing those waratah flowers. Australia had no pollinating bats early on, and the lack of any fossils at all implies that flying foxes entered recently from Asia; they chanced upon a nectar-rich system rather than participating at the start, although they add great value as pollinators. Australia lacks large social bees, and this was probably one factor favouring birds and then bats at flowers.[43]

In the Northern Hemisphere birds are considered 'young' pollinators because hummingbird flowers often show clear descent from those visited by insects. A 30-million-year-old-hummingbird fossil found in Germany was claimed in *Science* to be the 'earliest evidence for nectarivory in birds'. But the Victorian pollen is more than twice that age, and Western Australia has produced a 40-million-year-old banksia cone that matches the banksias that birds pollinate today.[44]

Where the Big Birds Are

The Orians and Milewski theory is limited by the fact that bird size and soil infertility don't always match. The Andes has the word's largest hummingbird, the giant hummingbird, without having especially poor soils. But at 24 grams the giant hummingbird is still comparatively small, falling below the size of more than thirty species of honeyeaters and lorikeets, including yellow wattlebirds, which are more than ten times its weight (reaching 260 grams). Asia's largest nectar bird, the spectacled spiderhunter, is more substantial (49 grams), but the Borneo rainforest I watched them feuding in was fertile enough to support orangutans in large numbers. With good reason are these birds described as 'exceptionally quarrelsome'.[45]

In Australia itself size variations don't match infertility and sunshine. Tasmania's yellow wattlebird occupies far richer landscapes than does the western wattlebird, half its size. And New Guinea and

New Zealand have large aggressive honeyeaters living where the soil can be rich and sunshine limited.[46]

While Africa abounds in sunbirds, her largest flower bird, at 43.5 grams, is something different. The Cape sugarbird *is* bound to infertile soils, but as well as this it depends on one of the plant groups at the heart of this story, the Proteaceae. Sugarbirds are protea specialists. South Africa's Cape Province is famous for producing the world's best wildflower displays, and the similarities to south-western Australia invite wonder. Each region boasts vast numbers of extravagantly flowering plants bound to extremely poor soil in a Mediterranean climate – although only 300 or so species in the Cape fynbos (heathlands) are visited by birds, compared to three times that number in Australia's south-west. Too infertile for elephants or other big mammals, the fynbos has a convincingly 'Australian' look. A sugarbird on a protea looks like a honeyeater on a banksia. The males are brazen brown birds with super-long tails that flail behind them when they fly. As Africa's biggest nectar birds, they fit the Orians and Milewski theory by living where the soil is extremely poor.[47]

But much larger nectar birds once graced an unlikely place – Hawaii, whose lush islands became the bird extinction capital of the world. On a visit to the back rooms of Honolulu's Bishop Museum I was shown a large drawer full of departed 'o'os, mamos, ula-ai-hawane, and many more, the multicoloured victims of a complicated plight involving introduced disease. Some began disappearing as soon as they were found. Only four skins remain of the kioea, the giant of the nectar birds. I stared hard into the glass eyes of the only specimen kept in Hawaii, wondering about its past. It looked so much like a wattlebird that it was thought to be a honeyeater of Australian origins, until its DNA proved otherwise. It was far larger than any nectar bird found in Africa, Asia or the Americas. Hawaii was the one place far from Australia where the dominant trees, 'ohi'as, fed nectar to a substantial proportion of birds, including some that were unusually large. The volcanic soils range from fertile to phosphorus-deficient.

Avif of Laysan etc

CHÆTOPTILA ANGUSTIPLUMA (PEALE)

The extinct Hawaiian kioea was universally believed to be a giant honeyeater with Australian ancestors, until a 2008 study showed a relationship to Northern Hemisphere waxwings, which look completely different. It is a striking example of evolutionary convergence, which can be explained by the observation that it 'frequented the flowers of the Ohia', a tree related to Australia's eucalypts.

I met biologists who had seen and heard the last 'o'os in the 1970s and '80s, including the very last bird, and they had poignant stories to tell. Here is naturalist Robert Perkins, reminiscing more than a hundred years ago about the scenes that were lost:

> the Great Ohia trees, at an elevation of 2,500 feet, were a mass of bloom and each of them was literally alive with hordes of Crimson 'Apapane and scarlet 'I'iwi; while continually crossing from the top of one great tree to another, the 'O'o could be seen on the wing sometimes six or eight at a time . . . The picture of this noisy, active, and often quarrelsome assembly of birds, many of them of brilliant colors, was one never to be forgotten.[48]

I witnessed fragments of that scene when I roamed through Hakalau Forest, on the side of a giant volcano, watching rows erupt among the tiny surviving birds. 'Apapanes and 'i'iwis look like scarlet honeyeaters, although they have finch-like ancestors. A biologist broke down and cried when we discussed their future. 'The animals you study are like your children, they are not supposed to die before you do.' The mosquitoes responsible for fatal diseases had just reached this altitude, boosted by rising temperatures. I helped inspect the buckets stationed in the forest to detect them.

The main tree in Hawaii, 'ohi'a (*Metrosideros polymorpha*), belongs in the same family (Myrtaceae) as Australia's eucalypts and paperbarks, possessing fluffy red flowers that could easily pass for eucalypt blooms. It thrives on rich as well as poor soils. DNA evidence shows 'ohi'a to be descended from a coastal New Zealand tree (pohutakawa, *M. excelsa*). Wind, or migrating waders, took the minute seeds from island to island across the Pacific.[49]

Hawaii and Africa both share with Australia something unusual. One had giant nectar birds using the Myrtaceae, and the other has its largest nectar birds relying on the Proteaceae. I assume that these two plant families gave Australia its nectar giants, since they are the main nectar sources of Australia's parrots and large honeyeaters. Australia is dominated by the Myrtaceae, and the Proteaceae, by number of species, is the third-largest family.[50]

Lending weight to this interpretation is New Caledonia, an island east of Queensland that was reached by Australian plants and birds, including grevilleas, paperbarks, honeyeaters and parrots. With the Myrtaceae standing out as its largest plant group, and the Proteaceae prominent and plentiful, it has rainbow lorikeets as its main parrots and honeyeaters dominating its shrublands. Especially fierce is the New Caledonian wattlebird, which I saw driving native starlings out of tall rainforest trees beside a river. [51]

Accessible Nectar

Biologists once thought each kind of flower fitted its own special callers, like sockets to plugs, illustrating nature's perfection. 'Butterfly flowers' matched butterfly tongues, and 'bird flowers' fitted hummingbird beaks like red stockings over taut legs. Hummingbirds do attend many red tubes with nectar concealed at the base where insects can't get at it. I have read in American texts that bird flowers are usually red and cylindrical, but the situation in Australia is far more complicated.[52]

It has shrubs with red tubes attended by small honeyeaters, but the flowers that incite noise are usually brush-like or fluffy, typically white or cream (although sometimes orange, red or yellow), and crowded together in masses. The nectar is usually accessible to any caller – bird, mammal, butterfly or bee – in what is a promiscuous system of pollination. I once worked as an entomologist collecting native bees, and by mounting a net on a set of poles I could sweep up all the insects lured to lofty eucalypt blooms. All the flies, wasps, beetles and bees were astounding, coming in a kaleidoscope of colours and shapes, reminiscent of a coral reef community. I keep emphasising birds, but insects and bats are important pollinators too.[53]

In eucalypts, bottlebrushes, and most related trees the flowers are cups surmounted by crowds of stamens that animals push past for the nectar. Proteaceae honey is even more accessible when on the large heads of banksias, waratahs, grevilleas and hakeas, although it's hidden down corollas on some (*Adenanthos, Lambertia*). Nectar in South Africa is always hidden. Sugarbirds by necessity have longer and finer bills than honeyeaters of matching size. I have watched Cape white-eyes, with their short beaks, lap nectar from Australian silky oaks (*Grevillea robusta*) growing in South African gardens, but they avoid their local proteas – plants in the same family – except when they can pierce the flowers.[54]

Parrots may account for the difference in accessibility, since their beak shape means they can't get nectar from tubes unless they tear open and ruin flowers. I suspect that, to limit damage from parrots, Australian flowers evolved readily accessible nectar, and this liberated honeyeaters from the need for specialised bills. Parrots are missing from South Africa's heathlands, which lack trees and the hollows they might nest in.[55]

African proteas are grown for the flower trade in Western Australia, and on some farms parrots have become a serious problem, with one grower complaining of losing half his blossoms. Banksia flowers are not ruined by parrots seeking nectar, because it's so easy to reach; I often suck it from the flowerheads when I'm passing by. Black-cockatoos do damage the flowers when these are infested with moth grubs, but by eating these the birds help the plants.[56]

Most honeyeaters have much stouter bills than hummingbirds and sunbirds, plus broader tongues. Stout bills work better as weapons and for handling a wide range of items, so hat while nectar remains important to honeyeaters it does not stop them eating very different foods. There are mangrove honeyeaters hunting snails and baby crabs, spiny-cheeked honeyeaters swallowing berries and leaves, and crimson chats living largely on the ground. More than a few species prey on chicks in nests, and some take lizards and frogs, or are shot in orchards for pecking fruit. Blue-faced honeyeaters have been

The noisy friarbird is a typical large honeyeater in possessing a much stouter bill than is seen on nectar-loving birds on other continents. It is apt to jab aggressively at unfamiliar birds it meets in flowering trees.

seen eating pet food. The author of a highly regarded book about honeyeaters, Wayne Longmore, told me he liked them because 'they do everything'.[57]

That can't be said about hummingbirds, even though some of them do specialise on sap or insects rather than nectar. It can't be said about Africa's sugarbirds either; I've seen one leap adroitly from a protea head to snatch a passing gnat, but their slender bills, like those of hummingbirds and sunbirds, restrict what their owners can do.

I said that a sugarbird on a protea matches a honeyeater on a banksia, but there is one key difference that leads to another: concealed nectar requiring a more specialised beak.[58]

Something else to factor in are the different tongues. Those of honeyeaters and parrots have brushy tips, ideal for mopping up nectar from large flowers. A wattlebird has something like 120 bristles on its tongue. Hummingbirds and sunbirds have slender tongues with forked tips, which limit the sizes these birds can reach and still imbibe fluid efficiently. Sugarbirds have brushy tongues but there are only two species in one corner of Africa, compared to scores of sunbird species, showing that outside Australia a brushy tip does not buy success.[59]

Australia's open nectar allowed honeyeaters to become the continent's largest family of birds, numbering some seventy-five species, with many more on nearby lands, especially New Guinea. Evolution has taken these birds in some unusual directions, which don't always involve nectar. The gibberbird is a honeyeater that runs like a lark over desert gravel, snatching insects and pecking up grass seeds. It will sprint down rabbit burrows when alarmed. The open flowers allow many birds that don't specialise on nectar to take it when it is plentiful, including crows, magpies and trillers.[60]

As for parrots, most species take some nectar, and the lorikeets and swift parrot are nectar specialists. Parrots have such short beaks that pollen dusts their feathers when they dine at flowers, and some of this must travel vast distances when lorikeets roam in fast and noisy flocks. Some species (the lories) have spread far into the Pacific, where coconuts have open flowers to maximise pollination on remote isles where insects may be limited.

A possible scenario for the rise of Australia's nectar systems goes like this: in the rainforests that dominated long ago, the two plant families Myrtaceae and Proteaceae prospered in the infertile soils around swamps, sandstone outcrops and quartz beds. Their floral architecture suited visits from birds, and so did the difficult sites they grew in, since the sparser canopy on harsh sites made flowers easy to find and defend. As Australia dried and the soils leached, the plants and birds multiplied together. High investment in pollination brought eucalypts and related trees immense success, by fostering evolution for different soils and changing climates. Infertile soils produce bigger birds, as Orians and Milewski say, but there was more going on than just plants overproducing carbohydrate. One reason to suppose that birds mattered early on is that many Proteaceae once had pollen so fine it must have been spread by wind alone, but this pollen vanished from the fossil record early on.[61]

An unusual feature of the leaves and sometimes the flowers in this family is their toughness. Banksia flowers have such wiry stamens that the old flowerheads have served as brushes. Two Arakwal elders from Byron Bay, Lorna Kelly and Linda Vidler, told me how as children they had brushed their hair with the scented flowers. Leathery (sclerophyllous) leaves resist collapse during drought, while stiff flowers survive large, argumentative birds. The genes for tough leaves may have produced tough flowers as well, recommending these plants for birds. The Myrtaceae have dense wood as well as tough leaves, a trait that aided the switch from rainforest to fire-prone woodlands.[62]

Australia does have other important flowers, of which the parasitic mistletoes stand out. When they infest eucalypts they often bloom out of phase with their hosts, the sap they steal funding flowering seasons that can last six months or more. Mistletoes turn eucalypts into duel nectar suppliers. Offering flowers or fruit in every month, grey mistletoe (*Amyema quandong*) has been found to feed spiny-cheeked honeyeaters all year round. On the acacias it attacks it can average more than five clumps a tree. Australia abounds in mistletoes, with more than sixty species, in every setting from mangroves to desert. The dense clumps formed by their stems are ideal hiding places for nests, and some 200 bird species take advantage of this. But there's a drawback in that large honeyeaters wanting nectar may add to the menu any chicks they find nearby.[63]

Mistletoes do well on paddock trees in full sun. Capertee Valley, north-west of Sydney, is a hotspot for endangered regent honeyeaters, and mistletoe helps explain why. When regent expert David Geering showed me around I saw the mistletoes thriving on river oaks left by farmers in paddocks to protect creek banks, but missing from oaks in nearby shady forest.

Mistletoes evolved from plants in the ground, and the proof resides in the Christmas tree (*Nuytsia floribunda*) which instead of growing on other plants attacks them underground. It can cause blackouts when its roots attack buried powerlines instead of shrub roots. The spectacular orange and brushy, festive-season displays attract birds in the same part of Western Australia that brims with banksias. The DNA of this tree identifies its line as the oldest in the southern mistletoe family (Loranthaceae). If, as thought, this family has Cretaceous origins, it adds to the evidence that beaks were probing corollas in the age of dinosaurs, since bird pollination is almost universal in the family and presumably always was. Other flowers that nourished birds early on probably included heaths and kurrajongs.[64]

Wattles (*Acacia*), Australia's main shrubs and small trees, flower heavily but offer no nectar, relying on pollen-feeding insects to fertilise their flowers. But many species do secrete something like nectar, in tiny glands (nectaries) on their leaf margins, to attract ants, which then attack the leaf-eating insects that blight the wattle. Birds sometimes take this substance as well. In golden wattles (*A. pycnantha*) it appears only on leaves close to flowers and mainly when they open, serving like floral nectar to feed pollinating birds.[65]

Australia is a place where birds and plants responding to each other have produced not only the world's biggest flower-feeding birds, but the world's tallest flowering plants. The blossoms that open highest above the ground are those of the mountain ash (*Eucalyptus regnans*), a tree that has honeyeaters as visitors, and which contributed so much fuel to the horrific 2009 bushfires, when 173 Victorians died. One of these giants can throw out a million flowers in one season and a thousand flames the next. Grown widely for wood, they now lend an imposing air to European lands, and may assume a role as bio-fuels. The mountain ash is the biggest thing alive to need fire (to provide ashy clearings for its seedlings), making it one of the world's more dangerous plants. Eucalypts rely on fire to exclude many competitors, just as honeyeaters rely on aggression. I could say that birds helped make this tree, but the point about Australia is that plants invest heavily in animal pollination, not just pollination by birds. Mountain ash honey attracts insects and possums as well, and insects are useful pollinators.[66]

Pollination is a fascinating topic but not one I can linger on, because there are other sources of sugar to consider for their role in inciting aggression.

Two

Forests that Exude Energy

Before the modern age of plastics, much value was placed on sub-stances called exudates that ooze from trees. Ships hauled cargoes with names that sound fanciful today: tragacanth, sandarach, caoutchouc, gamboge, mastic, balsam, gum arabic, dragon's blood. Shellac secreted by sap-sucking bugs was made into gramophone records, and gutta-percha from rainforest trees insulated the deep-sea cable that connected London to Sydney. Other exudates found their way into medicines, cosmetics, glue, paint, rubber, even chewing gum and golf balls.[1]

Australia was colonised at a time when interest in exudates ran high. The proof is preserved in the name of the dominant trees: 'gum trees'. Their sticky secretions served well against dysentery, becoming one of Australia's first exports, sold in London as 'Botany Bay Kino'.

By sniffing and licking anything that oozed from trees, colonial botanists found that Australia abounded in interesting exudates. There was talk of making rubber from gum tree sap and explo-sives from grasstree resin. Wattle gum was sold in Sydney stores as chewing gum, although in texture it is more like boiled toffee. The banyalla tree excited interest for having rancid-tasting resin that triggered nausea and headaches.[2]

Two substances that aroused comment in those times remain of interest today for their importance to birds. Here is Lieutenant-Colonel Godfrey Mundy, in 1852, describing one:

It sounds strange to English ears, – a party of ladies and gentle-men strolling out in a summer's afternoon to gather manna in the wilderness: yet more than once I was so employed in Australia. This substance is found in small pieces on the ground under the trees at certain seasons, or in hardened drops on the surfaces of the leaves; it is snowy white when fresh, but turns brown when kept like the chemist's drug so called, is sweeter than the sweetest sugar, and softer than Gunter's softest ice-cream.[3]

Others who tried manna were reminded of sweetened flour, sugar-plum, a delicious sweetmeat, and icing on a wedding cake.[4]

As well as the old blue gum mentioned in chapter one, Hobart has a small patch of old-growth forest lingering unusually close to office blocks in a large park, the Queen's Domain, on the edge of the city's heart. In 1855 this area was mentioned at an exhibition in Paris for the manna that was 'formerly collected largely, and eaten freely, by the aborigines of the island'. The biggest trees there today are manna gums (*Eucalyptus viminalis*), one of which is wider across her trunk than I am tall. I suspect her of being Hobart's oldest and biggest living thing, and that by kneeling before her to collect her harvest I did what people were doing centuries ago.[5]

Manna is almost effervescent, with an intensity that only increases as it melts away. About 60 per cent sugar, it has an intriguing flavour. Its appeal was strong enough to bend 'proper' manners, gentlefolk condescending to pluck lumps from the ground to eat. Manna could be bountiful on and under trees when there was no rain to dissolve it. There are reports of upwards of 20 pounds being procured from one tree, of 'many bushels' gathered 'in a short space of time', of Aborigines collecting a pound in a quarter of an hour.

I have not heard of yields like that in modern times.[6]

The manna in the Old Testament can't be identified with certainty today, but could well have been the sweet, congealed fluid secreted by tamarisk trees after attack from bugs. Iraqi Kurds identify it as a sacred secretion of mountain oaks, and they sell it today in sweet shops. Another 'manna' is the dried sugary sap of the Mediterranean manna ash, a traditional laxative.[7]

Australian manna is soft and white, appearing after insect attack on the foliage or trunks of particular eucalypts, especially manna gum and brittle gum (*E. mannifera*). The insects responsible have not been identified with certainty, and what was said of the matter in 1849 remains true today: 'the question of its origin becomes of very great difficulty'. A chemist in the 1960s could not coax manna from trees even after 600 attempts, including 'punching holes in leaves, scarifying twigs, cutting leaves in half, drilling a hole in the trunk or cutting a blaze in the tree'.[8]

The other exudate of interest to birds as well as people is lerp. A stockman stranded in mallee groves north-west of Melbourne in 1845 fed happily upon it for a day or two. 'Lerp is very sweet', wrote Robert Cay afterwards, 'in size and appearance like a flake of snow, it feels like matted wool, and tastes like the ice on a wedding-cake.' Tracts of scrub wore it like coats of snow, he said, and the 'natives' became fat after it appeared.[9]

In a charming piece written for the *Journal of the Microscopical Society of Victoria* in 1880, 'How the Lerp Crystal Palace is Built', W. Wooster explained what lerp was. Tiny insects 'in the larval state protect themselves from the sun and their enemies by building over themselves little tents, or rather crystal palaces, composed of a gummy and sugary secretion, which is exuded in a semi-liquid state from the tube at the hinder end of the body'. Lerp is in fact modified bug excrement, produced by aphid-like bugs called psyllids. It contains starch, a compound rarely synthesised by animals. The word 'lerp' is Indigenous, from the mallee lands of Victoria.[10]

Australian lerp and manna became minor marvels of the age, described in detail in academic journals in Europe. At the Universal Exhibition in Paris in 1855, manna from Hobart was displayed along with wattle gum, sperm whale teeth, and the cranium of a Tasmanian aborigine. There were visions of lerp industries providing sugar for the table, malt for brewing, and starch for stiffening shirts. Plantations of manna gums were proposed. All of this is long forgotten today, but lerp and manna remain important foods for millions of birds.[11]

Awareness of this was late in coming. Because biologists in Europe and America never saw birds eating exudates, this escaped comment in Australia. Lerps are usually less than 6 millimetres wide and difficult to see up in trees. Many observers did mention birds eating 'scale' but attached no importance to this.

Special Foods

In the 1970s an astute PhD student, David Paton, noticed that the honeyeaters he was studying were often busy among leaves or bark. This was nothing unusual, experts having decided long ago that many honeyeaters fed largely on insects, but David saw that it was sugar and starch that attracted these birds, not meals on legs. In one manna gum stand he counted six species taking manna. Some were eating nothing else, others hardly anything else. David also saw honeyeaters taking honeydew, a sweet fluid secreted by tiny bugs hiding in bark cracks, on branches that were sometimes sticky to touch. Birds feuded when these foods abounded.[12]

In what proved a watershed article, David announced in 1980 that manna, honeydew and lerp were important honeyeater foods, 'but the significance of this has been ignored'. It is recognised today but remains underappreciated. Lerp and honeydew exist because there is far more sugar than amino acids in tree sap, forcing sap-suckers to imbibe more sap than they need in order to meet their protein needs. The excess flows from their anuses as honeydew, or is sculpted into lerp 'palaces'. These often look like limpets, occurring singly or in dense clusters. In Victoria and South Australia, David came across exudates everywhere: 'In all areas I found honeydew, manna or lerp within ten minutes of searching for them and often saw birds collect them.' A student was seeing what all before had missed.[13]

Well, not everyone. In a book published in 1842 I found a telling mention of manna. Perth pioneer George Fletcher Moore explained that birds achieve 'excellent condition during the season when it abounds'. Aboriginal hunters told him that birds were plumper and tastier when manna flowed. In 1835 the *Hobart Town Almanack* mentioned manna as something 'greedily devoured' by birds, and a 1951 article about honeyeaters mentioned several hundred devouring 'scale insects', in trees that appeared to have more birds than leaves.[14]

Other biologists have helped join up the dots. In the 1980s another PhD student, John Woinarski, concluded that pardalotes – those tiny birds with monotonous calls – specialise on lerp and manna. Pardalote beaks are unusually stout, for levering sweets

Pardalotes have stubby beaks and nest insides holes. Their beaks are designed to help them scrape lerp off leaves, and their holes are designed to protect them from attacks by lerp-eating honeyeaters. These are striated honeyeaters, one of the four species.

from leaves. By controlling psyllid outbreaks they 'play an important role in maintaining the health of the ecosystem', wrote John. When lerp becomes scarce they can form roaming flocks of hundreds that sometimes wander out to sea and drown, or which die of starvation. In one flock I followed, the clicks of little beaks sounded like rain. The white spots on three of the four species suggest to me that evolution has produced birds that wear pictures of their foods.[15]

Australia's smallest bird, the weebill, also has a stubby bill and likes lerp. And there are honeyeaters (in genus *Melithreptus*) with short beaks and unusual braced jaws designed, it seems, for levering lumps from leaves. One member of this group, the white-naped honeyeater, is doing poorly around Adelaide, where koalas are defoliating local manna gums. Larger honeyeaters molest it more easily when the foliage is sparse.[16]

John Woinarski ran one study in which newly caught birds were caged for twenty minutes with lerp-infested foliage. Of the twenty-nine species he tested, which included thornbills, whistlers and silvereyes, an amazing twenty-two took some lerp. Every one of the small foliage birds ate it. Indeed, any small bird seen among gum leaves may have lerp in mind. It contains more starch than sugar, hence its broad appeal. Birds either eat the bugs with the lerp or leave them behind.[17]

The tiny psyllids that produce lerp are the most plentiful insects on eucalypts in southern Australia, one study found. Leaving the National Library in Canberra one day, after consulting old documents about lerp, I found it on the first tree I walked to, and also on trees outside Parliament House. It occurs outside Australia, but only in a few situations. In southern Africa explorer David Livingstone saw villagers eating 'a sweet gummy substance' on the leaves of mopane trees. His 1857 book, *Missionary Travels and Researches in South Africa*, quoted an expert comparing this to 'New Holland' lerp.[18]

In 2011 I drove for two days through mopane stands in Kruger National Park searching for this food. At last, beside the Letaba River,

I found a troop of baboons, twenty or more, sliding mopane leaves through their teeth or pulling them close to nip off lerp. Through binoculars I could see the white flakes scattered over their fur. An hour passed before I saw two bulbuls drift through the grove taking lerp. Outside Australia I have found only one article – from Botswana – about birds (starlings, doves, bulbuls, francolins) eating this food. It looks and tastes just like Australian (*Glycaspis*) lerp, but the psyllid responsible (*Arytaina mopane*) is only distantly related and must have come to produce lerp independently.[19]

At Mopane Camp in the national park a fuel attendant, Thomas Mathebula, accepted a leaf I showed him, plucking off and eating the lerp, calling it 'maboda'. 'You can make your stomach full if you eat many,' he said about this treat he'd often eaten as a child. In Namibia it sells in markets.

In Hawaii I once chanced upon a flock of house finches eating fallen lerp. They were outside a military base, under river red gums imported from Australia. In this globalised age insects keep crossing the seas on vessels and planes. Lerp must be inviting to Hawaiian birds, having been detected on the islands only three years before my visit. Researchers in California found that this same lerp (*G. brimblecombei*) became important to birds within three years of its arrival in 1998. In one infested red gum near Santa Monica I watched hooded orioles feuding over the foliage, and hummingbirds fighting over the flowers. One birdwatcher in Sacramento recorded twenty-nine species taking lerp in his garden, including crows and jays; birds discovered the food 'almost immediately' after it appeared in his street.[20]

California now has a lerp epidemic. Stricken eucalypts drop leaves and drip sticky liquid on parked cars. Wasps lured to the sugar make matters worse. Tiny parasitic wasps have been brought from Australia to attack the psyllids – just as Australian ladybirds were imported in 1888 to save California's oranges from another Australian bug, cottony cushion scale, in what became the world's first big biological control success.[21]

The scientist who imported the wasps confessed to liking lerps: 'I eat them all the time,' said Donald Dahlsten in a newspaper interview. Another American, Christopher Nyerges, author of *How to Survive Anywhere*, has lauded lerps, and I also find them appealing. They can be as sweet as confectionary or very bland, depending on the psyllid species, the season and the weather. Honeydew is sometimes present as well, if the bugs imbibe more sap than they can deploy in their shelters, hence America's sticky cars.[22]

Lerp feeds an amazing range of Australian birds, although no list has ever been compiled. Many parrots eat it, including swift parrots, as well as finches, bowerbirds, trillers, currawongs, choughs, apostlebirds, woodswallows, shrike-tits, moorhens, crested pigeons, and the small birds that John Woinarski tested. Reed-warblers, which otherwise keep to reeds, and purple swamphens, will seek it up in trees. 'They look pretty awkward but they can do it,' remarked David Paton about the swamphens he has seen.[23]

Lerp was sometimes prolific in the mallee lands of south-eastern Australia, a dry region once clothed in small, multi-stemmed eucalypts (mallees), most of which were bulldozed for wheat and sheep. On some mallee leaves in South Australia I have seen such large clusters of lerp I could have subsisted on it. The one biologist I found who has witnessed an outbreak, Joe Benshemesh, told me of fallen lerp looking like drifts of light snow. For some months a malleefowl he was studying obtained half its sustenance from this windfall. 'Everything grows fat on lerp when they outbreak,' he told me. 'It would be hard to overestimate the potential importance of such events in the ecology of mallee.' He suspects that malleefowl raise young only in years when something like lerp provides a boost.[24]

The river red gum (*E. camaldulensis*), Australia's most widespread tree, is another good source, as American birds have found. David Paton has seen 30–50 lerps per leaf, enough to fill a cereal bowl in half an hour. The African mopane lerp can be this plentiful, but it matters more to baboons than to birds. The comprehensive leaf drop of

These lumps of lerp on mallee leaves are unusually large because the shelters built by many psyllid bugs have stuck together. Lerp seldom reaches such high densities on leaves, and can do so only when few birds are taking it. Massive outbreaks sometimes occur in mallee woodlands, a dry infertile habitat in which birds tend to be sparse.

mopane trees in the dry season rules out Africa having lerp specialists. In Australia, lerp varies in abundance by season, location and tree species, but some birds enjoy a year-round supply.[25]

Manna appears to be a food of birds only in Australia. A search of the global database Web of Science brings up scores of journal articles about it, but only those from Australia mention birds. Honeydew *is* eaten elsewhere, ornithologists in the 1990s voicing surprise that warblers and hummingbirds were consuming it in Latin America, describing this as an unusual phenomenon 'most often reported from Australia'. Honeydew is important in New Zealand as well, where large parrots (kakas) and honeyeaters obtain it from beeches (*Nothofagus*). On the South Island, in dark forests heavy with a honey smell, I have watched New Zealand bellbirds chase silvereyes from this food, although it was too plentiful to defend with success. It drips in such volumes that trunks and branches blacken from the mould it attracts. In Latin America, Russell Greenberg described the interspecific territoriality it inspires, marked by attacks between species, as a rare phenomenon globally 'restricted to a few ecological situations'.[26]

Honeydew provides a reason why honeyeaters, more than most nectar birds, investigate bark. Eucalypts have unique epicormic strands that run from the bark deep into their wood, ready to sprout new leaves after fires of any intensity. Smooth-barked eucalypts can for this reason afford to invest less in protective bark than most woodland trees. Thin bark allows them to perform water-saving photosynthesis through their trunks and branches, explaining why very small bugs can feed from large limbs, since the sap is close to the surface. Smooth-barked eucalypts shed their aged bark each year rather than letting it accumulate on the trunk, and when partly detached this shelters honeydew bugs and many edible insects.[27]

Exudates also appeal to Australian mammals: gliders and Leadbeater's possum eat large amounts, and flying foxes like lerp. Outside Australia, Madagascar is the only place where exudates are important foods for mammals (lemurs).[28]

Why is Australia different? In their 'Ecology of Australia' article Orians and Milewski portray sap and other sweet liquids as something cheap and superfluous that eucalypts can afford to waste, because nutrient-poor soils hamper their use of the sugars they produce. Where Europe and America have many aphids, Australia instead has larger bugs (psyllids and coccids) imbibing tree sap at higher rates. They are better than aphids at processing nutrient-poor fluid, having a bypass filter in their gut to quickly remove water and sugar while retaining amino acids and other rare nutrients. Unlike aphids, they hide behind loose bark, or under their lerps, thereby escaping the drying winds that gust through flimsy eucalypt foliage. Australia's evergreen trees allow psyllids to feed and breed year-round, and bug numbers sometimes explode, creating feasting and breeding opportunities for birds. Aphid honeydew, wherever in the world it is found, is mostly taken by ants.[29]

Honeydew shows up mainly in cool places, perhaps because sugar plays a role as antifreeze in sap. On Ben Lomond in Tasmania I have admired towering mountain white gums that were glittering from all the congealing honeydew catching the afternoon sun. It was behind almost every bark strip, sweet to lick and honey-thick. In snow gum forest on Mt Wellington I watched crescent honeyeaters on feeding runs, dabbing their beaks at points on small branches where I could peel back bark and find minute bugs beside glints of liquid. In mallee woodlands I watched a yellow-plumed honeyeater feeding in the same assured way, visiting limbs it had obviously visited before. The bugs may benefit from birds removing their waste, which otherwise might drown them or attract moulds. Ants remove it as well. But very little is known about this resource, which has never been properly studied and is so easily overlooked.

The ragged Australian woods were something early visitors, including a travel-weary Charles Darwin, saw fit to complain about. Why did the trees shed not their leaves but their bark? We now know that trees in very poor soil invest their scarce nutrients in leaves that are tough and long lived rather than soft and deciduous. Smooth-barked eucalypts (those called 'gums') have been found to stay cooler during fires than species like stringybarks, which retain their old bark as a shaggy, and flammable, coat, suggesting a reason why bark is often shed in such a fire-prone land. Birds benefit twice over, from leaves that can supply lerp year-round and from honeydew (and edible insects) found behind slowly shedding bark. When deciduous trees drop their leaves they rid themselves of insects, and the price of keeping leaves is to accumulate their enemies.[30]

The implications of exudates are profound: these foods can sustain more birds than would nectar alone. Manna, like honeydew, is found mainly in southern Australia, the region with the largest and most prolific honeyeaters. When birds come to flowering mountain ashes, exudates are part of the attraction. Experts have remarked that when studied closely most honeyeaters are found to consume some 'alternative carbohydrate foods'. Birds need food security when they breed, and exudates fill gaps in the flowering calendar.[31]

Eucalypts thus fund bird aggression on a scale that would not apply if nectar was the only sugar source. Where there is lerp, honeyeaters pursue pardalotes until they sometimes kill them. When honeydew flows, honeyeaters turn on and drive away treecreepers and other bark birds. Even mild birds that eat nothing sweet are set upon when exudates increase.[32]

Two more secreted foods deserve a mention before I dive deeper into the social side of this. Birds will lap at sap bleeding from trees damaged by possums or boring insects. On a dry ridge in central Queensland I once found a Lewin's honeyeater, away from its usual rainforest haunts, guarding a wattle tree. When it was not repelling other birds it was probing damp places where bark strips had been torn off, apparently by something with teeth. I returned at night to find a sugar glider in my spotlight beam. Days later I found the arrangement unchanged, with the honeyeater in charge by day and the glider at night, like shift workers manning a fuel depot, although neither would have known the other existed. In north Queensland, fights arise at incisions made by yellow-bellied gliders when the daytime visits of rosellas, lorikeets, bowerbirds and honeyeaters coincide. In the dry season in inland north Queensland, I have found honeyeaters in large numbers, and lorikeets as well, targeting congealing sap on damaged buds of boxes (*Lophostemon grandiflorus*), which seemed to be the main sugar source at the time.[33]

In Tasmania birds once came to cuts made by stone axes. Holes hewn by Aborigines into cider gums (*E. gunnii*) filled with up to half a litre of slowly fermenting sap each day. In 1826 there was recorded 'a great eucalyptus cider orgy' when the brew cellared inside trees was drunk. Rock slabs were placed over holes to stop thieves such as wattlebirds. Cider gums are Australia's most cold-adapted trees, with sap that thickens into a honey-like syrup you can lick from the tree. In the Northern Hemisphere, maple sap only becomes sweet when it is boiled down to syrup.[34]

Other continents, instead of gliders, have woodpeckers such as sapsuckers piercing trees for sap, and hummingbirds exploit this. A river red gum I found at the edge of a Californian wood had hummingbirds on its flowers and other birds taking lerp; it also had neat rows of holes on an upper trunk, the work of a red-breasted sapsucker. Nothing in Australia punctures bark like that. Having drawn birds away from their native vegetation, this planted tree spoke volumes about eucalypts as an energy source.

Ringneck parrots in Western Australia's blue gum plantations have taken to tearing open stems to lick sap, ruining young trees. They are the only Australian birds I know of that open trees for sap.[35]

The final secretion to mention is the 'nectar' produced by glands at the base of leaves of certain wattles. After Cyclone Larry in 2006, when other foods were scarce, I saw birds as large as helmeted friarbirds taking it (from *Acacia mangium*) in north Queensland. Whether sap and wattle nectar qualify as important foods of birds in Australia is difficult to say because they have never been properly studied, but the continent is certainly well endowed with birds that can exploit them.[36]

The Forest Dominators

Studying birds in the Chiltern forest in Victoria for his PhD, Barry Traill found communities among the boxes and ironbarks sharply divided between haves and have-nots. Most of the small birds – robins, whistlers, weebills, silvereyes, thornbills, twenty-two species in all – kept to infertile ridges. The richer slopes and gullies were controlled by small honeyeaters, which shared their space only with birds too big to expel or which fed near the ground.

'There was a line no wider than a road dividing the two systems,' Barry told me. 'Any small bird that crossed that line was instantly attacked.' The small honeyeaters tried to evict larger birds but lacked the might. Barry found the same kind of division in central Victoria, and similar dynamics occur elsewhere. Although they are smaller than sparrows, fuscous honeyeaters, white-plumed honeyeaters and yellow-tufted honeyeaters may have once controlled the best woodlands over large areas of south-eastern Australia. White-plumed honeyeaters are the characteristic birds of river red gums along inland rivers – the prime outback real estate.[37]

Barry found that each pair of fuscous honeyeaters kept to a tiny breeding territory but helped neighbours evict intruders, including unfamiliar fuscous honeyeaters. A honeyeater banded and released 50 metres from its own patch was immediately 'whacked' by others of its kind, but closer to home its neighbours were tolerant. These birds eat nectar and lerp, and sometimes wattle 'nectar'. Like all honeyeaters they also take insects. The exclusion of other birds that take the same foods must be very useful.

Nowhere else in the world are birds excluded like this. Territoriality, the world over, is usually selective. Robins drive off other robins, sometimes of other species, they don't harry fantails. Birds will often repel members of their own genus, but seldom birds from different families. When they do so it's usually a predator, or a bird that has strayed near a nest. Different species are chased away from flowers, and sometimes from bunches of fruit, but it's only concentrated food sources which are defended, not a whole territory. The main exceptions I know about, apart from honeyeaters, are birds that feed on short grass or bare ground, notably lapwings and Australian magpies. Their defence works because their two-dimensional territories are easy to survey and guard.[38]

When Chiltern was gripped by drought one year, none of the trees flowered and exudates were probably scarce as well. Most of the honeyeaters left and gentler birds took their places. Rain a few months later brought the honeyeaters back and the placid times ended.

But territorial possession can be more extreme than this. Two honeyeaters of a larger size – bell miners (also called bellbirds) and noisy miners – practise the most intense resource defence of any birds on earth. For this reason they are targets of intense research.

The trees that bell miners live among usually sicken and eventually die, because instead of removing harmful insects the miners guard them from other birds. They live mainly on lerp, levering it off leaves with their tongues, usually without harming the bug beneath, leaving it free to secrete another starchy cap. Heavily infested eucalypts often shed their leaves, producing new growth especially appealing to psyllids. These reach such high densities that bell miners can live in colonies numbering in the hundreds, although they can also be small. Said to farm their food, they do so unsustainably: the high numbers of psyllids cause the trees to die, which is like a farmer letting his cows kill the grass. Wherever bell miners occur in south-eastern Australia their tinkling chimes sound constantly from dawn to dusk, 365 days of the year, in what is surely the most saturated sound broadcast by any animal. Their acoustics turn harsh when their territories are broached.[39]

That is a recent picture of these birds. To generations past they were cheery denizens of ferny grottoes whose dulcet chimes charmed wayfarer and poet. Henry Kendall's beloved poem 'Bellbirds' (1869) portrays them as silver-voiced darlings of daytime, 'softer than slumber, and sweeter than singing'. Sing they do not, but their bell-like notes are charming in modest doses. Most people fail to register their call as a warning of harm to other birds. Rural real estate with bell-birds attracts a premium, despite the promise of dying trees.

The control that bell miners exert became known in 1982, after Victorian biologist Richard Loyn, hearing of a landholder whose sick trees came good after miners were illicitly shot, removed these birds from a defoliated forest near Melbourne. Even before the last few were caught a suite of newcomers had arrived: rosellas, pardalotes, thornbills, white-naped honeyeaters. They consumed psyllids and lerps at a combined rate of 650 items per minute on the 3-hectare site. The released trees flushed new growth.[40]

'Bell miners are true farmers,' said Richard. 'Their territoriality turns the trees into producers of bird food via psyllids, but at the forest's expense.' The densities they reach – up to thirty-eight birds per hectare, or fifteen per acre – show how dependable lerp can be. One of Richard's articles on the subject found a place in *Science.*[41]

His trial removal of bellbirds was repeated in 1993 by Michael Clarke and Natasha Schedvin of La Trobe University, who witnessed 'large influxes' of the same newcomers after killing 189 miners near Melbourne. Crimson rosellas plucked leaves and slid them through their beaks, much as African baboons do, leaving a forest littered with green leaves. Pardalotes on the ground took the last lerps from these. Six weeks later most of the lerp was gone and most of the new birds had moved on.[42]

Bell miners are reckoned a problem since their numbers keep rising and trees over vast areas are dying – though most people don't know this. Biologist Paul Meek, once employed to manage what is called Bell Miner Associated Dieback (BMAD), showed me Toonumbar National Park, New South Wales forestry land that became a national park in 1995, shifting a problem from one department to another. We saw glades of dead and ailing trees, in place of the tall timber foresters had expected to rise after they last cut over this land. From an original epicentre in one stand of gums on shallow soil, bell miners had spread out like an infection along a ridge and down a slope to dominate miles and miles of increasingly unwholesome forest. We watched them on one ridge in spotted gums, a habitat they had previously avoided, and which has different psyllids from those in damper forest.

'We're going to have to rebuild a whole national park,' said government ecologist John Hunter, who joined us for the day.

'A *whole* national park?' I asked.

'Well, a significant part of it.'

The Bell Miner Associated Dieback Strategy (2004), commissioned by the New South Wales government, talks hard. 'Extreme degradation' is mentioned, along with 'Major disruption in ecosystem function', 'trees stressed and dying', 'increased weed invasion', and 'reduction of diversity and abundance of threatened fauna and flora'. At stake as well are tourist revenues, water quality, flood intensity, and 'timber supplies, honey production, shelter belts and forest-related lifestyles'. All from a bird and some bugs.[43]

Bell miners have always interested naturalists and travellers, but those who wrote about them seldom mentioned dying trees until recently. Only in the 1990s was a problem identified, after the New South Wales government, by halting rainforest logging, increased the pressure on eucalypt forests. About the precise causes researchers disagree. Foresters reject accusations of poor management. Thinned forest suits bell miners, but problems also arise in places that have not been recently logged, such as Cunningham's Gap in southern Queensland where, over the past decade, trees by the score have died along a stretch of highway.[44]

Lantana (*Lantana camara*), a weedy shrub, is often part of the problem, sheltering miner nests and possibly increasing the available soil nitrogen, resulting in richer eucalypt sap for bugs.[45]

One response to miners is to cull them. Mandeni Resort in southern New South Wales was promoting birdwatching as an activity, which did not stop the manager obtaining a permit to kill thousands of the pretty green miners. Local bird-lovers were unfazed. 'We are all irritated by bell miners,' said Barbara Jones of the local South Coast Birdwatchers Club, whom I phoned about the cull. She complained at length about miners invading and driving other birds away. Other birds that used to breed in her garden seldom even visit today. She confessed to sometimes shouting at miners in exasperation. 'I can't get away from that call!'

Culling of bell miners is deemed necessary to protect Victoria's state bird, the endangered helmeted honeyeater (a subspecies of yellow-tufted honeyeater), which is down to 130 or so birds. Dam building and other changes to the land may have helped bell miners take over by producing stressed trees ideal for lerp. To protect state forests, many trees are now injected with systemic insecticide.[46]

At Toonumbar young rainforest trees were rising through the lantana and in these I saw whipbirds, bowerbirds and monarchs.

By handicapping eucalypts, bell miners sometimes help rainforest. Given all the past rainforest loss, that is not altogether bad. Much of Toonumbar may end up with rainforest birds rather than bell miners, but since lantana stifles rainforest, saplings will often need help to push through.

The Ultimate Aggressor

The noisy miner – the other problem miner in Australia – elicits even more concerns, since it reaches much higher numbers. It dominates drier eucalypt forests and woodlands over much of eastern Australia, keeping small birds out of vast areas. Noisy miners exacerbate habitat loss by rendering the remnants left behind unsuitable for small birds.

A 2012 paper in *Ecology* speaks of 'despotic aggressiveness over sub-continental scales', mentioning fifty-seven bird species being suppressed. Researcher Martine Maron describes their rise as 'one of the most important mechanisms through which habitat fragmentation and degradation threaten populations of eastern Australian woodland birds'. She wants them culled, as cost-effective restoration. Colleague Michael Clarke described them in a journal article as a 'mess we have created' and a responsibility we should accept. They have been nominated under federal law as a Key Threatening Process.[47]

Noisy miners take defence further than any bird. Bell miners tolerate other birds in the understorey, but noisy miners live where this is sparse enough to impose control from the ground to the canopy. An intruder can find itself mobbed by up to a hundred snapping and scolding miners. They will turn on almost anything: koalas, cows, bats, pigs, snakes, lizards, people (sometimes), and birds big and small – including bell miners, which are slightly smaller. They break eggs and kill the chicks of other birds. Their aggression is often expended on animals of no consequence, such as ducks. Their twelve different predator calls include one for snakes and another for birds of prey. Most of the snakes I see in my garden are announced by noisy miners. Once their call drew me outside in time to see a tree snake lunge at a large water skink before tumbling with it down some stairs – an exceptional scene. Another time I saw two miners have sex during a mob attack on a bird of prey.[48]

Where forest tracts are thinned by logging, grazing, fire or roads, noisy miner colonies can extend for kilometres with no obvious breaks. They now dominate hundreds of thousands of hectares of Barakula State Forest in Queensland. Eucalypt patches in paddocks also prove highly defendable, as do sharp boundaries between forests and pasture, noisy miners being a classic 'edge' bird. In fauna surveys they count as evidence of degradation, a box to tick alongside those for overgrazing and weeds.[49]

Cities and towns suit them as well. One expert noted in 1969 that noisy miners 'are not a garden bird in towns and cities', but today they occupy Hyde Park in Sydney's CBD, and in Brisbane I have watched them over the years become the dominant bird, helped along by the planting of urban eucalypts. The sparrows of my childhood are so rare today I often go a year without seeing one. Cats attract blame for bird loss around cities when noisy miners are a more tangible explanation.[50]

There are probably more of these birds alive now than ever before, and Australia probably sounds harsher because of this. Their original habitats are poorly known, but tall open woodland on relatively fertile soil was likely favoured. Around Sydney, noisy miners spread from the Cumberland Plains into the sandstone country when bird-attracting shrubs were planted in gardens, providing a good reason, where there are miners about, not to plant nectar-rich shrubs.[51]

A biologist who migrated from Canada (and a former lecturer of mine) was the first to divine how unusual noisy miners are. The title of Doug Dow's 1977 paper was 'Indiscriminate Interspecific Aggression Leading to Almost Sole Occupancy of Space by a Single Species of Bird'. He surmised that 'successful defence of an area for the whole year against encroachment by virtually all other species of birds appears unique'.[52]

But not all birds are evicted. Butcherbirds, which are partly carnivorous, enjoy special status. Larger than miners, they join their attacks, striking with far bigger beaks. I have seen possums in great distress not so much because miners were heckling them for showing themselves by day, but because butcherbirds, incited by the commotion, were inflicting injury. Such dramas unfold in gardens every day. Butcherbirds drape the small birds they kill on twigs before dismembering them to eat – hence their name – but when they live among miners they know not to eat miner young. The noisy miner nests in my garden, with many adults attending the noisy chicks, are very obvious. My butcherbirds benefit from miners driving off the Pacific bazas (hawks) that come for the same grasshoppers butcherbirds eat. Butcherbirds may at times *need* miners. On the Atherton Tableland, where miners are scarce, grey butcherbirds occur only where they do. In Victoria butcherbirds disappeared when one miner colony was disbanded.[53]

Pied currawongs will also forego meals of miner chicks to win acceptance. I have seen one land near a miner nest in my garden without harming the young, but currawongs are hounded out of some miner colonies. Also tolerated at times are crows and magpies, which join the attacks, and kookaburras and parrots, which resist eviction.

The society of noisy miners is as extreme as their defence tactics. Their colonies can number in the hundreds if not thousands of birds, with individuals keeping to small areas within the colony. Males befriend neighbours, while females shun females but mingle widely with males, granting sexual favours to some in what is a very promiscuous as well as violent society. Males greatly outnumber females, sometimes by five to one. They fight among themselves from an early age, sometimes pecking their sisters to death. This may explain the gender bias, which is not present at birth. Adults will kill chicks, and brawls often arise between different sectors of a colony.[54]

When they are not fighting, the males help at nests. Up to twenty-two may feed one brood of chicks. The highest ranking and most diligent is the nominal father of the young. He may bring food to fourteen nests. Many birds around the world breed cooperatively but never on such a scale. Helpers are often unrelated to those helped, and may be impressing females to gain favours when top males die.[55]

Trees guarded by noisy miners do not die, although they can suffer more insect attack, improving in health when miners are removed. As for the birds that come into harm's way, three of them are endangered species. A noisy miner was seen to drive away a pair of regent honeyeaters after destroying their nest. The regents, whose gold, black and white embroidered design befits a Versace accessory, have gone through a collapse which, like that of America's extinct passenger pigeon, could never have been predicted.[56]

This species made an impression on Gould: 'It generally resorts to the loftiest and most fully-flowered tree, where it frequently reigns supreme, buffeting and driving every other bird away from its immediate neighbourhood; it is, in fact, the most pugnacious bird I ever saw . . .'

Gould encountered a 'great abundance' of regents in brushes where precious few remain today. When the red gums bloomed the Melbourne Cricket Ground was one haunt. Small groups commandeered rich-flowering trees, defending them against all comers, just as I have watched friarbirds do inside a noisy miner colony near my house. One regent was reported seeing off other birds at the rate of thirty-two pursuits per hour, and up to seventy when building a nest, a frequency unmatched by any other bird. Lerp and honeydew feature in their diet, as I saw near Armidale when I watched a pair ply sticky lerp to their young.[57]

The regent honeyeater has been squeezed out by the loss of the best trees, those on fertile soil, and by a boom in larger miners and wattlebirds. It may be the wrong size for the times – not big enough to profit from its aggression. Its future is very uncertain.

Regents and swift parrots are 'rich patch nomads' that roam widely until they find sugar hotspots. Able to carry pollen immense distances, they have almost stopped doing so, having become two of Australia's three most endangered pollinators. The noisy miners contributing to their demise are probably Australia's least mobile pollinators; their female-activity ranges near Brisbane measure 52–95 metres across, with male core home ranges averaging 129 metres across. Bees are far more mobile than that.[58]

In climate change reports for governments I have warned that noisy miners will handicap eucalypts by reducing the mobility of their pollen. To produce seedlings with a future, trees will need pollen from drier and hotter places, not pollen from the next tree. Droughts that thin forests will aid miners. Lorikeets, red wattlebirds and flying foxes will assume more importance in the future since they spread pollen widely, little deterred by miners.[59]

Efforts to restore land often worsen the miner problem. Eucalypts by the million have been planted on denuded farms, and millions more will be grown for carbon storage, providing ideal miner habitat – namely trees without shrubs, and trees adjoining paddocks. Small birds face violence if they use the corridors that link forests across such paddocks. When degraded areas of buloke (*Allocasuarina luehmannii*) are restored, fast-growing eucalypts are often planted instead of slow-growing buloke, and miners move into a habitat they otherwise avoid, with predictable outcomes for other birds.[60]

One proposal is that corridors be made wide enough – at least 600 metres – to include a core without miners. Another is that leafy wattles be planted to discourage them, although these ruin corridors for the ground-feeding birds that most need them. In large forest tracts miners can be disadvantaged by removing fires and cows. A brochure put out by La Trobe University, with funding from the Australian, Victorian and New South Wales governments, nominates culling as 'the most humane, practical, cost-effective' method of easing their impact. In cities, though, next to nothing can be done to loosen their hold. I would rather something else in my garden but cannot help admiring their bravado and social complexity. Miners are not troublesome in every city: in Canberra they dominate many parks but stay out of gardens, and around Adelaide, at the dry western edge of their range, they are milder in disposition, probably because there is less food to defend.[61]

There is a third species, the yellow-throated miner, that has begun occasioning concerns. It occupies vast tracts of inland woodland but seems to be most combative in Victoria, along easily defended road-side strips.[62]

The Dawn of Group Defence

With some seventy-five honeyeater species in Australia, why should the behaviour of two of them be so extreme? While bell and noisy miners are much larger than nectar birds on other continents, they're smaller than wattlebirds and friarbirds, which rarely attack in groups.

A major driver of evolution in Australia has been the long-term drying of the continent. Twenty million years ago the land was largely covered in rainforest, and even a few million ago there was more than today. Many birds found among eucalypts have close relatives living inside rainforest, suggesting descent from rainforest ancestors. DNA trees of honeyeaters show that miners have their nearest relatives in the mountains of New Guinea. These are the melidectes (*Melidectes*), several of whose nine species are loud and argumentative, although they don't live in groups. Having kept to rainforest they have proba-bly changed less over deep time than miners. They could be very close to what miners evolved from. That none has been studied became my reason to spend more than a week in moss forest in the Central Highlands, prime melidectes habitat.[63]

From a high window I could follow a Belford's melidectes on its daily visits to the flowers of a melastome tree, epiphytic orchids, and a small tree in the Myrtaceae family. After wiping its beak on a favoured perch it would survey its small domain. Its main call sounded so like an aggrieved noisy miner I sometimes forgot where I was. A tree with curled leaves provided protein, which it extracted by opening its beak inside the curls to reach hidden grubs. Nothing else went near that tree.

From other melidectes as well I built up a picture of regular belligerence, enacted at times for no clear reason. Smoky honeyeaters and Papuan lorikeets seemed unfazed by melidectes' lunges, the smoky honeyeaters (which are large and take fruit) casually dodging them, and the lorikeets, fiery red birds with golden streamers, all but ignoring them. The melidectes preferred open places, feeding in trees that pushed above the canopy – a setting suited to defence because intruders can be seen and chased.

A bird guide told me that melidectes 'is a bully bird', mentioning wallabies being harassed in clearings. I saw one attack black sittellas, small bark birds that take completely different foods. One nature resort tried killing melidectes to help other birds. At a lower elevation I watched a yellow-browed melidectes drive superb birds of paradise and a honeyeater from a fruiting tree, in which it then showed no further interest. Noisy miners are often like that, as if control is a point of principle, although in this case the melidectes was probably protecting food for the future.[64]

These two melidectes species look much like noisy miners if you go past their faces (see colour section). They share with them grey, pale-edged feathers that give a scalloped look, and olive on the wings and tail. There is black and white on the faces and coloured skin round their eyes, but organised differently. My garden birds looked faded when I returned home, miners being paler. The young melidectes beg with the same loud call – a sound in my garden that probably arose in rainforest millions of years ago. If honeyeater evolution went from rainforest to drought-prone land, the sequence would have gone from melidectes, or something like it (in rainforest), to bell miner (wet eucalypt forest), to noisy miner (dry eucalypt forest). We can suppose that noisy miners did not evolve directly from bell miners, but from their recent ancestors, which were probably grey rather than green. Miner DNA implies that bell miners did diverge first. Leading biologist Dick Schodde believes that melidectes evolved in Australia then spread to New Guinea when the climate dried, and that seems very plausible. The sequence implies a change from a diet of mostly nectar to mostly lerp, then to one that includes both.[65]

About noisy miner diet something interesting has been noted: they are big for what they eat, and their young are fed small items 'at an exceptionally high rate'. Small insects are plentiful because small birds that might eat them are kept away. Two miners raising chicks alone might have trouble supplying tiny items fast enough, but there are many helpers to share the burden. An eye for tiny foods could have been inherited from a lerp-eating ancestor.[66]

In some regions miners are tied to rich sources of nectar or lerp, such as yellow gums (*E. leucoxylon*) on fertile flats in Victoria and ironbarks in central Queensland. A century ago they may have been nectar and lerp specialists that took small insects as well. Now they occupy places such as Hyde Park where sugar is scarce. In some parks they take bread and butter, even meat. Fruit is pecked at in orchards and kangaroo and emu dung is flipped over for the insects underneath. On rare occasions frogs and lizards are swallowed. Population expansion is driving them into ever more habitats, including some with no eucalypts, showing that food preferences matter less than vegetation suiting defence.[67]

Looking around the world for analogues to miners, the closest I've found are some thrushes in Europe and Asia, fieldfares, which nest often in colonies, call harshly and mob predators, sometimes killing hawks with a barrage of faeces. Impressive though they are, the only competitors they displace are other thrushes. I watched one challenge a woodpecker with a lunge that was as brief as it was ineffectual. America's acorn woodpeckers practise group defence, but only around the few trees in which they store acorns in thousands of small holes. Forest fragmentation has helped American cowbirds replace other birds on a vast scale, but as nest parasites rather than aggressors. I watched a young one being plied with eucalypt lerp by its duped foster parent, a junco. None of these birds comes close to dominating large tracts of forest at the expense of other species.[68]

Miner evolution probably required an unusual resource to drive it. Rich and reliable lerp promoted group breeding and group defence in birds that inherited aggression from a diet of nectar. Lerp promoted an interest in small foods while encouraging defence of foliage as well as blossoms. That is my reading: a progression from nectar to lerp that was possible only in Australia.

Miners have much in common with us. They live, often unsustainably, in complex cooperative societies dominated by males, who accept most of the responsibility for defence. They displace other species. Noisy miners even practise a division of labour, some males helping with food and defence while others do just one or the other. These birds tell of something larger: a continent with such influential birds that whole forests can change if one species multiplies. Introduced Asian mynas are widely disliked, but despite an international reputation for aggression they are no match for the native birds named after them. Research in Sydney shows that miners, wattlebirds and magpies perpetrate far more attacks on other birds.[69]

We should keep in mind that miners represent only the pinnacle of Australian aggression. In a blunt article, 'Why Have Birds in the Woodlands of Southern Australia Declined?', wattlebirds, friarbirds and small honeyeaters came up as well when Hugh Ford and three colleagues warned that interspecific aggression was a growing problem.[70]

Three

The First Song

As well as harsh-sounding birds, Australia has many sensational songsters. In 1955 film star Katharine Hepburn became obsessed with one of them, visiting Sherbrooke Forest near Melbourne again and again with ballet supremo Robert Helpmann to view performing lyrebirds. Her biographer says she found their shows a 'thrilling, rib-tingling experience, like a sexual, exciting play'. Helpmann dedicated to her his lyrebird-inspired ballet, *The Display*, featuring an extravagant lyrebird costume designed by Sidney Nolan.[1]

Lyrebirds have attracted many admirers. French composer Oliver Messiaen transcribed their song into a score, on which he based a movement of his last orchestral piece. American philosopher Charles Hartshorne ranked them the world's finest songsters. A more recent American devotee, musician David Rothenberg, singled them out in his book *Why Birds Sing*. There are reasons why Australia has aggressive birds, and so it is with the lyrebird's song. Some history is needed first.[2]

Northern Orthodoxy

For most of the twentieth century it was thought that birds had their genesis in the Northern Hemisphere. Germany had yielded the world's first bird, *Archaeopteryx*, and Australia could only have received its birds from Asia. The most forceful advocate of this thinking was the redoubtable Ernst Mayr.

Mayr was one of the leading minds of his time. The 'Darwin of the twentieth century' was one accolade given him when he died in 2004. He gave the word 'species' an evolutionary meaning where before it had only been descriptive. He founded the journal *Evolution* and wrote a stack of important books. But this same man, when the evidence mounted against him, stuck to his view that Australia had never been more than a receptacle for northern birds.

His interest in the region was stimulated early on, when he was sent by Lord Rothschild to New Guinea in 1927 to collect birds of paradise and other specimens. He spent the next two and a half years in the South Pacific collecting for the American Museum of Natural History, observing the same kinds of differences between islands noted by Darwin in the Galapagos, and Alfred Russel Wallace in Indonesia. Because each island is, in effect, a test tube with a unique mix of ingredients, they have special insights to provide. Mayr's seminal book, *Systematics and the Origin of Species*, which inspired a generation of biologists worldwide, had golden whistlers and scarlet robins on islands near Australia demonstrating evolution at work.[3]

In an article that appeared in 1944, Mayr explained that Australia had received its birds from Asia in waves. Arriving first were the forebears of emus, lyrebirds, honeyeaters, parrots and honeyeaters, followed later by robins, creepers, and other less distinctive forms. This was a version of *terra nullius*, an empty land filling with good things from the north. It was the obvious way to think before the notion of continental drift reared its difficult head. The Old World and Australia shared orioles, babblers and robins, and since evolutionary theory demanded common origins, this could only mean ancestry in the north. Australia was effectively a junior branch office of the Great Northern Lands.[4]

But when the chance came, the birds flouted their allotted roles.

Mayr visited Australia in 1959 to test the classification of the day, to see how well bird behaviour substantiated the nomenclature, the latter based as it was on skins and bones in museums. But his first treecreeper (*Climacteris*) shocked Mayr by dropping down and hopping on the ground. Real creepers (*Certhia*) never did this. 'There is little doubt in my mind that *Climacteris* is not in the least related to *Certhia*,' he concluded, 'and should at once be removed from the Certhiidae.' His shock would have sharpened had he seen one sipping nectar, as they occasionally do.[5]

A glimpse of a scrub-robin left him feeling much better: 'The bird behaved entirely like a thrush. It was characterized by the quick runs and sudden stops of a thrush, and in every other aspect it also acted like a member of that family . . . it certainly is a thrush.'

A companion recalled that Mayr had spent half a minute eyeing the bird before announcing, 'That's not a robin, that's a *Turdus* [thrush].'[6]

But Australia's scrub-robins are neither robins nor thrushes. What Mayr did not suspect was that unrelated birds, to procure the same food, might evolve similar mannerisms.

The real challenge to Mayr's convictions came with the acceptance of continental drift as a paradigm to explain global distributions. I remember in the 1970s my university lecturers struggling to integrate it into their thinking. Australia had been part of a vast supercontinent before it became an island. Some bird groups could have started out in the south, before Gondwana broke apart and Antarctica froze over. This could explain why Australia and South America were both rich in marsupials.

Mayr was unconvinced. He rebuked Australian biologist Dom Serventy for proposing at a conference in 1971 that some birds might have southerly origins, insisting that evidence for an original Australian bird fauna was 'slight, if non-existent'. But another American soon came to Serventy's aid. Noting a lack of giant flightless birds on northern continents, Joel Cracraft in 1972 proposed a Gondwanan origin for ratites (ostriches, emus, cassowaries, rheas, kiwis), the birds deemed the most primitive. He mentioned two other groups found entirely (penguins) or mainly (parrots) below the equator. He chided Mayr for 'misleading and erroneous statements' about plate tectonics.[7]

Serventy had also mentioned some tiny feather fossils found in a road cutting in Victoria a few years before. Unbelievably old – more than 100 million years – they dated back deep into the age of dinosaurs. Only Germany's *Archaeopteryx* was older. Birds, it was now clear, had been living down under for a very long time.[8]

Ernst Mayr should have backed out of the debate, but he soldiered on. At an international ornithological congress in Canberra in 1974 – the first ever held in the Southern Hemisphere – Mayr was dismissive of those fossils, one of which happened to be on display. Yes, he explained, Australia had birds right from the start, but its birds today are more Asian than Gondwanan, so the original birds 'must have become entirely extinct'. Cretaceous floods left them susceptible to replacement by newcomers from Asia.[9]

But Australian biologists did not think much of Old Testament-style deluges flushing away the unworthy, and the notion of a partly Gondwanan bird fauna took hold, helped along by a paper from another American, Pat Vickers-Rich. She did not find her subject easy. In a chapter she wrote in 1982 with Jerry van Tets, by which time she was an Australian resident, she mourned the loss of certainty: 'It is difficult to give up a theory such as that of Mayr, which offers one solid solution concerning Australia's avifaunal origins. It doesn't feel good giving it away for the uncertainty of multiple hypotheses.'[10]

But for those with no loyalty to past ideas, origins had become very interesting.

Forward Thinking

The way forward was forged by another giant of twentieth-century ornithology, Charles Sibley. Recognising that flight, by severely constraining the forms birds take, conceals the evolutionary distances between groups, Sibley turned to chemistry, to molecules. At his labs in Yale he boiled up sections of double-stranded bird DNA until the molecules cleaved in half, then he allowed them to cool down and rejoin. Complementary strands with the same genetic instructions were attracted and formed strong bonds. When DNA from two species was heated together and left to cool, the amount of hybrid DNA measured the species' genetic similarity. High similarity implied a recent common ancestor, and justified placement close together on the classificatory tree.[11]

Sibley and his assistant Jon Ahlquist were nothing if not ambitious. Over ten years they tested 1600 bird species from around the world. Sibley's grand opus – his 'tapestry' – was a whole new classification of birds, one that is still widely cited, although the better genetic methods used today show his conclusions were often wrong. He was the grand alchemist, offering endless shocks and surprises. Dabbling with mammal DNA, he became the first to show how close humans and chimpanzees are.[12]

In his later years he recalled his findings about Australia with special pleasure, his interest in the area dating back to birds he saw during World War II service in New Guinea and the Solomon Islands. Sibley showed what others could only guess at: that Australia's robins, flycatchers, warblers and babblers were not what their names suggested, but part of a vast Australian songbird radiation. Just as marsupials had developed into Australian forms, some of which had analogues elsewhere, so had songbirds, but on a much larger scale. Their DNA strands kept clinging to each other, not to Asian birds.[13]

But there was more than this, and Sibley's ultimate conclusion so offended northern sensibilities it was widely ignored. Not only had Australia evolved its own songbird clade, it had been an exporter on a grand scale. Many of Europe and America's feathered finest – including magpies, jays and shrikes – were part of the Australian radiation.

They are closer to Australian birds than to most northern birds.[14]

The incubating strands showed songbirds falling into two groups, the Passerida – which Sibley decided had radiated in Africa and the Northern Hemisphere, to produce such familiar birds as sparrows, swallows, starlings and thrushes – and the Corvida, which had poured out of Australia. Sibley was not saying jays and shrikes evolved in Australia, but that one Australian ancestor reaching Asia long ago had spawned these groups.[15]

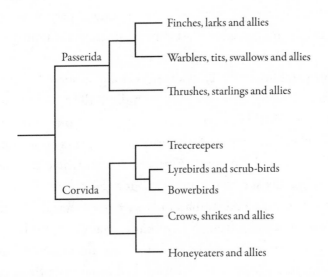

The first inkling that Australia played a major role in bird evolution came when Charles Sibley's pioneering DNA studies suggested that the world's songbirds split into two groups, one with the world's crows, shrikes and their kin appearing as relatives of Australian birds, implying a remote Australian ancestry for them. Sibley used various names for these two groups, including Passerida and Corvida. (Diagram simplified from figure 1 in Sibley and Ahlquist 1985.)

I should clarify at this point what songbirds are. All the world's birds have been arranged into more than forty orders, but more than half of all species alive today, and a large majority of individuals, fall into just one – that of the perching birds, order Passeriformes, also known as passerines. They are the outstanding success story among birds, taking their name from the sparrow, which is *passer* in Latin.

(They are not to be confused with the smaller groupings that take their name from the sparrow: the Passerida; and the sparrow family, Passeridae.) All the petite little birds that forage in foliage are perching birds: robins, warblers, flycatchers and finches, as well as much larger crows and magpies. Not included are parrots, pigeons, cuckoos and owls, each of which constitutes an order of its own. Good at perching these birds may be, but they differ in other ways that indicate different ancestors. Birds that feed in water – gulls, herons, ducks and the like – are almost never passerines. The water surface marks the boundary of their success. Something that sets them apart from nearly all other birds is an opposable hind toe with a tendon that operates independently of the tendons in other toes.[16]

Most perching birds (80 per cent) belong in one suborder, that of the oscines or songbirds, and they include all the birds renowned for singing. Songbirds learn part of their song repertoire, while the suborder of suboscines, which make up about 20 per cent of all perching birds, found mainly in South America, are hardwired for their calls, which are usually short and simple. (But a few suboscines apparently do learn their calls.) Australia's only suboscines are three rainforest pittas, which came in from Asia recently. Many songbirds are not to our ears musical at all – crows and sparrows, for example – and they are best thought of as birds with sophisticated calls that are partly learned.[17]

Songbirds are often lauded as special. Here is Alfred Russel Wallace in 1856:

The Passerine order comprises at once the most perfect, the most beautiful, and the most familiar of birds. The feathered inhabitants of our fields, gardens, hedge-rows and houses belong to it. They cheer us with their song, and delight us with their varied colours. Their activity and elegant motions are constant sources of pleasure to every lover of nature. They are the birds with which from our infancy and boyhood we are most familiar, and we therefore involuntarily derive from them that ideal or typical form of animal life with which we connect the general term, Bird. And thus doing, who can doubt but that we are correct? The lightness, activity, elegant forms, brilliant colour and harmonious voice by which birds as a whole are peculiarly distinguished from all other animals, find in this group their fullest expression and most complete development.[18]

Sibley was daring to say that many of these birds had Australian roots. Europe's rooks and nutcrackers are closer to whipbirds and whistlers than to Europe's thrushes and starlings. Sibley liked to be confrontational, and ire was soon forthcoming. Adelaide biologist Jeremy Robertson, a young postgraduate in 1980 when Sibley unfurled his tapestry at a meeting in Arizona, remembers less about the talk than about the arguments afterwards. 'There was a huge resentment towards this idea that such an important group of animals as the Corvida could not come from the Northern Hemisphere,' Jeremy told me. One combatant was a young Stephen Jay Gould.

Ernst Mayr was nicer, praising Sibley's drive and innovation, and welcoming his division of songbirds into two clades, although he would not buy an Australian role in anything. 'There was,' he insisted, 'no escape from the conclusion' that the Passerida gave rise to the Corvida, and it was 'quite clear' that the Corvida had arisen in Asia.[19]

The next step was taken by a young Australian, Les Christidis. An undergraduate at Melbourne University with a penchant for genetics, Les wondered why it was always mammals that were used in lectures to explain genetic principles, and when he discovered that nothing was known about bird DNA he chose finches for his PhD. Their DNA was easily obtained from the blood of pet shop stock.[20]

He then teamed up with the CSIRO's Dick Schodde to do a test of Sibley's findings. Dick had befriended Sibley while supplying samples for his DNA work. Where Sibley had compared bird DNA,

Les decided to compare proteins in the organs and muscles of different birds, to see if these told the same story. Les collected the data and
Dick wrote it up.[21]

Sibley welcomed this test of his tapestry, but the proteins did not
show a division of songbirds into two large groups. The Passerida
formed a cluster amongst the Corvida, and lyrebirds stood apart from
everything else. 'We got lyrebirds at the base of everything,' Les told
me in 2004, about the phylogenetic tree that emerged. A tree is drawn
by placing closely related birds on adjoining twigs, to show recent
separation from a common ancestor. The more the proteins differ,
the more distant the relationship, and the deeper the separation on a
branch. Les was saying that lyrebirds have the most divergent DNA,
indicating the largest evolutionary distance, justifying placement on
the lowest and longest branch.[22]

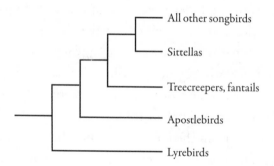

Tests conducted by Les Christidis showed that lyrebirds and several other Australian
birds were clustered at the base of the songbird tree, indicating an Australian origin
for all songbirds. This tree is a simplified version of one that emerged from his data.
We now know that treecreepers belong on a deeper branch than apostlebirds.

Lyrebirds have as their closest relatives two rare birds not tested by
Les – the scrub-birds – which belong with them on that lowest and
longest branch. Lyrebirds and scrub-birds were recognised as far back
as 1880 as 'the most anomalous forms of passerine birds yet known'.
Biologists in the nineteenth century, whose interest in comparative
anatomy ran deep, found that lyrebirds and scrub-birds deviated from

all other songbirds by having, for example, fewer vocal muscles and no wishbone. Max Fürbinger in Germany declared in 1888 that all the world's perching birds were close enough to go into one giant family, except for Australia's two lyrebirds and scrub-birds, which he placed in another. Richard Bowdler Sharpe in London went far further, giving lyrebirds and scrub-birds their own order, which was to say they stood out from all other birds as much as do penguins and parrots. Australia's first popular bird guide, *An Australian Bird Book* (1911), adopted this approach.[23]

English zoologist Alfred Newton had in 1893 portrayed the lyrebird as the 'nearly sole survivor of a very ancient race of beings' destined for extinction. In a very detailed anatomical paper in 1974, Sibley linked them to bowerbirds and birds of paradise. Experts on three continents have pondered these birds. The proteins tested by Les Christidis, and the anatomical oddities, suggested they had features shared by the world's first songbirds, which all other species (except scrub-birds) have lost.[24]

Lyrebirds and scrub-birds might be called 'living fossils', the most 'primitive' surviving songbirds, but these terms have lost standing today. Evolution has no trajectory that bestows inferior status on early lineages. Cockroaches and termites might also be called primitive, but like superb lyrebirds they are not drifting towards extinction. We do better to talk in terms of early and recent features than of primitive and advanced birds. Lyrebirds and scrub-birds are modern birds that retain early songbird features. They are the 'sisters' of all other songbirds. The first songbird produced two daughter species, one giving rise to lyrebirds and scrub-birds, and the other to everything else. Because the term 'sister' does not put one branch above another, it has become popular today.[25]

With the Passerida forming one massive branch within the crown of the songbird tree, lyrebirds and scrub-birds forming the lowest branch, and other Australian birds on low branches, an Australian origin is implied for every songster in an English country garden, for all the chickadees, cardinals and jays in America, for bulbuls, babblers and sunbirds in Asia, and weavers, whydahs and bush-shrikes in Africa. All songbirds, in other words, have Australian roots. This went far beyond what Sibley thought.

When Les and Dick submitted their work to international journals they were rebuffed. After four years or so their paper finally surfaced in the British journal *Ibis*, in 1991, its conclusions muffled to avoid offence. In his office at the Australian Museum Les showed me the wording imposed on them: 'Thus it is not beyond reason to draw attention to the possibility of a Gondwanan origin for the order . . . Although purely speculative at present, this hypothesis does warrant testing . . .'[26]

But instead of anyone testing their conclusion, it was ignored. Ignored as well was a 1982 American study of lyrebirds that argued from anatomy that 'the entire order Passeriformes is of southern origin'.[27]

Some years passed before the next page in the story was written.

In rainforest in New Zealand I remember hearing a thin, high squeak that at first I took for an insect. But so scarce are forest birds in New Zealand today – a legacy of introduced rats, carnivores, possums and deer – that I scanned the foliage hoping for something better than a cricket, until finally I spied two midget birds, greenish-brown with petite bills and no tails to speak off, flicking tiny rounded wings as they skipped up the axis of a mossy tree. They were riflemen, members of a remarkable but fast-disappearing family of perching birds called the 'New Zealand wrens' (Acanthisittidae), although they are unrelated to anything else called a wren. Of the four species known in the nineteenth century only two remain, the other being the rock wren, a dumpy little denizen of mountain scree. Both are rare.

Alfred Newton had in 1893 described New Zealand's birds as 'extraordinary forms – the relics of perhaps the oldest Fauna now living'. He voiced fears for their future the year before one of them went

The New Zealand 'wrens' posed an enigma to 19th-century naturalists, and not until 2002 was it realised that they are lone survivors from the first flourishing of perching birds. Their ancestors may have come from Australia. In the 140 years since this plate was produced, riflemen (the top birds) have dwindled, the rock wren (lower left) has become vulnerable to extinction, and the bush wren (lower right) has vanished, not seen since 1972.

extinct. The Stephens Island wren was once the world's only flightless perching bird, and also the tiniest flightless bird. Obliterated from most of New Zealand long ago by Polynesian rats, it was in 1894 removed from its last outpost, a 1.5-square-kilometre island, by feral cats. Had Newton been heeded, the fourth New Zealand wren species, the bush wren, might have escaped a similar fate. It was last seen in 1972, a victim of rats.[28]

Like lyrebirds, these 'wrens' had always proved awkward to classify by seeming to lie outside or between the two groups of perching birds. They have some oscine and some suboscine features, plus some of their own. In 2002 two DNA studies published by the Royal Society of London, drawing on newer and better methods, finally found a place for them – below the other groups. The ancestors of the 'wrens' had branched off before songbirds and suboscines existed.[29]

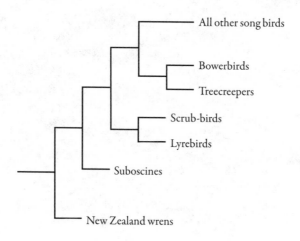

The first division in the perching birds separates the New Zealand wrens from all other perching birds, which make up more than half the world's bird species.

But a paper published three years later really turned the world over, by putting Australian birds on top. In 2004 a news agency wrote it up. 'A DNA study has found the northern hemisphere's much-loved songbirds actually evolved in Australia 45 million years ago', the news article began.[30]

American biologist F. Keith Barker was quoted as saying 'all songbirds trace their origins to Australasia and New Guinea'. Barker may have thought he was saying something new, but the article went on

to quote a vindicated Les Christidis. 'When we first suggested this . . .
we got laughed at by the Americans,' he said. 'Australia doesn't have
that many birds relative to the rest of the world, so how could it be the
centre of everything?'[31]

Here at last were Americans talking up Australian birds. Ernst
Mayr remained silent. Any number of recent DNA studies have since
confirmed these findings, and Australia's status as the original home of
songbirds is no longer in question. To those biologists in Europe and
the US who work on origins, this has become a central tenet, the pivot
on which many papers turn.[32]

Songbirds make up 47 per cent of the world's bird species but there
are more of them than this figure suggests. Surveys in Britain indicate
that eighteen of its twenty most plentiful species are songbirds, which
go to make up 90 per cent of all its birds. The most abundant bird
on earth, Africa's red-billed quelea, is a songbird whose population
(about 1.5 billion) exceeds that of all the world's seabirds combined
(about 1 billion). These and other statistics imply that the majority of
the world's birds have ancestors from Australia.[33]

So how was it that Australia came to have the first songbirds?
If we go back to a time when Gondwana was still partly intact, the
evidence came to be interpreted like this: the 'wren' lineage evolved
first, surviving in isolation in New Zealand when that land – as part
of a larger landmass called Zealandia – drifted away from Gondwana
82–85 million years ago. Wren relatives remaining on what was left of
Gondwana, notably Australia, Antarctica and South America, then
evolved in two directions, to produce the songbirds that thrived in
Australia and the suboscines in South America.[34]

The DNA evidence is so strong and consistent it carries a lot of
weight, and there are bits of fossil bone to back it up. Scientists could
claim a northern origin for most things because most fossil hunting went on in the north, skewing the fossil record. But songbirds
are missing from the same quarries in Europe that have yielded up
early remains of ducks, pheasants, rails, cranes, swifts and many other

groups. Songbirds were for this reason considered a young group, which suited the notion of the best species (namely humans) appearing last, like Adam on the seventh day of Creation, or royalty at the end of a procession.[35]

All this changed in the 1990s when three tiny bones found near Kingaroy in southern Queenland, about 54 million years old, reached the hands of the Australian Museum's Walter Boles. He identified them as fossils of perching birds, the oldest ever found by a margin of some 20 million years, a point subsequently made by Walter in the journal *Nature*. The bones don't match those of songbirds and could be closer to New Zealand wrens. When I saw them I was shocked by how tiny and insignificant they seemed. The smallest was so amorphous to my eyes it could almost have passed for a mouse dropping. Had Walter dropped it I'm not sure we could have found it on the dirty museum floor.[36]

In the years since then doubts have grown about the identity of these fossils, and the best we can say is that they come from the ancestors of perching birds or a related lineage. They remain extremely important as the oldest evidence of anything close to a perching bird. The site they came from has also yielded Australia's first known marsupial (*Djarthia*), bat (one of the world's oldest), frog, and Australia's only known salamanders. The window it provides on the past is priceless.[37]

If we turn to identifiable songbirds, Australia has produced the oldest remains of living genera by an amazing margin. Lyrebirds (*Menura tyawanoides*) and logrunners (*Orthonyx kaldowinyeri*) were scratching aside leaf litter in the early to mid-Miocene (at Riversleigh, the famous fossil site in north-west Queensland), at which time the perching birds in Europe did not belong to living families, much less living genera. As for Africa, the rich deposits of bird bones from Miocene Kenya and Namibia lack songbirds altogether, although some have been found in Morocco. Australia is called the land of marsupials but its claim over songbirds is stronger. Marsupials evolved in South America from mammals whose ancestors came from China.[38]

Ernst Mayr had been captive to a fallacy that goes back to the *Origin of Species*, in which Charles Darwin wrote about the spread southwards of northern plants:

> I suspect that this preponderant migration from the north to the south is due to the greater extent of land in the north, and to the northern forms having existed in their homes in greater numbers, and having consequently been advanced through natural selection and competition to a higher stage of perfection, or domineering power, than the southern forms.

Early Musicians

Because Australia is portrayed as a land that was colonised, the truth about its birds has been difficult for Australians to digest. I don't hear it talked about much, even though any last doubts have been laid to rest. The silence was a motive for my writing this book. Recent textbooks don't help because they are out of date. *The Speciation & Biogeography of Birds*, produced by Academic Press in 2003, mentions very few birds from Australia going anywhere. It cites Mayr, 1944, in support.[39]

If lyrebirds and scrub-birds are half-faithful facsimiles of the first songbirds, as their anatomy suggests, their coming heralded a new dawn for planetary acoustics. Never before had the world thrilled to notes so liquid, pure and powerful. A lyrebird's song can reportedly reach a hillside 3 kilometres away; up close, it hurts the human ear. A startled chick can scream loudly enough to make the nest difficult to approach.[40]

The mimicry of lyrebirds is world-renowned. As well as copying birdsongs, they have broadcast with uncanny accuracy the wing beats of pigeons, the howling of dogs, the nocturnal honks of wandering swans, the pleading of young magpies, the coughing of a smoker, the siren of an ambulance, koalas grunting, trees creaking, parrot feathers rustling, kookaburra bills snapping, cockatoos tearing wood. A lyrebird chased up a tree by dogs barked for three weeks afterwards. Pet lyrebirds have imitated rattling chains, violins, pianos, saws, the creaking of a horse and dray, a child crying, and the screaming of slaughtered pigs. Albert lyrebirds, the most northerly of the two species, can sound like tractors starting up, frogs in chorus, a bird landing with a thump on a branch, and even garbled human conversation, with phrases such as 'Hey Bill' thrown in. Baby lyrebirds start calling when still inside the egg. Tests suggest that grey shrikethrushes are fooled by simulations of their calls. (Because lyrebird stories are so often exaggerated I have taken all my examples from the seven-volume *Handbook of Australian, New Zealand and Antarctic Birds* – the ornithologist's bible.)[41]

Lyrebirds do have their own calls as well, but imitation rules the repertoire, which increases with age. Males sing for hours each day to impress. Females mate with the most skilled male, then rear the chick on their own. One male has been heard singing from dawn to dusk with only a three-hour break in the middle to feed. Lyrebirds learn largely from other lyrebirds, so their sounds survive past their life spans. Thirty years after lyrebirds were taken from Victoria and freed in Tasmania in 1934, their descendants were imitating whipbirds the island does not have, providing a compelling example of culture in birds, of one generation passing knowledge to another.[42]

Scrub-birds, of which there are also just two species, match lyrebirds for gusto. Smaller than starlings, they too are deafening up close, leaving ears ringing. Found in rainforest on ranges running southwest of Brisbane into New South Wales, the rufous scrub-bird will sing for hours in season. Lyrebirds are one subject of its mimicry. This bird is one of the few to occupy Antarctic beech (*Nothofagus*) groves, Australia's oldest surviving forest type.[43]

The second species, the noisy scrub-bird, has a call described as 'astonishingly sharp and far-carrying'. It is a splendid ventriloquist with a sweet, descending crescendo, but is not much of a mimic. I have spent days in wind-buffeted heathlands in Western Australia creeping up on calling males without really seeing one. Instrumental in saving this rare bird from extinction was Prince Phillip, who in the 1960s spoke out against plans to turn its last refuge, Two People's Bay, into a housing estate. Fewer than a hundred individuals then remained. Numbers are higher today, thanks to strict fire control and the founding of new populations, including on an island.[44]

The American philosopher and bird-lover Charles Hartshorne swooned over the lyrebirds he heard as a touring Fulbright scholar in 1952. 'Here is almost a Shakespeare among birds, giving one everything from the clown's laugh (the Kookaburra imitation) to the delicate love song (the Pilot Bird's lyric).' Hartshorne was reputably the first person since Aristotle to know as much about birds as he did about metaphysics. In one article, 'A Foreigner's Impressions of the Lyrebird's Singing', he rated the lyrebird's song as 'one of the great ones'.[45]

'This style is one of boldness, tingling vitality, a challenging, proud, imperious air,' Hartshorne declared. It was more operatic than the nightingale's, though less neat and refined. It was more original and coherent than the mockingbird's, less prone to 'slightly boring' repetitions, but not as mellow. In his book about birdsong, *Born to Sing* (1973), Hartshorne ranked the superb lyrebird first in the world out of 194 'superior singers'. Coming second was the Albert lyrebird, and in seventieth place the nightingale.[46]

In *The Life of Birds* David Attenborough presented a lyrebird right after a nightingale, and the song he praised as possibly the most elaborate, complex and beautiful in the world was not that of the nightingale.

One may wonder if, vocally, songbirds went backwards after the lyrebird. How could evolution create the best musician first, with thousands of younger lineages producing simpler and weaker sounds? I am not saying that lyrebirds themselves date back tens of millions of years, but the arresting similarities between lyrebird and scrub-bird acoustics, despite other differences that justify placement in different families, suggest they had a common ancestor with a commanding voice. This situation is even stranger than it seems, given that lyrebirds and scrub-birds have fewer throat muscles than other songbirds. They have been called 'syringeally primitive'. New Zealand wrens lack intrinsic muscles on the syrinx (vocal organ), suboscines have up to two, lyrebirds and scrub-birds have three, and other songbirds – all 4500 species – have four. Lyrebirds bend their necks to change their resonance. Parrots match them by producing skilful sounds (especially when they learn human talk) from a simple syrinx.[47]

I suspect the earliest songbirds, evolving long ago when there were plenty of birds but no songbirds, took over the airwaves, so to speak. Hartshorne was struck by the broad range of the lyrebird, declaring that the total pitch range 'is one of the greatest of any bird, greater than any song in Europe or North America'. Nightingales and mockingbirds, to avoid competition, often sing at night (hence 'nightingale'). Shakespeare, writing in *The Merchant of Venice*, understood why:

> The nightingale, if she should sing by day,
> When every goose is cackling, would be thought
> No better a musician than the wren.[48]

Lyrebird acoustics would not work in today's tropical rainforest, with all the competing noise. Their closest analogues in Asia, pheasants, rely on coloured trains to impress mates. One night in India I heard the passage of a leopard announced by throngs of wailing peacocks, which, like other pheasants, do not expect their calls to impress. The forests occupied by lyrebirds and scrub-birds lack a wide range of birds, and by calling mainly in winter lyrebirds minimise vocal overlap.[49]

The narrower repertoires of all the other songbirds help them coexist, by reducing vocal interference when numerous species choose to sing at once. With songbirds making up almost half of all bird species, many attempts have been made to explain their global success.

Differences in song promote speciation as well as cohabitation, by helping populations distinguish each other and breed true.[50]

How would the world sound today had songbirds never evolved? Would anything very pleasing rise above the caws and squeaks and rattles in a forest? Probably not. Although budgerigars are said to have a 'warbling song' it's not one that ever moved a poet. There is no music in the quacks of ducks, the arguments of gulls or screeches of owls. Cuckoos and kingfishers are masters of monotony. Insects and frogs can sound pleasing but there is no complexity to their calls, nothing that qualifies as song. Humpback whales do produce 'songs', but these reach our ears sounding like moans. All the sounds that resemble music come from songbirds.

This is one way of saying that human tastes for sound are closer to those of songbirds than to anything else. Darwin grasped this, observing that birds 'have nearly the same taste for the beautiful as we have. This is shewn by our enjoyment of the singing of birds, and by our women, both civilised and savage, decking their heads with borrowed plumes . . .'[51]

Birds have of course influenced human music. Aside from compositions with direct imitations of birds, such as Beethoven's *Pastoral Symphony*, Joseph Haydn's *The Seasons* and Messiaen's *Réveil des Oiseaux* (based almost entirely on bird calls), there is surely another debt that has gone unacknowledged. Mozart was so attached to a pet starling that he conducted a funeral when it died, and eight days later composed a divertimento. One may ask whether human music would have reached the heights it has had that first songbird not sung in an Australian rainforest.

Striking parallels have been drawn between birdsong and music. Patricia Gray and colleagues wrote in *Science* of birds that pitch their songs 'to the same scale as Western music, one possible reason for human attraction to these sounds'. The American wood thrush follows our musical scale 'very accurately', while the interval in a ruby-crowned kinglet's song is often a full octave.[52]

And here is Hartshorne:

> . . . birdsong is recognizably musical by all basic human stand-
> ards. It has nice bits of melody, charming rhythms, even bits of
> harmony (for birds, unlike us, can sing contrasting notes simulta-
> neously); it has obvious examples of theme with variations, neat
> examples of accelerando and rallentando, crescendo and diminu-
> endo, interval inversion, even change of key and tempo contrasts,
> as, for instance, when the same pattern, e.g., a trill, is given at half
> or double speed.[53]

If the comparisons are valid they may tell of birdsong influencing the evolution of human acoustic perception, and in particular our
sense of what sounds pleasing.

Complicated calls and intelligence seem to go together. Of the
three groups of birds that learn rather than inherit their calls – songbirds, parrots and hummingbirds – the first two count as the most
intelligent of birds, some proving better at problem solving than most
mammals.[54]

Intelligence is probably also on show when songbirds forage in
trees and shrubs. Nearly all the birds that search among leaves and
stems for hidden insects are songbirds. Birds of prey, kingfishers and
suboscines may snatch up insects when they spot them, and cuckoos
sometimes search assiduously, but most of the detailed inspections of
foliage are made by songbirds. They have able brains in small packages, and good gripping feet that equip them to search complex
environments for concealed prey. Fossils in Europe show that many
older birds vanished after songbirds arrived, probably because they
could not compete.[55]

Songbird nests also display intelligence. Songbirds and their suboscine kin are almost the only birds to build sophisticated nests, some
of which even have a built roof. Eagle nests look impressive because
they're so big, but a child could lay out sticks in much the same way.

The nests of waterbirds, seabirds (such as they exist), kingfishers, and most other non-passerines are simpler again, with the striking exception of those made by hummingbirds and swifts. Parrot intelligence does not show through in their nests, which seldom amount to more than some chewed wood in a hollow. Perching birds build all the 'best' nests, all those clever baskets and neat cups, with strands of stem and hair and cobweb, disguised at times with dabs of lichen or moss. Cisticolas, once called 'tailor birds', draw leaves together with cobwebs to hide their inner nest. Perching birds freed themselves from the limited supply of tree holes and other cavities, and they are often small enough to hide their nests among leaves.[56]

The Australian Way

In a world now upside down, with the receptacle in the north, Australia's songbirds can no longer be thought of as oddities. If anything, they set the norm. A biologist in Europe or America wanting to understand robins or rooks should keep in mind an Australian origin. The aggression levels are unusual, but the fact remains that Australia has the oldest songbird lineages as well as the wealth of form and behaviour that goes with tens of millions of years of slow and steady diversification. The evolutionary distance between a lyrebird and, say, a currawong, is larger than that between any two songbirds in Europe or the Americas. The difference between lyrebird lyricism and wattlebird rasping provides one measure of distance, and the variety of breeding systems another.[57]

Australia has songbirds that breed in pairs, but it also has the white-winged chough, which comes closer than any non-human (apart from ants) to practising slavery. To raise a chick, a pair of these crow-like birds needs at least two helpers, which are usually their young from earlier broods. What marks them out as unique is that young choughs from nearby groups are sometimes 'kidnapped' to serve as extra helpers. Large groups assail small groups, and during the ensuing melee fledgling birds are lured away by wing-tail displays that command a following response. Kidnap victims are fed and preened and otherwise treated well, but even so, they often wise up a week or two later and return to their former homes or go somewhere new. But in what is a dishonest society, those that remain sometimes fake helping, placing a morsel in a chick's mouth then swallowing it themselves when the parents aren't watching closely.[58]

Songbirds in Europe and North America are comparatively conservative. The male uses spring song to stake his claim, then helps his mate rear their brood. Biologists saw in this the monogamous nuclear family idealised by Western society. Most of the world's mammals, by comparison, are promiscuous.

Australia's birds break every 'rule', and in every possible way. Male bowerbirds do nothing to aid the young they sire, instead pouring all their energy into boudoirs kept for sex. At the other extreme are large miner groups, where many males bring food to one nest. Fairywrens are highly promiscuous, males almost never siring the chicks in their own nests, inseminating instead the faithless wives of rivals. The highest levels of female infidelity ever detected in birds have been in Australian magpies (82 per cent in one population), followed by superb and splendid fairy-wrens.[59]

Cooperative breeding, where three or more birds help with parenting, excited interest in the 1970s when Australia was found to lead the world, in respect of songbirds especially, including some treecreepers, woodswallows, robins, miners and apostlebirds, and non-passerines such as kookaburras and magpie geese. Cooperative breeding is rare in Europe but common in Africa. First noted in Australia in the nineteenth century, the phenomenon lacked credibility until it was documented by an American, Alexander Skutch, in Central America in 1935, then summarised by him in 1961. In the 1960s it fitted the emerging theory of kin selection, of helpers advancing their genes by 'selfishly' feeding their younger brothers and sisters. Skutch thought it was something else – reproductive restraint to help the species – but we know today that animals are not motivated by loyalty to their species. A popular theory today is that helpers lack vacant territories to fill, and by acquiring parenting skills at home they position themselves to claim future vacancies.[60]

This justifies extreme infidelity, since a son that stays home may inherit his father's place, risking inbreeding. Indeed, male fairy-wrens often pair up with their mothers, and daughters with their fathers, but without having sex together.[61]

Why should so much helping happen in Australia? Is it to do with erratic rainfall, poor soils, or a lack of spring flush to deliver insects? Each of these theories assumed that the environment was wanting, but when the evidence was dissected each theory proved wanting. There is a weak relationship between cooperative breeding and variable rainfall, but that's all.[62]

Charles Sibley's new classification turned the thinking around, when someone noticed that group breeders the world over fall mainly into his Corvida, leaving the Passerida dominated overwhelmingly by pair breeders. Cooperation has an evolutionary basis. Pair breeding has no right to be considered the norm. The question went from 'Why do so many Australian birds cooperate?' to 'Why is it uncommon in the Northern Hemisphere?' Or, to quote expert Andrew Cockburn, 'the problem becomes to explain the absence of cooperative breeding rather than its presence'. He went on to call the Florida scrub-jay 'an old Australian bird carrying its cooperative legacy to a new home'.[63]

Group breeders may be less common outside Australia because they were less likely ever to leave it. Their young make poor travellers, since they often stay home to help rather than seeking somewhere new. In pair breeders, not only do the young travel, but two are enough to breed if they arrive somewhere new. Completely promiscuous birds are also held back because they seldom travel in pairs or flocks. The songbirds that left Australia would have been pair breeders.[64]

The birds in northern Europe and northern North America, where most ornithologists live, have unusually narrow habits, limited by severe winters. Territories advertised by song are held only while breeding lasts. Our sense of what the world's birds are like was skewed by these northern birds.[65]

This distortion was so strong that as recently as 1996 we can find an article in the *Journal of Avian Biology* depicting most of the world's birds as unusual. American author Thomas Martin declared, 'Tropical and southern hemisphere birds represent a particularly interesting and apparently paradoxical system for studying life history evolution.' Insights eluded him because his every question was inverted. 'Is food more limiting in the tropics and southern hemisphere?' he asked, not 'Is food especially plentiful in the northern summers?', which it is, since insects peak then. His most prescient point was that 'many general perceptions may be incorrect or misleading'.[66]

The thinking was a little clearer in a 2001 book, in which authors Bridget Stutchbury and Eugene Morton declared that tropical birds set the norm, and that 'song function in temperate zone birds must be viewed as specialized and atypical'. But they failed to distinguish the cold temperate north from the mild temperate south. Australia's temperate zone birds are not, as these biologists wrote, 'constrained by climate to breed in a short time span'.[67]

Songbirds from northern latitudes typically breed while young, their offspring leaving home early and seldom living long, thanks to harsh winters and long migrations. Summer peaks of insects allow for large clutches to compensate for brief lives. Fervent spring song helps pairs time their breeding to the narrow insect peak. Life in Australia is safer and slower. With a mild climate and evergreen trees supporting some insects year-round, most birds face no dire winter hardship. 'Life in the Slow Lane' was the title of a scrub-wren study in 2000. Comparing Australia, Europe and North America, John Woinarski found that Australian songbirds 'are more sedentary, and appear to have smaller clutches, longer incubation periods and fledging periods and longer nesting seasons. Also they may be more likely to live longer and show communal breeding.'[68]

The western wattlebird surprised Gould by laying 'but a single egg' and lacking a regular breeding time. The superb lyrebird can take two months to hatch its one egg – a world record for a songbird. By way of comparison, European starling eggs, which number four to six to a clutch, hatch in eleven to fifteen days. The noisy scrub-bird, a smaller bird, spends more than a month on its one egg. Young choughs are fed for eight months after leaving the nest, as part of a four-year adolescence. Lyrebird males can take nine years to grow complete tails. Birds in Europe and the US are seldom if ever this slow.[69]

Perching birds with eggs and chicks that develop slowly have relatively large brains. Two biologists deduced this after measuring 10000 museum specimens of some 1400 of the world's bird species. Slow breeding probably allows more neurons – and neuronal connections – to form.[70]

Fast-and-fecund and slow-and-steady are strategies that both work. A factory producing a few good shoes can have as much footwear in circulation as one that puts out many cheap shoes. A point to note about songbirds is that some Asian groups of them have spread back into Australia – these include finches, swallows and silvereyes – and they are fast and fecund. Zebra finches can breed at the age of two months and fill a nest with seven eggs.[71]

Cooperative breeding fits life in the slow lane. Young birds have long periods of dependency on their parents, and have to be long-lived to inherit territory vacancies. They acquire social skills while they wait.

Play is something young mammals often engage in, but in birds it is rare and taken as evidence of unusual intelligence. When American biologists Judy Diamond and Alan Bond gathered together every example of play in birds that is known about, their global list of ten songbirds included three from Australia: the chough, apostlebird and magpie. These are Australia's three largest cooperatively breeding songbirds (although most magpies breed in pairs). Magpies are very playful. To quote one biologist, they 'roll around, breast bump each other, pull on wings, run after each other, hide, crouch, engage each other with legs, play fight, peck, grasp, jump and even play hide-and-seek'. These are exceptional behaviours for birds. Choughs can use tools, pounding old mussel shells against live ones to open them, a sign of intelligence. Overlooked by Diamond and Bond were chestnut-crowned babblers, which, according to one observer, 'spend much time in play, becoming very excited, and making much noise and chasing each other in a follow-the-leader game'. By racing around one shrub again and again, they formed a circular runway more than a metre across and averaging 16 centimetres wide and 9 centimetres deep. Babblers too are cooperative breeders.[72]

Helpers at the nest often assist territorial defence. Group breeders can afford to be noisy and assertive, there being more beaks and claws to win an argument, providing another reason why Australia can sound abrasive. Gould found choughs to be 'harsh, grating, disagreeable and tart'. Their closest relatives, apostlebirds, sound like 'tearing sandpaper'. These two birds often roam together in very vociferous groups. Such birds have no need for spring singing because there are no pair bonds to build or reaffirm.[73]

Apostlebirds are so named because they are said to roam in parties of twelve, while another name for them, CWA birds, alludes to them chattering like a Country Women's Association meeting. They are a favourite of mine because they are so intelligent and curious and plucky. In her poem about these 'self possessed and clannish' birds Judith Wright declared them rude to strangers.[74]

Smaller and less assertive, Australian babblers go by such names as cackler, chatterer, cur-cur, barker, dog bird, chatterer, yahoo, fussy, happy jack, happy family and the twelve apostles – which allude to their rowdy calls or to their gregarious ways.[75]

Some years ago in Central America I was having daily confrontations with big brown birds in groups, brown jays. Whenever they saw me they would fly up and complain. Their calls have been described as explosive and obnoxious, with a loud screaming 'often repeated mercilessly'. They seemed so 'Australian' I guessed they must be cooperative breeders. I looked them up back home to find they were the very birds that convinced Skutch that cooperative breeding existed. Group breeding, strident cries and feistiness often go together, as noisy miners show.[76]

Two biologists who confronted thirty Australian songbirds with an owl, a predator, found that all of them voiced alarm at a lower pitch than European songbirds. High frequencies hide the caller's location. In Africa I saw an owl mobbed by small birds whose cries were so highly pitched I could barely hear them, as if the volume had been turned right down. Australia is never like that. Aggrieved birds, especially group breeders, let themselves be heard by making obvious noises when they are riled.[77]

The sizes of the birds contribute to this. Many visitors to Australia are surprised by all the big birds they see. I've been surprised myself after long trips abroad. Small birds are by no means lacking, but it's the large birds that are prominent, in suburbia especially, where plentiful food invites the big to expel the small and meek. Australia might be the smallest continent but it has thirty-two heavy songbirds (weighing 120 grams or more), compared to eleven in Europe and seventeen in North America (not counting Mexico and Central America). All the European heavyweights fit into one family (of crows and jays, the Corvidae), and the American birds into two (Corvidae, Icteridae). Australia's big songbirds, which include lyrebirds, bowerbirds, honeyeaters, figbirds, choughs, currawongs and birds of paradise, extend across nine families.[78]

Some 20 million years ago at Riversleigh, Australia had a songbird (*Corvitalusoides grandiculus*) bigger than any on earth today, apart from lyrebirds and the very largest ravens. Riversleigh had a lyrebird as well, and at least one more big songbird (weighing about 180 grams).[79]

The Australian magpie, much heavier than its northern namesake, is a striking example of a large songbird. In surveys, 85 per cent of

Australians admit to attacks from this domineering bird. So often are mailmen harassed (dive-bombed up to 200 times a day) that Australia Post has funded research hoping to find out why. In 2008 an elderly woman in New South Wales had her leg amputated after a jab from a magpie turned gangrenous. Horses, sheep, dogs and possums sometimes lose eyes to the stiletto beaks; koalas have been blinded, lambs killed.[80]

Magpies can distinguish kindly adults from scheming boys, a skill that fosters trust when they visit gardens for handouts of meat. Some of the best-fed birds show high cholesterol levels. Their confidence finds favour in a country founded on egalitarianism, on a convict disdain for subservience, for if any Australian animal is beholden to none, it is this one. Naturalist George Bennett in 1856 rose to language no marsupial ever inspired: 'It is a bird of much importance in its own estimation, struts about quite fearless of danger, and evinces, on many occasions, great bravery.'[81]

In New Zealand, where they were imported in the nineteenth century, magpies are notorious for harming other birds. The forty-odd species they attack include parrots, herons, harriers, seagulls, pheasants, chickens, ducks, and several threatened species. Small birds are often killed. They also attack hedgehogs, rabbits, possums, cats, dogs, sheep, cars and model planes. Attempts have been made to eradicate them from parts of New Zealand.[82]

In Australia magpies often join noisy miner attacks, injecting even more violence. With choughs they have a testy relationship, frequently driving them from the pastures they both feed on, and initiating fierce battles when their competitors refuse to leave. Choughs huddle into what has been described as 'a screaming piebald mass of black and white with a dozen or more crimson eyes, gaping bills and flashing white wing patches' – a response given specifically to magpies. Choughs, too, often fight among themselves, leading at times to eggs being pushed from nests, and nests from trees. Biologist Robert Heinsohn followed one destabilised group that 'went on a rampage', ruining the nesting attempts of six other groups before settling down to breed, hostility yielding to parental affection.[83]

Extreme behaviour in birds is more likely in Australia than anywhere else because its songbirds have been diversifying for so long. As Les Christidis put it to me, 'You can have a picnic at Ferntree Gully in Melbourne and see a lyrebird, a treecreeper and a bowerbird, and what you are seeing is something like 50 million years of bird evolution from your picnic table.' Magpies, wattlebirds, bowerbirds and lyrebirds represent four different outcomes when, given enough time, songbirds reach great size.[84]

Australia has fewer species of songbirds than many places but far more genetic breadth, ensuring large contrasts, with small, milder-mannered songbirds living beside big combative ones. English settlers, noticing more screeches than lyrebird refrains, romanticised what they missed. Of Australia's many charming songsters, ranging from dainty heathwrens and whistlers to larger butcherbirds and magpies, the rufous whistler charms me the most, because in spring, males on my forest block vie to produce the longest possible improvised whistle, a sublime song that goes on and on and on. Ecologist Hugh Ford chose this very bird as one whose breeding behaviour 'would be unremarkable in North America or Europe'. Pioneers settling in Brisbane called it the 'robin redbreast'. As for the Australian magpie, it is almost the only large songbird to offer pleasing songs as well as harsh attack calls. Kept around homesteads to announce intruders, it provided, in effect, the warble of the canary and the bark of the dog.[85]

Ernst Mayr never accepted he was wrong about Australia's songbirds, although he did praise Sibley for ushering in a new era in bird classification. But northern prejudices are falling now, and Australia is the fulcrum on which the new thinking turns.

Four

New Guinea: Australia's Northern Province

Just north of Australia lies a country better known for feathers than for anything else. Papua New Guinea is the land of birds of paradise, and of men who wear their plumes and demonstrate in a shimmering, quivering, prancing sort of way how deeply shared, by birds and people, a sense of beauty can be. The plumes adorning the male birds can look like skirts, ribbons, fans, streamers, fur, whips, even lace. They show how strongly female choice can drive evolution, and so do the extreme poses the males strike, which often seem impossible for anything made to fly. But there is no sensuality to complement their beauty, for as dancers they are jerky, monotonous, and too obviously keen to copulate.

New Guinea is central to this story: the island is not so much a neighbour of Australia as a core part of it, biologically speaking. To include Tasmania but not New Guinea is to let nationalism distort ecological thinking. Whenever sea levels have fallen New Guinea has joined the mainland for longer than Tasmania has, because the water that separates it is shallower. The line between northern Australian savanna and rainforest is more limiting to birds than is the Torres Strait. New Guinea's birds are part of the Australian bird fauna.[1]

This can be difficult to digest since 'Australia' has two poorly distinguished meanings: it is a nation when it votes, but a biological realm when it produces songbirds. That realm includes Papua (claimed as a province by Indonesia and once called Irian Jaya) and Papua New Guinea. The word 'Australia' serves well enough when deserts are the topic, but not when Australia is called the home of marsupials or the land of parrots and kookaburras, and New Guinea is left out. 'Australian rainforest' assumes a powerful meaning when the latter is included. I cannot keep talking about Australian birds without including New Guinea in the picture. Birds of paradise provide an opportunity to show why.

The Cape York rainforests harbour a bird darkly iridescent with a resonant call, the trumpet manucode. With riflebirds, this is one of the four birds of paradise on Australian territory. The nearest birds of paradise to Sydney – paradise riflebirds – are a mere 200 kilometres away in Barrington Tops. From the main street of Brisbane the same birds can be reached in fifty minutes by driving to Mt Nebo, and on the Gold Coast they are closer again. Clad in velvety black with a glittering turquoise chest and crown, a courting male sways in a taut, muscular way, flicking and weaving powerful wings like oriental fans. These birds of paradise, neither secretive nor rare, attract little attention today, although their feathers reached Europe in bygone times as part of the plumage trade. They were also shot as orchard pests.

The more spectacular birds of paradise in New Guinea sparked a sensation in Europe centuries before New Guinea mattered in any other way. The first skins arrived in 1522 with the expedition of Ferdinand Magellan, a gift from the Sultan of Tidore to the King of Spain. Others returned on Spanish, Dutch and Portuguese spice ships.[2]

The impossible splendour of the first skins – which had no legs attached – inspired tales about heavenly residence. Of hens drifting towards the sun sipping ethereal dew while cocks served as floating nests for the eggs. A later portrayal in a bestselling book had them living among spice trees, such that 'the groves which produce the richest spices produce the finest birds also'. They were appropriated into paintings of the Garden of Eden, the first accurate depiction of one appearing in a sixteenth-century prayer book. New Guinea hunters, by supplying the birds, were influencing the religious imagination of Europe long before Europe imposed its religion on New Guinea.[3]

In a trade that began before the time of Christ, paradise feathers found their way to China, India, Arabia, Nepal, Europe and North America. By Nepal's kings and generals they are still worn at special events. Western women once vied to enhance their beauty by flaunting the plumes on their hats. One court dress in Europe was reportedly trimmed with 150 Queensland riflebird breasts, before their export was banned in 1913. In Dutch New Guinea the plume trade continued until 1931.[4]

In New Guinea itself they became role models on a scale unmatched anywhere else, teaching men that male success came from displaying the best feathers with the most vitality. At one ceremony attended by David Attenborough the 'relatively fresh' feathers worn by 500 men showed that 'at least 10,000 birds had been slaughtered to provide the headdresses of that one dance'.[5]

Alfred Russel Wallace made his way to New Guinea in 1855, eager to apply a scientific eye to these birds. Sick at the time, and facing danger, he saw just two of the forty or so species in the wild, and one of these only poorly. His interest soon turned to acquiring skins to sell. Upon returning to Singapore he obtained at great cost a live bird, whose contortions back in England became a public spectacle.

In an article for the Zoological Society of London, Wallace vividly evoked their world:

> Nature seems to have taken every precaution that these, her choicest treasures, may not lose value by being too easily obtained. First we find an open, harbourless, inhospitable coast, exposed to the full swell of the Pacific Ocean; next, a rugged and mountainous country, covered with dense forests, offering in its swamps and

precipices and serrated ridges an almost impassable barrier to the central regions; and lastly, a race of the most savage and ruthless character, in the very lowest stage of civilization. In such a country and among such a people are found these wonderful productions of nature. In those trackless wilds do they display that exquisite beauty and that marvellous development of plumage, calculated to excite admiration and astonishment among the most civilized and most intellectual races of man . . . [6]

That picture of inhospitality endures to this day, inhibiting study of these birds. Much that is known about them is second-hand. Many specimens in northern museums came from Europe's millinery markets. The most detailed field study was undertaken in an odd location – Little Tobago Island in the Caribbean, after English bird-lover Sir William Ingram, fearing the plume trade would drive them extinct, purchased the island as a sanctuary for forty-seven greater birds of paradise that he freed there. More than sixty years later, in 1965, an American student showed commendable diligence by spending every day for nine months observing their dwindling descendants, which numbered seven. By 1979 one remained, a spirited male who performed before oropendolas, tropical American birds with similar colours and calls.[7]

To rebut critics, Charles Darwin had to explain how evolution could possibly produce the gaudy plumes of birds of paradise and pheasants. He answered that drab females drive evolution by choosing extreme outfits: 'As any fleeting fashion in dress comes to be admired by man, so with birds a change of almost any kind in the structure or colouring of the feathers in the male appears to have been admired by the female.'[8]

This works because birds of paradise are promiscuous, females obtaining nothing more from their mates than insemination. A gaudy father bringing breakfast to the nest might alert predators, but nothing this risky happens.

Jared Diamond has offered one sketch from New Guinea that suggests not survival of the fittest, but something from a fable: 'Male astrapias in convulsive flight, dragging a tail that is up to 1 m long, look as if they are about to plummet to the ground at each wing beat.' The white ribbons of these birds of paradise represent the longest tails, proportionally, of any bird on earth.[9]

One might expect such eye-catching birds to be secretive, in order to survive, but they are often noisy and obvious, even near villages. The most common large songbirds in New Guinea, they survived the plume trade because females and young males are too plain to attract a hunter's arrow. Many a showy male disports himself in full view of a main road, something I have seen for myself. High lead levels detected in their bodies in one 1992 study were blamed on exhaust fumes.[10]

They are not surviving everywhere. Around many villages they are eaten, and Papua a growing illegal trade has fuelled fears about their future. According to a 2001 *Jakarta Post* story, those involved include soldiers, police, government officials, professional bird traders and locals. Because they have guns, the police and military are often part of poaching rings. The animal-protection group ProFauna Indonesia cited one police officer reported for possessing three sacks of stuffed parrots and birds of paradise. The *Jakarta Post* quoted someone saying, 'Every time a navy ship leaves Irian Jaya, it is like a zoo of birds inside.'[11]

Plentiful food explains how the males can while their days away with display. It allows bowerbirds in the same forests to obsess over boudoir maintenance and display. South America's luridly coloured cotingas and manakins are also thought to benefit from abundant fruit. Biologist David Pearson compared tropical rainforest birds on four continents and found that New Guinea had the most birds specialising on fruit. This food not only provides 'much leisure time' (Diamond again) but allows for female success as sole parents. Some fruits seem to exist for these birds of paradise alone. The rich pulp ripens inside stiff capsules that only they can breach, as if the trees, needing their seeds spread widely, have evolved specific relationships with birds of paradise. American biologists Bruce Beehler and John Dumbacher asked why:

> What peculiarity of the Papuan region might permit the evolution of plants that rely on birds of paradise for seed dispersal? We believe the absence of most arboreal placental mammals that elsewhere serve as seed predators may have opened the door to the evolution of a foraging/dispersal system that in other regions would be evolutionarily precarious.[12]

Birds of paradise are products of forests lacking apes, monkeys, civets and squirrels. Stiff capsules in Asia are levered open by fingers, paws and teeth instead. Rather than lacking feet, as Europe once thought, these birds have very large ones, used at times to pin capsules against perches to open them, something few birds can do. New Guinea is rich in marsupials but, surprisingly, none specialises on fruit.[13]

Regrettably, the island is at serious risk of being overrun by feral monkeys. Long-tailed macaques are entering Papua as pets, brought in by Javan workers and migrants, and some have won their freedom near Jayapura. If monkeys spread across New Guinea they will change it forever. As well as fruits and seeds they eat lizards, crabs – and importantly – the eggs and chicks of birds. Taken to Mauritius hundreds of years ago, these same monkeys contributed to the dodo's demise. In Palau, another place of introduction, they reach densities of a hundred per square kilometre. With good reason they appear on the IUCN's list of '100 of the World's Worst Invasive Alien Species'. I have warned about them on national radio, since they could one day reach Queensland's northernmost islands, although my main concern is for New Guinea itself. Pleas to the Australian government to offer assistance for their control have gone nowhere. No amount of plume hunting could match the harm monkeys will do to New Guinea. Palm civets freed on the nearby Aru and Maluku islands pose another high risk.[14]

Fossil expert Mike Archer has told me that Australia once had two families of possums (pilkipildrids and miralinids) whose teeth indicate a possible diet of fruit. But they disappeared back in the Miocene, as Australia's climate dried and the rainforest declined. Had they survived in New Guinea it might have proved less congenial for birds of paradise as well as for feral monkeys.

Rainforest Finds a New Home

One of the many surprises to arise from recent DNA work is that the genetic distances between birds of paradise genera are relatively large, implying great age. One molecular clock suggests the ancestral species arose some 25 million years ago, well before the New Guinea landscape existed. There may have been some low islands in its vicinity but the cordillera on which these birds abound only began to rise 6–10 or possibly 14 million years ago, depending on the geologist you believe. The lower figures are more popular.[15]

New Guinea is in fact younger than much of its wildlife. The clock date given for birds of paradise is likely to be wrong, but not enough to alter the conclusion. These birds probably originated somewhere else. Their nearest relatives are New Guinea's melampittas and Australian choughs and apostlebirds, drab birds one and all.[16]

Australia 40 million years ago was strongly dominated by rainforest. The continent slowly dried as it wandered north, 7–8 centimetres a year, into the dry middle latitudes, forcing rainforest into a long retreat towards the coasts, until, by a couple of million years ago, when the Pleistocene ice ages had become a new force, it survived only in patches near the east coast and more marginally in the north. These became far too small, especially during ice age peaks, to conserve a full complement of rainforest birds.[17]

But there was a saving event – a collision with the Pacific tectonic plate. A giant 'bruise' erupted along Australia's leading edge, a cordillera that rose up 5000 metres. On high slopes under banks of mist the altitude answered for latitude by providing the equable wet climate that Australia elsewhere was losing. Dick Schodde proposed that the cordillera 'probably came just in time' to serve as a refuge for Australian rainforest birds. Birds of paradise are one group he believes relocated north.[18]

Everyone does not think like this. Often New Guinea is taken at face value, as having acquired its wildlife *in situ*. Schodde decided otherwise from work he did in the 1970s:

> We found that the birds and mammals of the montane New Guinean rainforests had their closest affinities not with those in the tropical rainforests of lowland New Guinea and Cape York Peninsula, but with others further south in Australia, from the Atherton Tablelands southwards . . . And so we interpreted the faunas of these forests as old, relictual and ancestral to the bird and mammal faunas that had radiated in Australia's scleromorphic flora.[19]

Fossils could confirm if birds of paradise originated further south, but bird bones are so light, to suit flight, they seldom survive as fossils. If we turn to mammals, their fossils support the idea of New Guinea as an Australian refuge. In the Mt Etna caves near Rockhampton, Scott Hocknull of the Queensland Museum has found remains, less than 300000 years old, of tree kangaroos, cuscuses, ringtail possums, striped possums and rodents, closely matching species found today in montane New Guinea. These mammals might have originated in New Guinea and spread south, but that can't be true of New Guinea's forest wallabies (*Dorcopsis*), known from remains found in Victoria. Tim Flannery has highlighted 'striking similarities' between 4.5-million-year-old Victorian fossils and New Guinea's mammals.[20]

By generously shedding leaves and pollen, trees leave far more fossils than birds or mammals, and these show that New Guinea served as a major refuge. It is exceptionally well endowed with southern beeches (*Nothofagus* species), trees that clothed Australia when it was younger and wetter. They belong to a group (subgenus *Brassopora*) that grew as far afield as Antarctica and South America, but which survives today only in New Guinea and New Caledonia. The variety of beech capsules in New Guinea compares with that in 30-million-year-old Tasmanian fossils, leading to the suggestion of 'an ancient and distant origin'.[21]

Conifers tell the same story. Celery-top pines (*Phyllocladus*) survive in New Guinea and Tasmania but nowhere in between. Other conifers surviving from Mesozoic times persist in Australia only in upland New Guinea. *Dacrydium* lasted in Queensland until 20 000 years ago, vanishing during the last and harshest ice age.[22]

While Queensland and New Guinea both carry lush rainforest, their differences are stark. The largest area of rainforest on the mainland today, the Wet Tropics of north Queensland, has 112 bird species, while New Guinea has more species than that (165) recorded from one square kilometre of rainforest. One-tenth the size of the mainland, New Guinea has nearly as many birds, more than 700. Dick Schodde suspects that the mainland's current bird fauna, reduced by aridity, may be little more than half its past size.[23]

The fossils at Mt Etna imply this. By having more similarity to mammals in New Guinea than to those of north Queensland they show that extinctions must have occurred in the Wet Tropics as well as Mt Etna. Plant fossils argue for this too. We can suppose that many of the bird groups confined to New Guinea today were once found further south, something also indicated by bird DNA.

Refuges are usually portrayed as old and stable, but New Guinea is somewhere new and unstable where old things survive.

What the Birds Reveal

When it fluffs up to impress, a male crested satinbird could be wearing a bright orange wig on its back. Keith Barker's pivotal DNA study in 2004 showed that satinbirds are not the birds of paradise they were thought to be, but a distinct family of three species confined to montane New Guinea. Berrypeckers and longbills stand out as another family (Melanocharitidae) unique to the island. Molecular studies show both families to be old.[24]

A paper appearing late in 2013 helped clarify their positions. They are sisters to the Passerida, the world's largest grouping of songbirds. Along with Australia's robins, slightly higher up the tree, satinbirds and berrypeckers can be called basal Passerida. They are very distant cousins of the sparrow. My encounters with satinbirds and other oddities – ifritas, melampittas, painted berrypeckers – in the vicinity of ancient conifers (*Phyllocladus, Dacrycarpus, Dacrydium, Podocarpus, Papuacedrus*) fit the idea of a walk up a New Guinea slope being 'like a walk back into time'.[25]

That these lineages are old is beyond dispute, and the only alternative to a mainland origin is one on islands that formed in the vicinity of New Guinea and were then absorbed into it as Australia pushed north. A few geneticists believe this, hence the claim that 'proto-Papua' provided a major songbird radiation that 'subsequently invaded all other continents'.[26]

But the islands that were not absorbed – New Britain, New Ireland and the Solomons – are remarkably poor in relict birds, a 2014 paper noted, undermining the notion of an island radiation. The mainland, being older, more stable and much larger, is far more likely to have been the evolutionary crucible. New Guinea probably spawned young groups such as fantails but not older ones.[27]

Take orioles. In the time of Aristotle, golden orioles were thought to issue out of funeral pyres, but today's thinking is that orioles came out of Australia. They are the only songbirds in Europe in a genus 'made in Australia', all of the others belonging to genera that evolved after an ancestor left Australia. The evidence about their past fits together neatly. A 2010 DNA study found that Australia has the world's oldest oriole lineage. The only fossil species has been found in Australia, at Riversleigh. The oriole family has three genera, all of which are found in Australia, compared to only one (or none) on other continents. New Guinea is the world's oriole centre, the only part of Australia with all three, although the fossil (*Longimornis*), which dates back more than 15 million years, implicates the mainland in their early evolution.[28]

Quail-thrushes are birds of dry places, of spinifex and saltbush, mallee and mulga, with one species using woodland near the coast: that at least is the depiction in Australian books, the version I grew up with. But New Guinea has a quail-thrush inside rainforest. I was transfixed when one fluttered down noisily near me, paced over wet litter, then paused on a buttressed root. It was like seeing a lyrebird in a desert – the setting seemed wrong. That bird looked nearly the same as the cinnamon quail-thrushes I have seen camouflaged against hot inland sand, bearing similar colours and facial markings, although it was darker. Its sister position on a genetic tree encouraged scientists to conclude of quail-thrushes that 'ancestral forms that inhabited tropical rainforest or moister Australian forests' gave rise to inland species. If the molecular dating can be trusted, quail-thrushes go back more than 30 million years, to long before serious aridity struck Australia. They went from rainforest to desert without a change of clothes, so to speak. Where birds of paradise have embraced visual novelty, quail-thrushes are plumage conservatives.[29]

It yaps like a dog, but the rufous-bellied kookaburra is a genuine kookaburra (genus *Dacelo*), confined to New Guinea. Much smaller than its famous relative, it is darker as well, to suit its rainforest domicile. Its habitat has been described as unusual, but rainforest was surely the first kookaburra home. It is the only habitat used at times by all four species, three of which occur in New Guinea, two nowhere else. Kookaburra calls often feature in Hollywood jungle scenes, in anything from *Tarzan* to *Raiders of the Lost Ark*, with some validity. These birds are kingfishers, which in the tropics live mainly in rainforest, often along rivers. The popular species in England, the original 'kingfisher', itself qualifies as a rainforest bird, having a vast range that extends all the way to New Guinea. At the edge of tall Malaysian jungle I watched one snatch up small fish sheltering at a river's margin to escape the strong flow. Most kingfishers are tropical and this species probably originated near the equator before spreading to Europe.

More than a few birds from open places have close relatives inside New Guinea rainforest – fairy-wrens, babblers, woodswallows, sittellas, and the well-known magpie-lark. In a dirt-floor school in the Central Highlands I paged through my field guide and watched the fingers of lively Grade 2 boys (no girls were enrolled in that grade) converge on the torrent-lark plate, with the enthusiasm children further south might show for the magpie-lark or 'peewee', its close relative. One lives beside rainforest streams, the other on sparsely grassed flats, well out into the outback. Ernst Mayr evinced surprise at Australia's savannas and semi-deserts having birds whose nearest relatives were 'in the steaming lowland or the misty mountain forest of New Guinea'. Queensland rainforest lacks torrent-larks today, but what of the past? As for fairy-wrens, New Guinea's emperor fairy-wren is the sister to all the rest.[30]

The New Guinea connection deepens when we compare related genera, denoting relationships further back in time. The rock warbler of the sandstone terrain around Sydney, in its own genus, has New Guinea mouse-warblers as its closest relatives. Like the Wollemi pine and southern leaf-tailed gecko, its ancestors survived the demise of rainforest by associating with sandstone; I watched one carry a feather to its bulky nest in a recess under a boulder. Inland Australia's crested bellbird also has its nearest relative in another genus in New Guinea, this being a vivid green bird found up to altitudes of 3600 metres, the rufous-naped whistler. A close relationship means having an ancestor in common, which, given the youthfulness of the arid zone, presumably occupied rainforest. Species that stayed inside rainforest probably changed less over time than those that moved out, since their world changed less, hence the notion of rainforest as a refuge. Mouse-warblers may have pecked at the feet of Wollemi pines in gullies where Sydney now sprawls, while rainforest in Queensland probably had kookaburras, quail-thrushes, and a range of birds of paradise, although I may be mistaken about this. Fossils that prove the point may materialise one day.[31]

Placing New Guinea

New Guinea was included in the Australian region by Alfred Russel Wallace, the father of biogeography, and by the great Baldwin Spencer, who in 1896 carved Australia into three bioregions: the arid Eyrean, southern Bassian and northern Torresian. His Torresian region reached from northern New South Wales to northern New Guinea. His divisions remain in use today, but regrettably New Guinea is often left out.[32]

One reason for this was a growing belief that New Guinea's wildlife evolved in isolation from Australia's, to produce a twin fauna from the same ancestral stock. But recent DNA studies, not just on

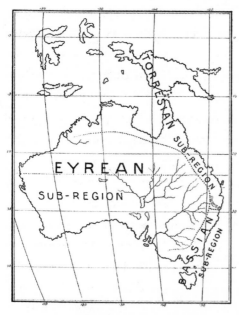

FAUNAL SUB-REGIONS OF THE AUSTRALIAN REGION.

One of Australia's leading 19th-century scientists, Baldwin Spencer, was in no doubt that New Guinea was biologically part of Australia, and he produced this map for an important report published in 1896. It implies that northern Australia has more in common with New Guinea than with the rest of Australia.

birds but on gliders, rats, snakes, fish, crayfish and plants, show multiple links between north and south, as if the sea were never there. A major paper on honeyeaters and their allies found no evidence for separate radiations of New Guinean and Australian members. Another study, on whistlers and their kin, offered this take-home message: 'The emerging picture is one where Australia and New Guinea share a single avifauna with complex connections that increasingly involve the dry central areas of Australia rather than the east-coast rainforests.'[33]

There was a sea barrier for much of the past, but it proved less of an obstacle than the cordillera. New Guinea's southern rivers share sixty-one freshwater fish species with the mainland, which is far more than they share with the island's northern rivers, a situation mirrored by many birds. Northern New Guinea, for instance, has a different cassowary from the one shared by southern New Guinea and Queensland.[34]

When precision is needed in discussions about Australia and Papua New Guinea, biologists variously say Australia-New Guinea, Australo-Papua, Meganesia, Sahul, or Greater Australia. Richard Dawkins was so underwhelmed by these choices he concocted his own word, 'Australinea'. By their very number these names show that none has caught on, that something crucial has been missed. From here on, when I use 'Australia' in this book New Guinea is included. Every biologist I have asked about this has agreed, often with great emphasis, that New Guinea *is* Australia.[35]

Tom Heinsohn and Geoffrey Hope describe New Guinea as 'an integral part of the Australian continent', one that forms its 'youngest geological province'. Talk of a province frees the mind from the shallow water that currently separates it. Biogeography should reflect deep time, not a recent sea level rise. Heinsohn and Hope point out that from 100,000 years to about 11,000 years ago it would have been possible to hike 5000 kilometres from northern New Guinea all the way to southern Tasmania.'[36]

New Guinea is most obviously Australian in the drier woodlands found on some of its southern foothills and plains. The Sogeri Plateau near Port Moresby, when rainbow lorikeets shout from blue gums (*Eucalyptus tereticornis*) rising above kangaroo grass (*Themeda triandra*), looks just like the Gold Coast hinterland. The New Guinea woodlands have magpies, wedge-tailed eagles, brolgas, boobooks, corellas. They have wallabies, taipans, death adders and frilled lizards as well.[37]

The Cape York rainforests are the flipside of these, dominated by New Guinea birds. These include eclectus parrots, palm cockatoos, manucodes and, eyeing you with curiosity, white-faced robins, living among cuscuses and green tree pythons. Like the Mt Etna fossils, they tell of times when rainforest ran south from New Guinea without hindrance. Everyone interprets them that way. Recent spread south was possible sometime between 80000–120000 years ago, and perhaps less than 10000 years ago. At such times there must have been more species than survive today in the Cape's patchy rainforests.[38]

Did Papuan hornbills once spread their hefty wings above the canopy? They colonised the Solomon Islands, which had no past bridges to New Guinea. In his book *Life in the Cape York Rainforest*, Robert Heinsohn argued that these huge, long distance fliers 'must have penetrated into northern Australia'. Their Queensland reign would have ended a few thousand years ago, when a drying climate shrank the scrubs below sizes that could support such needy birds. White-faced robins need less forest to survive. Heinsohn wondered why Queensland lacks king birds of paradise – resplendently red birds with ridiculous tails – and twelve-wired birds of paradise, when each resides 'only a stone's throw away' in New Guinea. His surmise – that the fruit supply is inadequate – might not have applied a few thousand years ago when the rainfall was slightly higher.[39]

About Queensland's tropical rainforests much gets said, but very little about there being the two kinds, one flavoured by New Guinea and the other, the Wet Tropics, boasting upland birds and mammals found nowhere else (although often with sister species found on New Guinea mountains). The division shows that for rainforest birds in the recent past, spread south from New Guinea was easier than spread north from the Wet Tropics. In other words, Torres Strait proved less of an obstacle than the woodlands just north of the Wet Tropics.

Australian Birds shown in New Ways

When New Guinea is included, Australia's songbirds reveal themselves in new ways. One of them, the hooded pitohui, is orange and black like a giant wasp, a likeness it carries through by being poisonous. Hunters complain that its meat burns the throat. Sneezes and watery eyes are the reward for handling one. Walter Boles acquired a headache by inhaling an old museum specimen, something he warned me not to try.[40]

A penchant for toxic beetles may explain how several New Guinea birds, not just hooded pitohuis, have chemicals in their flesh like those in the poison arrow frogs of Amazonia. The high doses in breast and belly feathers may help protect eggs in nests from snakes and rats. The hooded pitohui is in the same family as orioles, some of which, in Asia, have the same bold black head and wings (but not the orange), making it possible that a warning pattern developed in Australia survived as a uniform thousands of kilometres away.[41]

The visual mimicry on show in New Guinea is also unusual. Young brown orioles disport themselves like helmeted friarbirds, adopting similar poses, movements and calls, substituting black cheek feathers for friarbird black skin. Being large honeyeaters, friarbirds are apt to jab at strange faces in busy trees, so natural selection has given orioles a demeanour that invites respect.[42]

There are also roaming flocks of big brown songbirds, including rusty pitohuis, cuckoo-shrikes and bird of paradise females, which look similar and forage together. Uniformity is thought to improve group responses (hence uniforms on soldiers). An attacking goshawk stands out as the only non-brown bird in these flocks, which are led by babblers – or rather, the babblers are followed by the others. As noisy group breeders, they suit this role. Queensland may have had pitohuis and large brown flocks living among those 'New Guinea' marsupials at Mt Etna.[43]

If we look past what is different about New Guinea – the climate, the mountains, the people – there is much that matches. The same bird groups prevail: parrots, pigeons, honeyeaters, robins, whistlers. The honeyeaters display the versatility I mentioned before. MacGregor's honeyeater – long thought to be a bird of paradise until its DNA revealed it to be a honeyeater, the world's largest – lives on the highest peaks eating *Dacrycarpus* fruit. There is aggression. Studying nectar birds in rainforests, an American biologist found the 'quiet, even orderly, aspect of group feeding' in Panama to pale against the 'veritable riots of inter-individual aggression' in New Guinea. Colourful language like this often comes up in Australia.[44]

New Guinea's songbirds are often large. Melidectes honeyeaters are twice the size of nectar-feeding spiderhunters, and birds of paradise tend to be larger than fruit-eating songbirds in Asia, where mammals eat much of the fruit. I have birded in montane moss forest in Malaysia six weeks after visiting New Guinea, and the bird sizes seemed very different. Small insectivores proved as plentiful in outer tree foliage in Malaysia as they were scarce in New Guinea. They are scarce as well in Australia's northern savannas, a situation blamed on honeyeater aggression.[45]

New Guinea is part of the Australian story. Torres Strait is not a reason to doubt this, when Bass Strait is wider and deeper. Nor is the fact that New Guinea is mostly wet where the mainland is dry – this only makes it truer to Australia's rainforest past. Narratives about a drying continent leave out this part of the story. New Guinea is now the main home of Australia's rainforest birds, the place where they retain an evolutionary future, the front from which future invasions of the world will occur. So small and impoverished are the mainland rainforests that they are almost evolutionary dead ends. New Guinea is made different by its people, but anthropological lines should not define biological realms. The social differences in any case are less than they seem, and many anthropologists talk about northern and southern Sahul rather than New Guinea and Australia, having decided that their peoples are more similar than different. As for understanding Australia's birds, we can't get far if we leave half of them out.[46]

Five

Land of Parrots

The nineteenth-century Englishman, contemplating life in Australia, had books he could turn to for advice. One that was popular, *Sidney's Australian Hand-book: How to Settle and Succeed in Australia* (1848), must have given readers pause over this: 'The woods swarm with birds good for the pie and spit, especially parrots, pigeons and black cocka-toos . . .'

Parrots and cockatoos were exotic pets back then – kept by gentlemen in parlours and by publicans in inns – not ingredients in pies. In Sidney's day, thousands of parrots were hawked each year on London streets, six shillings buying a cheap cockatoo. The wili-est dealers – the bird-duffers – touched up their wares, daubing paint on plumage and varnish on beaks and legs to contrive what nature could not. Her Majesty's sailors were permitted to bring back parrots from abroad, since teaching them words lessened shipboard bore-dom. Good talkers fetched high prices. The reformist writer Henry Mayhew learned all of this by roaming London streets befriending and interviewing the working poor.[1]

At a time when inns and coffee houses had parrots in place of the television sets in pubs today, the prospect of eating the entertainment must have struck an odd chord. So would the notion of living among such birds, yet Australians do just that. The average suburb has more parrot diversity than most tropical countries. New South Wales, with thirty-three species, is almost as rich in parrots as all of Africa (twenty-two species) and mainland Asia (fourteen) combined. If you include its forested fringe, Sydney boasts more parrot species than any country on those two continents – apart from Indonesia, whose islands gained parrots from Australia in the past.[2]

Animals such as rats succeed in cities by living discreetly, but that hardly applies to parrots. With their noise and colour they intrude more on Australian lives than most animals. Tourists from abroad soon notice lorikeet screams or cockatoos leering from trees. An American article in *Living Bird* magazine, 'The Land Down Under', opened by recounting the 'awful racket' coming from rainbow lorikeets outside a hotel window.[3]

These birds help Australia sound harsh: no parrot was ever accused of bursting into song, and an aggrieved cockatoo makes one of the most abrasive animal sounds there is. A flock of them creates a cacophony, as naturalist Price Fletcher found when he reached a remote waterhole in 1878: 'What a babel of tongues, what incessant screeching, what a whirling, flying moving mass of noise; 50,000 cockatoos all screaming at once! Just for one moment try and realize it, reader, and you will involuntarily put your fingers up to stop your ears.'[4]

Gould was awed by lorikeets: 'The incessant din produced by their thousand voices, and the screaming notes they emit when a flock of either species simultaneously leave the trees for some other part of the forest, is not easily described, and must be seen and heard to be fully comprehended.'

In modern times, biologist James Shields has told of the screaming flock of sulphur-crested cockatoos (white cockatoos) that for three deafening hours followed him over 5 kilometres of country, ruining the wildlife survey a colleague and he were conducting. Few animals ever draw attention to themselves with such commitment.[5]

The beaks of parrots are relevant here. The two muscles that work them (Musculus ethmomandibularis and M. pseudomasseter) are unique to parrots. The lower jaw is deep and strong, the tongue muscular, and the upper jaw hinged flexibly to a thick skull, allowing for some independent movement. The head is large. Parrots are the only birds that can really crush and chew.[6]

A pigeon taking grain just pecks and swallows. Its bill resembles a set of forceps or tongs. A pelican swallowing a fish, or a kookaburra a snake, again bring tongs to mind. The light bones needed for flight preclude most birds from having crushing jaws. Parrots are one of two exceptions (finches are the other), and their mouths are like Swiss Army knives – used at times to crack kernels, husk seeds, pulp fruit, tear wood, hold nuts and climb inside trees. Corellas dig, and New Zealand has a parrot that stabs sheep.

Parrots can damage other birds, and it helps to have this known. When cockatoos come screeching into trees the message sounds like 'Don't mess with us.' Sulphur-crested cockatoos can crush olive stones, while palm cockatoos do to pandanus nuts what people need axes to do. Australia's peregrine falcons outdo those on other continents by having especially big feet and a massive beak – for subduing biting parrots.[7]

Noisy miners are sparing in their assaults on parrots, even when both birds take the same foods. In a land of bickering songbirds, parrots, except the very small ones, can hold their own. 'Extraordinarily pugnacious' is one description of the rainbow lorikeet, the parrot that most imposes on noisy miners by eating the same foods. While honeyeater calls are loud so as to protect resources, those of large parrots probably help signal the harm they can do. Small parrots with weak beaks sound meek.[8]

More than most birds, parrots have a distinctive bauplan (body plan), one characterised not just by the beak but by the feet, with two toes running forwards and two back, equating to four fingers and four thumbs. Almost alone among birds, they can hold what they eat. When they grasp nuts and fruits to nibble they become the squirrels and monkeys of the bird world. This helps explain why Australia is a land of parrots – because a land of squirrels and primates it is not. Alfred Russel Wallace noted that the global overlap between parrot beaks and squirrel teeth is minimal. New Guinea has more than fifty parrot species (with twenty-nine of those recorded in a single location) and no squirrels or monkeys, while the Malay Peninsula has twenty-five squirrels, seven monkeys and just five parrots. Squirrels and monkeys are usually easier to find than parrots in Malaysia.[9]

Parrots are unusual among the world's seed-eating birds in often taking nectar. A majority of Australian parrots eat nectar; in fact they are second only to honeyeaters as consumers of sugary foods. They have broader tongues than other birds, which, when brushy-tipped, make good mops for dabbing up nectar. Their slicing beaks are perfectly shaped for scraping lerp from leaves, which they can hold up while they feed. Australia's parrots fall into two families – cockatoos, and all other parrots – and the second of these is the largest bird family in Australia after the honeyeaters. That the two largest families are the main nectar birds is another indication that sugar plays a special role in Australia.[10]

Eucalypts are pivotal to parrot success, offering lodgings as well as food, namely the holes most parrots are reared in. Although eucalypt wood is hard, Australia has termites that eat heartwood softened by fungal attack. A storm snaps off a limb, rain soaks into the base, fungus grows, termites arrive, holes form. Fires play a part too, by providing the entry points termites need. Termites and fires have helped Australia become a land of parrots.[11]

Termites outside Australia seldom produce hollows, and when eucalypts are grown abroad they do not form them. In Zimbabwe I walked along the trunk of a giant fallen eucalypt that lacked even the smallest recess, while on the River Murray, by way of contrast, I once found a giant river red gum grove with seven parrot species loitering around the tree holes. In branches that groaned in the wind like creaking doors a red-rumped parrot chick reached out of one hollow during a feeding bout, while corellas in the next tree were annoying regent parrots by peering down their holes. Eucalypts are often hundreds of years old before deep hollows form, and the demise of large trees in farmland raises concerns about future parrot success. That said, parrots sometimes do without trees. In Canberra, en route to a ministerial advisory meeting, I've seen red-rumps enter building vents at the Department of Foreign Affairs.

Aptitude with Attitude

Parrots have unusually good brains to work their jaws, tongues and feet. An African grey parrot named Alex, trained by Massachusetts professor Irene Pepperberg, was a star performer in experiments, doing much to undermine the assumption that mammals are more advanced than birds. He could reply in English when shown mixed items and asked, 'How many blue?' He could identify substances (paper, wood, cork), objects (keys, toy trucks), and knew more than a hundred words. Numerically he rated near chimpanzees and very young children. He talked to himself:

'Snap, snap, snap, snap. How many? Four.'

'You be good, gonna go eat lunch, I'll be back tomorrow.'[12]

Alex rated a *New York Times* obituary when he died in 2007. Another parrot made world news that same year with the 'first genuine evidence of animal dancing'. A sulphur-crested cockatoo handed in to a pet shelter in Indiana amazed neuroscientist Aniruddh Patel by showing that 'a nonhuman animal can synchronize to a musical beat'. Snowball became a YouTube sensation, attracting 200000 hits in one week. TV spots and advertising contracts followed. Snowball bobbing and swinging his feet to the Backstreet Boys is something to see, although as Patel noted, he showed limited reliability and tempo flexibility.[13]

Parrot playfulness says something about their minds. While many young mammals engage in play, it is known in birds only among parrots, songbirds and hornbills. Galahs will slide down and swing beneath telegraph wires, and keas skid repeatedly down tents. In his book about galahs – a serious work with statistics and graphs – CSIRO researcher Ian Rowley admitted: 'Galahs give the impression of enjoying their lives to a much greater extent than most other animals.' (I should mention that galahs, like corellas and cockatiels, are cockatoos in all but name.) A cockatoo pet of Konrad Lorenz, the famed Austrian ethologist, played such tricks as biting the buttons off Lorenz's father's trousers while he slept, and flying round and round a tree unravelling balls of wool. 'Our visitors used to stand in mute astonishment before this tree,' Lorenz recalled, 'and were unable to understand how and why it had been thus decorated.'[14]

The smartest parrots and songbirds outdo apes, dolphins and elephants on some tasks. To quote one neurobiologist, they exhibit 'cultural transmission of tool design, theory of mind, and Piagetian object permanence to a high level'. In relative brain size and intelligence they match apes, leaving pigeons, chickens and quail more at the level of rats and mice.[15]

As I have mentioned, parrots are the only birds apart from songbirds and hummingbirds that learn part of their calls. The sounds of all other birds are innate. From what we currently know, a capacity for vocal learning has evolved eight times, the other occasions being among humans, elephants, seals, dolphins and whales, and some bats.[16]

The gulf between people and birds is more than 300 million years deep, that being the period elapsed since our last common ancestor was alive – an amniote in the Carboniferous, which lived long before the first hint of a dinosaur. Humans are the most intelligent species to arise among mammals, and a parrot may be our avian equivalent.[17]

The slur 'birdbrain' dates back to nineteenth-century anatomist Ludwig Edinger, the father of comparative neurology. Finding them to lack a neocortex, the seat of mammal intelligence, Edinger never considered that bird intelligence might reside somewhere else. He had no reason to, for men of his era believed that nature, like society, had a hierarchy, going from birds to mammals to reach a pinnacle in man. Edinger coined terms for brain regions that entrenched the notion of birds as instinctive, not thinking, creatures. New names were the goal of a 2002 avian brain forum at Duke University, which over three days brought world experts together, at first to argue, then agree. This talk-fest advanced bird rights by bolstering the case that birds have good enough brains to feel misery and pain.[18]

Australia's largest parrots – palm cockatoos and black-cockatoos – bear confinement poorly and have not been tested as Alex was. But Queensland's palm cockatoos are renowned for making tools, something few birds can do. Small branches are severed and trimmed of bark to make drumsticks, with which the males, to proclaim their territories, proceed from tree to hollow tree pounding out messages up to a hundred beats long. They stamp their feet, tap their bills, and also drum with nuts. To house their massive jaws they have much bigger heads than other Australian parrots, providing space for lucid brains. Big nuts, by requiring giant jaws to open them, may have promoted their intelligence: the thick kernels presumably evolved to deter cockatoos, which could mean the birds drove the evolution of their own brains by attacking ever larger nuts. Parrots having bigger brains, relatively, than most birds, their jaws are implicated in two ways: they need good brains to operate their sophisticated mandibles, and large skulls for the jaws to attach to.[19]

Australia's budgerigars and cockatiels aren't renowned for intelligence but they are studied in universities the world over for the simple reason that they are cheap and easy to keep. Among the most interesting investigations carried out are on their calls. Each individual budgerigar knows not only the cry of its mate but the sound of its group when flocks combine. Parrots are garrulous because everyone talks to strengthen group membership. Calls converge as familiarity grows. Three budgies housed together will adopt a common contact call that only they know. After drenching outback rains produce a bounty of native grain, budgerigars coalesce into flocks so tight they can cast vast shadows. Branches collapse under their weight. By hearing extremely well in the 2–4 kHz range, pairs within flocks of thousands can keep together. A bird that gets separated searches for the dialect of its subgroup. These skills, probably common to all parrots, are keenly developed in budgerigars since their flocks are the largest. The words learned by pet parrots are the shared calls of the 'flock' they form with their owners.[20]

Whether parrot colours denote intelligence I cannot say, having found no critique of their purpose beyond the finding that red flags fitness. Nesting in dark holes, parrots are less likely than most birds to betray their young to enemies, a circumstance that freed up the parrot palette, something Wallace wrote about. Blocks of colour in certain places may help tight flocks coordinate rapid travel, and also promote social harmony through pleasure, much like colours on human clothes. Parrots are more social than many birds, forming groups that include unrelated birds, and parrot sociality is probably something else that promoted their intelligence. All their colours are not, however, ours to see. Courting budgerigars have a UV-reflecting cheek patch bounded by fluorescent yellow. To mark changes in mood, including anger over invaded personal space, other parrots can have facial skin that flushes red, or crests that flash colour when raised. These signals reduce the incidence of harmful biting.[21]

Parrots have a discerning palette, with somewhere between 300 and 400 taste buds, more than chickens (250–350) and many more than pigeons (37–75) and bullfinches (46), though far fewer than humans (9000) and catfish (100000).[22]

Parrots as Problems

The intelligence of parrots makes them formidable as pests. Indeed, Australia might also be called the land of parrot problems. The first crops grown in mainland Australia, by Sydney's convict settlers, were raided by king parrots and rosellas. Sulphur-crested cockatoos earned ire as crop marauders far and wide.[23]

'Picture to yourself a white tablecloth thrown over a newly-planted field of wheat,' suggested Henry Berkeley Jones in 1853, 'and you see it as it appears with a flock of these devourers upon it.' Another flock on a crop was compared to a snowdrift. Cockatoos were difficult to thwart as they seemed so cunning. Biologists talk about how, when a flock feeds on the ground, 'one or more birds are deployed on sentry duty in nearby trees'. Early traveller Godfrey Mundy found them to be so wary 'as to preserve an absolute impunity from gun or snare'.[24]

The problems they pose have never abated, a parliamentary inquiry in Victoria in 1995 hearing a torrent of complaints about ravaged nut crops, oranges dismembered for their pips, and potatoes and newly sown grain dug up and eaten. Some farmers who planted sunflower seeds found they were only feeding birds. To make matters worse, parrots, like humans, can be wasteful, often ruining more than they eat.[25]

Often, too, their damage seems wanton. Holes dug in bowling greens, race tracks, golf courses and tennis courts make sense when you know that cockatoos like to chew grass roots on dry days. But what about snipping off and dropping bunches of grapes, decapitating flowers in flower farms, ringbarking almond trees by the hundred, pulling nails from roofs, breaking lights at ovals, 'vandalising' verandahs and picnic tables, blacking out houses by severing cables, and tearing holes in walls? A fire was blamed on a corella chewing a cable, while a winemaker lost 800 out of 1400 newly planted vines. These were some of the problems conveyed to the Victorian inquiry. On revegetation plots, cockatoos hinder efforts to arrest salinity and over-clearing by uprooting seedlings. They roost near houses and cry out far too early in the morning and late at night, in Canberra ruining the sleep of senior public servants. To stop them breaking expensive light bulbs on the giant spire of the Melbourne Arts Centre, a major city attraction, a trained eagle and falcon were brought in.[26]

Conflicts with people may go back tens of thousands of years. George Caley, who became Australia's first resident white naturalist when he arrived in Sydney in 1800, reported that cockatoos cut off branches of geebungs (*Persoonia*), a plant with juicy edible fruits, 'to the great injury and vexation of the natives'.[27]

Modern parrot damage flows largely from two needs: brains that need engaging and beaks that need honing. With crops providing easy food, cockatoos have long days to while away, and exploring roofs and farm infrastructure gives them something to do. 'Cockatoos may be attracted by newly-dug soil, or may investigate where they have seen people working,' the Victorian inquiry concluded. 'Whatever the motivation, cockatoos visit newly-planted trees and either pull them out of the ground or bite them off.' Cockatoos only harm seedlings that are planted, not those that grow naturally. They don't eat them, they just pull them out.[28]

Holes appear in houses when a nearby resident doles out seed, a problem readily resolved by withdrawing the food. An acquaintance in Canberra said she stopped feeding cockatoos when they started to 'eat' her trampoline. The temptation to pull out loose roofing nails can be removed by putting in screws instead. Some problems are harder to solve. Suggestions by the Victorian government include running powerlines underground and carrying a shotgun in the golf buggy. When all else fails, the government's 'trapping and gassing team' takes out problem birds. Many cockatoos are killed by farmers and government agents around Australia.

Bell miners don't kill trees directly, but cockatoos do. I've seen the results of their work in the Tolga Scrub, a pocket of endangered Mabi rainforest in north Queensland, where sulphur-crested cockatoos roost after feeding in nearby fields. Their fights over sleeping arrangements were deafening. When they flew in one evening I saw that all the level branches were claimed first, by birds spaced out individually, each one beyond biting range of the next. Softening this scene were a few pairs roosting and grooming together. Newcomers kept aiming for spaces between perching birds, only to be repelled by screeches on both sides. Every so often a bird would fly at another and displace it as neatly as a ball in lawn bowls replaces another. Affronted cries carried on hours into the night. The flying foxes roosting on branches below them were far more subdued. The feuding seemed pointless when nearby trees remained empty – I knew they had been used in the past since, like many of the trees in contention, they were dead.

Returning at dawn I saw why. Parrot bills need trimming by constant use. Crops offer such soft easy fare that bills go without sufficient wear. I watched as tufts of leaves were shorn away and beaks honed on branches. These trees faced death by thousands of cuts. So serious is this problem that the Atherton shire has called for cockatoo culls to save the rainforest. Of the many threats faced by rainforest, parrot beaks may be the strangest. Cockatoos are probably the only birds in the world that kill trees by direct action. (Waterbirds sometimes kill roost trees with their excrement.) In southern Queensland, on long stretches of the Condamine River, they attracted blame for eucalypt deaths during drought in the 1990s. On farms, valued paddock trees die when galahs gnawing bark around their nest holes ringbark whole trees. In Victoria blue gum plantations are destroyed when cockatoos tear open saplings for witchetty grubs. The destruction is often greatest where food is plentiful, but forest loss concentrates birds into a small number of trees. The Tolga birds were targeting peanuts left in furrows after harvest. They floated past in a cloud so dense they hid the hills behind.[29]

New Zealand has big green parrots, keas, in her mountains which create their own category of problems. One stole my soap when I left my backpack beside a high trail. Other travellers have come off far worse, with windscreen wipers torn from cars, and motorbike seats vandalised. One Scottish tourist lost his passport to a kea that rifled through the luggage on a bus. In a separate incident, another Scottish tourist had $1100 stolen from the dashboard of his campervan.[30]

One problem seems scarcely believable – kea attacks on sheep, but these acts have been caught on film. A kea will land on its victim's rump, pluck away the wool, then stab with its great mandible to reach the flesh or kidneys. Most victims are disabled or dying sheep, and most keas do not partake in such attacks.

Testimonials published over a century ago make compelling reading:

> One day I came suddenly upon two or three keas, busy picking at the loin of what I supposed to be a dead sheep. There was a hole right through the sheep's back, and the birds were putting their heads right through to the inside of the sheep and pulling out portions of the intestine, but I cannot say if they ate them or not. I then went over, and to my surprise I found that the sheep was not dead . . .[31]

And this: 'A snow-slip carried some sheep with it. I found the sheep stuck in the snow, where it landed, still alive, with its hind leg eaten to the bone, and half a dozen keas tearing away at him.'

New Zealand has sheep-killing parrots. Keas are one of the world's most intelligent birds, possessing far more acumen than their victims, which sometimes leap over precipices in wild bids to escape stabbing beaks.

Shorn sheep escaped injury as their bare rumps offered no purchase. Under a bounty scheme to protect mountain flocks, 150000 keas were culled, and by the time this stopped in the 1970s keas were rare. They remain an occasional problem, resolved by relocating rather than killing the scarce remaining birds. Sheep attacks probably arose from a habit of scavenging dead animals, including moas killed by giant eagles, birds that met their end when the Maori arrived.[32]

Parrots as Pets

Parrots evoke such strong emotional responses that they are as loved as pets as they are deplored as pests. Their capacity for language, along with their intelligence and colour, gives them unique advantages as companion animals. Anthropologist Patricia Anderson noted the comfort of having a pet that yells, 'Daddy's home!' (without understanding the words). In any pet shop, parrots have far and away the best animal minds. They are the only birds to offer affection, and almost the only birds to enjoy being touched. An early naturalist, K.H. Bennett, had a free-ranging Major Mitchell's cockatoo that, on hearing him arrive home after nine months away, flew in from outside and settled on his shoulder, 'rubbing his head against my cheek and all the while emitting a low gurgling note expressive of gratification'. The larger parrots easily outperform dogs and cats on problem-solving tests. They are the birds most like humans: intelligent, talkative, playful, curious, dexterous, sociable, argumentative, and open to learning all through their long lives. Comparing them to children, Anderson noted that they are sometimes called 'fids' (feathered kids). I am put in my place when a tame cockatoo sidles up, gives me a knowing eye, and tries out its 'Hello, Cocky' routine.[33]

Known before the time of Aristotle, talking parrots influenced the human relationship with nature by coming closer than anything else to talking with us. 'Australian' parrots were part of this engagement well before Europe discovered Australia, because cockatoos, millions of years ago, spread to the Spice Islands, and trading ships, from the fifteenth century onwards, conveyed them to Europe. Maluku – to give the Moluccas or Spice Islands their modern name – is biologically an outpost of Australia, one that provided Europe with its first taste of Australia. Like birds of paradise, but while still alive, parrots became regular trade items. The word 'cockatoo' comes from the Malay language, reflecting the fact that many cockatoos in Mayhew's London came from the Moluccas. In New Guinea Wallace saw traders from Indonesia procuring vast numbers of parrots, some of which probably ended their lives in Europe.[34]

Parrots became one of Sydney's earliest exports. Gould saw draymen and stock keepers carting galahs to sell, and it was he who introduced the budgerigar to the world, taking some back to England in 1840. Years later he was proud to see 2000 budgies chattering in an English dealer's room. So easy were they to breed that by 1859 they cost less to buy in London than in Sydney. A yellow colour form was bred in Belgium by 1872.

Today they qualify as the world's most popular cage bird. The World Budgerigar Organisation boasts 120 member bodies on four continents. England has groups for particular breeds: the Lutino and Albino Budgerigar Society, the Spangled Budgerigar Breeders Association, and so on. The colour forms they dote on go by such names as slates, fallows, lacewings, spangle cinnamons and Texas clearbodies. Budgerigars have everything parrots can offer – intelligence, engagement, words and colour – in a package small enough to exclude grating screeches. Proud owners even describe the warbling they produce as song.[35]

The other Australian parrots to achieve global popularity are cockatiels and cockatoos. But as you might expect, parrots are the birds most imperilled today by the pet trade, the birds that fetch smugglers the highest prices. Indonesia's cockatoos have fared badly ever since *Baretta*, a 1970s TV show with a pet cockatoo, drove up demand in America. Smuggling is rife in Australia, according to an exposé on the ABC's *Background Briefing,* but the parrots of interest are foreign imports or stolen aviary stock, rather than wild-born Australian birds.[36]

The caging of intelligent birds raises ethical as well as conservation issues. Anyone who has seen gossiping budgerigar flocks in riverside trees has to wonder about the lone bird in its indoor cage. One vet noted: 'Most bird cages on the market are designed with the comfort of the owner, not the bird, in mind.' In Florida, in a little cage above the entrance to a mall, I saw a macaw, its life's purpose to provide colour for passing shoppers. In Sulawesi, where trapping is rampant, I watched a newly caught lorikeet throw itself against its tiny cage again and again, unreconciled to its fate. The species it belonged to was one I searched for in the wild in vain.[37]

Writing in the nineteenth century, naturalist Gracius Broinowski felt nothing for the caged galah: 'as a pet it is stupid and uninterest-ing, possessing neither docility nor intelligence, is most destructive in its habits, tearing and gnawing everything within its reach, and has a most unpleasant odour, while its screaming propensity increases in captivity, unless a pair is kept, in which case it is found that conjugal companionship has a soothing effect'.[38]

Children were taught to perceive parrots as stupid. An 1838 book of instruction, *The Rudiments of Knowledge,* offered this:

The Cockatoo

There is a bird of plumage rare,
In gilded cage exposed to view,
Procured with cost, preserved with care –
We mean the *pretty cockatoo.*

He is a foreign bird of fame,
And talks as parrots often do;
For if we ask him what's his name,
He'll say 'tis *pretty cockatoo.*

Yet in these words repeated o'er,
Does all this scholar's wisdom lie;
For to a thousand questions more
He only gives the same reply

If asked who made his gilded cage,
Or who his masters' portrait drew,
Who was in Greece the wisest sage?
He'll say 'twas *pretty cockatoo*.

Everyone did not think like this. When Madame Marzella's parrot troupe reached Australia in 1903 audiences found themselves entranced by cockatoo feats that included somersaulting, waltzing, bicycling, playing dead, and spelling names by picking out letters.[39]

Trees that Store Seed

More than most birds, parrots face divided futures, with some species thriving while others slip away. The only bird lost from mainland Australia since 1788 was a parrot, and two more – the orange-bellied and swift parrot – seem likely to join it. Parrot futures were something to ponder during the 2006 Melbourne Commonwealth Games, since the games mascot, Karak, was modelled on an endangered subspecies of red-tailed black-cockatoo. 'Karak' is a colloquial name taken from their call, much simpler than their proper name, so I will use it here.

Karaks are bound to one substrate – a vast sand sheet in western Victoria and nearby South Australia that formed when beach sand blew inland many thousands of years ago. These birds only have three foods – the tiny seeds of two stringybarks (*Eucalyptus arenacea* and *E. baxteri*) that favour this barren domain, and those of a she-oak called buloke found just to the north. One stringybark flowers every two to four years, the other every three years, and only when one follows the other do karaks have food enough to breed well. If this sounds precarious, that's because it is. Other subspecies of this cockatoo have many foods.[40]

Stringybark seeds remain on the trees for a couple of years, in capsules that become harder to breach with age, obliging karaks to spend all day feeding. Buloke trees provide better fare, but few remain today as they grow in rich soils valued for grain. Biologist Richard Hill, who showed me these birds to help publicise their plight, discovered their simple diet.

How could such a specialisation arise? The answer has to do with fire. Instead of shedding seeds when these ripen, stringybarks keep most of theirs in their crowns, awaiting fire, a practice called serotiny. Only when flames have turned the thick undergrowth and litter into fertile ash do their seeds have much chance of sprouting. Although a smouldering strip of stringybark captured by an updraft can spark a fire miles away, fires are local enough that karaks can always find unburnt areas to feed in. Their diet has been viable in the past because they live among trees that store seed. It is because their habitat has dwindled that fewer than a thousand of these birds now remain.

Mainland Australia is exceptionally flammable. It is dry and flat with few impediments to fire. To conserve scarce nutrients in the highly infertile soils, plants produce long-lasting leaves, protected from herbivores by high levels of fibre and often tainted with aromatic oils and phenolics. These defences are cheap for plants to produce because the key ingredient is carbon fixed during photosynthesis, rather than, as in most toxins, scarce nitrogen extracted from the soil. The oils and lignin in these sclerophyll plants, as they are called, burn readily, resulting in leaves that are often flammable when green. Orians and Milewski stressed that plants are flammable where soils are poor.[41]

Plants living where fire is inevitable but unpredictable often practise serotiny. Australia has more plants that store seeds than anywhere else. The flowers liked by birds – those of eucalypts, banksias, waratahs, paperbarks, hakeas, bottlebrushes – often proceed to pods that hold seeds until fires release them, sometimes by melting the binding resin. Banksia seeds often last ten and sometimes as many as seventeen years encased in wood. Eucalypts shed and replace seeds more often, with or without fire. Serotiny also shows up in South Africa and in some Northern Hemisphere pines, where poor soils and fierce fires go together.[42]

Serotiny provides remarkable food security for some parrots. Besides all the sugar in trees there are seeds as well, which only parrot beaks can reach. The storing of seed is especially prevalent in the south-west of Australia, where the hot dry summers guarantee periodic fires. Gum nuts almost as big as egg cups hang like Christmas baubles from the dark crowns of marri (*Corymbia calophylla*), a dominant eucalypt. With their long slender bill tips, Baudin's black-cockatoos and red-capped parrots tease out the seeds, their main food (see colour section). Red-caps look like circus clowns, with purple vests, lime cheeks, and other colours that don't quite match. The local red-tailed black-cockatoos also target marri, while Carnaby's black-cockatoos feed more widely, choosing banksia and hakea seeds as well, and the grubs found inside these. Flocks of 200 cockatoos can meet their daily needs from marri in an hour. Ringneck parrots feed below them on what they drop.[43]

As well as these parrots, Western Australia now has corellas from eastern Australia, escaped pets which have learnt to lift up ripening capsules with their heads held back and mouths opened to catch the falling seeds. Marris cast shade that for eucalypts is remarkably dark, and have seeds that are for eucalypts unusually large (1.5 centimetres long). To help the seedlings grow in shade, seed starch replaces sunshine to fuel early growth. The shade cast by marris is ultimately responsible for the parrots they attract, since food that is meant to help a seedling grow in shade often feeds a parrot instead. Cockatoos like marri nectar, and tear away the bark to get at grubs. This keystone tree (it provides lerp as well) was a major target of woodchippers

before a ban was imposed in the 1990s following a concerted conservation campaign. A recent concern is the Forests Products Commission's
poisoning of thousands of marris to make room for jarrah, a better
timber tree.[44]

Stored seeds are exceptionally nutritious, so as to help seedlings
get started in destitute soils. Their protein content can be 40–60 per
cent, two to three times that of soybeans. Because the soil is so much
poorer, Australia's serotinous seeds have almost five times the phosphorus of Africa's serotinous seeds, and they are high in iron as well.[45]

South Australia's Kangaroo Island has a bird with the narrowest
diet possible. She-oak seeds make up its every meal, and the island
only has one she-oak species. The glossy black-cockatoo spends up
to 60 per cent of each day extracting minute seeds from a hundred
or so cones. It is left-footed, holding every cone with the same limb.
The seeds look too small for such hefty birds but they are nutritious.
These birds vanished from mainland South Australia long ago, many
a mighty she-oak going to fuel paddle-steamers on the River Murray.
'Glossies' remain widespread in the eastern states, where most populations have two or three serotinous she-oak species to exploit. The
fossil record tells of more she-oaks in the past, which helps explain
how such specialised birds evolved.[46]

Serotiny has its own architecture. Conscripted into souvenirs and
bush art, the weird capsules, pods and 'woody pears' have shaped
aesthetic perceptions of the Australian bush. In May Gibbs' stories,
Snugglepot and Cuddlepie hide inside marri nuts while banksia men
plot through serotinous lips. Gibbs the child wandered among these
trees. Their pods show that seeds are much easier to protect from fire
than from birds: the thickest capsules are those targeted by parrots,
not those facing the greatest fire risk.[47]

The nexus between parrot, seed and fire is old. A seed valve missing from a 40-million-year-old banksia cone from Western Australia
persuaded two scientists to posit that a prehistoric parrot tore it off.
I doubted this until my attempt to remove a valve from a fresh cone

ended in its complete destruction. The valves are too firmly attached to come off by decay.[48]

Serotiny is all but lacking in northern Australia, where grasses that rise high in the wet season fuel regular fires in the dry, ignited by lightning or by people, providing regular openings for germination. There are fewer parrot species, all with broad diets. The most obvious serotinous plant is a small tree (*Grevillea glauca*), whose big nuts are the ones used by palm cockatoos for percussion.[49]

The Northern Hemisphere pines that store seed are grown in Australia, and yellow-tailed black-cockatoos have learned to exploit them. Landholders are bulldozing stringybarks to grow Monterey pines, a shift from one serotinous tree to another that lets one black-cockatoo replace another. No karak would look to a pine tree for food, but yellow-tailed blacks are more open about what they eat, and they have learned that pine cones conceal rich seeds. They have learned this in many places, and pine plantations now feed these birds across eastern Australia, supporting flocks up to 1500 strong, something no forest could do today. Whenever I see black-cockatoos floating over fields I can usually spy pines nearby. These birds float as much as fly, remaining aloft with little effort, as if they were big black butterflies.[50]

In Western Australia foreign pines now feed Carnaby's black-cockatoos, which have lost most of their habitat – partly to pines. Like Juan Fernandez firecrowns, they are endangered birds that rely today on foreign trees. Harvesting of pines is listed as a threat to their survival.[51]

On Eyre Peninsula in South Australia the last few yellow-tailed black-cockatoos (down to nine in 2008) depend on another exotic seed-storer, the aleppo pine. One large pine yields a thousand times the seeds of a native hakea. Pines are invading rare woodland remnants, their aromatic needles stifling the growth of rare plants, bringing bird and plant conservation into conflict. Monterey pines are also becoming weeds in karak country. In their native lands pine cones are attacked mainly by squirrels, with small birds, crossbills, extracting some seeds as well.[52]

Australia often has parrots in place of mammals, even when seeds are not the rewards. Long-billed corellas have drooping bills like witches' noses and bare skin around each eye suggesting a mask, making them look almost sinister. Their 'noses' are long because, like many mammals, they grub up juicy roots – or used to. They once ate murnong, a daisy whose tubers sustained Aboriginal people, until the sheep of the squatters grubbed it up, bringing 'deprivation, abuses and miseries', as one missionary wrote. Two ways of life folded at once. The Aborigines turned to handouts of flour, the birds to weeds and fields of ripening grain. Tubers are not typical bird fare. Orians and Milewski observed that instead of the digging rodents found on other continents, Australia has corellas 'with pickaxe-like beaks'.[53]

Other mammal replacements are New Guinea's pygmy parrots, which, like some squirrels, eat lichens and fungi growing on bark, and the New Zealand kakapo, the world's largest parrot, which walks out at night to nibble grass, stems, fruits, seeds, and the roots it digs up with its bill. It has a good sense of smell. It cannot fly.[54]

I said before that parrots thrive away from squirrels and primates, and this raises the question of why marsupials eat so few seeds. Rodents target seeds, but they reached Australia too late – just over 5 million years ago – to hinder parrot success. South America once had rodent-like gnawing marsupials (the argyrolagids), and Australia, up to a couple of million years ago, had the ektopodontids, which, expert Mike Archer told me, had 'remarkably rodent-like molars'. But they were never common and went extinct as Australia dried.[55]

Keas and kakapos did well in New Zealand, a land devoid of mammals. I assume that parrots command certain foods in Australia because they turned to them first, or harvested them with more efficiency than marsupials. This was possible since parrots are another group with an Australian signature.

Southern Origins

Scientists are so fond of the world's parrots that five big DNA studies were published between 2007 and 2011, and these revealed the existence of three groups – separate families – of very different sizes. Three large parrots in New Zealand – the kea, kaka and kakapo – are very different genetically from all the rest, showing that their line (family Strigopidae) diverged first. The other parrots on earth fall into two groups: the cockatoos of Australia and nearby lands (Cacatuidae), and everything else (Psittacidae). The two larger families have their deepest roots in Australia, indicating that as well as songbirds, Australia gave the world parrots.[56]

Gene trees show New Guinea's vulturine parrot to be the sister of Africa's grey parrots, and of macaws and all the other parrots in the Americas, whose forebears could have crossed Antarctica before it froze. South America became a second land of parrots, ending up with more species than Australia but far less variation in lifestyle

Because they had a head start in Australia parrots have succeeded in almost every habitat. Rock parrots can make a living on beaches, saltmarshes and other coastal habitats. They nest on the ground, sometimes under rocks on flat ground.

and form. Parrots have left Australia at least a dozen times. New Guinea tiger parrots are sisters to African lovebirds, Pacific lories, Australian lorikeets, rosellas, budgerigars, and many more. These vivid green birds add to the evidence that New Guinea is Australia's main refuge.[57]

Parrots and songbirds are far and away the world's most intelligent birds, and Australia gave the world both. Why should this be?

A possible answer came in a 2006 DNA study which indicated that parrots and perching birds – the order that includes songbirds – are close relatives. A larger study published in *Science* two years later took this further, concluding that parrots and perching birds are each other's closest relatives. In academic corridors the world over, everyone was caught out, because nothing about the build of these birds suggests a relationship. Their differences include perching birds often defending territories, which parrots rarely do, and making more complex nests. Charles Sibley's early work placed parrots with pigeons, while perching birds had been aligned, on evidence that was always thin, with anything from cuckoos to woodpeckers.[58]

Genetic techniques often don't work well at revealing extremely old relationships, but three genetic tests done more recently have shown the same link between parrots and perching birds, including one that used completely different genetic markers, so the relationship seems to be real, even though a couple of studies failed to find it.[59]

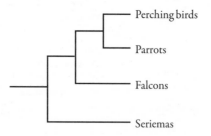

Perching birds evidently have parrots as their closest relatives, followed by falcons and seriemas. DNA analyses point to an origin in Gondwana for them all.

It shows that one early bird gave rise to four living groups, the seriemas (two large South American ground birds), falcons, parrots and perching birds. The last two have the closest relationship, which means they shared an ancestor more recently. Falcons turn out to be like owls, in having come to hooked beaks and talons independently of hawks and eagles, which no longer qualify as their relatives. Asked about this, falcon expert Penny Olsen told me that parrots and falcons smell similar ('slightly nutty'), and that hawks and eagles stand out by squirting rather than dropping their faeces. An American who at seven museums examined 4500 parrots and falcons found a unique pattern of moult in their wings, a sequence of feather loss and renewal shown by no other birds. They also share hooked beaks, and some fossil birds identified as early parrots (the Eocene messelasturids) were birds of prey, judging by their capable beaks and talons.[60]

All this physical evidence would fit the genetics better if parrots leaned towards perching birds as well as falcons. What does suit is anatomical evidence that perching birds had as their closest relatives the zygodactylids, small extinct birds whose feet approached those of parrots. What fits as well is a genetic tree pointing to South America as the source for the world's falcons.[61]

Parrots and perching birds are physically very different, chalk to cheese, with aptitude their most obvious shared attribute. Is this a plesiomorphy – a trait inherited from a common ancestor? It could mean birds became intelligent during the time of dinosaurs, since a common ancestor would almost certainly go back that far. It would mean they reigned for tens of millions of years as the most intelligent beings on the planet.

But their vocal systems are different (parrots use their tongues), and so too are the regions of their brains tasked with calling and hearing. These show that vocal learning evolved twice from an ancestor that merely possessed the preconditions for intelligence. All parrots learn their calls, but this skill evidently evolved in songbirds *after* the New Zealand wrens and suboscines had hived off. This could mean the parrot–perching bird relationship is spurious, although it doesn't show that at all. There was an explosive radiation of 'modern' birds after the loss of the dinosaurs and early birds, resulting in groups that, while related, have little to show for it because they diverged so fast.[62]

When birds reached the levels of intelligence they show today is an interesting question, about which parrots can teach us the most. To turn for a moment to primates, there is far more aptitude in young branches (humans, chimpanzees) than old (lorises, lemurs, tarsiers). The only wild monkeys to use tools, certain capuchins, evolved less than 3 million years ago, according to molecular dating. The nearest relatives of primates appear to be Asia's flying lemurs. In Borneo one glided past me at dusk, while in Java a mother and clinging baby stared down from a jungle trunk at night. They eat leaves. Nothing about them suggests acumen.[63]

Bird intelligence is more evenly distributed, turning up in all three parrot families. This suggests it is old, a plesiomorphy. If I used the language from a decade ago I could call the kea one of the world's three most 'primitive' parrots – based on its 'ancestral' anatomy as well as its DNA – yet the Department of Cognitive Biology at the University of Vienna, which keeps a captive colony, calls it one of the most intelligent of all birds. It is the feathered version of a Neanderthal savant. When some keas were pitted against New Caledonian crows – the world's best tool-makers after *Homo sapiens* – the keas came up trumps, soon feeding themselves snacks by shoving balls down holes while the crows remained hungry. Keas may fall behind grey parrots in aptitude – we don't know as they haven't been trained or studied as intensively – but they still rank in the top one per cent of all birds. They count as one of only three species known to engage in all four types of social play, beside apostlebirds and Australian magpies.[64]

Their nearest relatives, kakas, are smaller and less impressive, but they also like to play. In captivity their antics include jumping on the stomach of a companion lying on its back. As for the third member of their family, the kakapo, a nature book written a century ago, from an

The white-winged chough (ABOVE) is the only bird in the world to practise slavery, and one of the few to use tools. It has extreme social displays during which the eye conjunctiva engorge with blood. Along with apostlebirds (BELOW), its closest relatives, it is one of the intelligent group-breeding birds for which Australia is famous.

Africa's largest nectar bird, the Cape sugarbird (ABOVE), and mainland Australia's largest honeyeater, the red wattlebird (BELOW), are both attracted to the flowerheads of family Proteaceae, which they defend fiercely. But red wattlebirds are almost three times the size of sugarbirds, have much stronger beaks, reach far higher numbers, and live alongside many other big feisty nectar birds. Nectar-motivated aggression is thus a major force in Australia, but of minimal significance in Africa (and on other continents). The banksia flowerhead supporting this wattlebird has obviously evolved to sustain the weight of sizeable birds.

The blue-faced honeyeater (ABOVE) has features that recur in a related bird one-fifth its size, the black-chinned honeyeater (INSET). These birds indicate something unusual about honeyeaters – an impressive capacity to change size over geological time.

BELOW Swift parrots are endangered, while the Tasmanian blue gums they depend on have never been so plentiful, owing to mass plantings on six continents.

Many Australians have noisy miners (INSET) in their gardens, without knowing that New Guinea has closely related birds, including Belford's melidectes (ABOVE), which shares scaly grey plumage, black on the head, green on the wings, and coloured skin beside the eyes.

BELOW The form of the spotted pardalote can be interpreted as a response to lerp. At one end is a scoop-shaped beak, at the other, a very short tail that limits chances for lerp-feeding honeyeaters to catch and kill these tiny birds. The white wing spots resemble lerp.

ABOVE LEFT Australian manna is a sugary substance that congeals on the leaves of some eucalypts.

Vast tracts of dead trees in Toonumbar National Park (BELOW) are blamed on the bell miner (ABOVE RIGHT) barring other birds from reducing infestations of debilitating psyllid bugs. Before their premature deaths, these eucalypts were regenerating after logging.

ABOVE Australia leads the world for attacks by bird species on each other, and the Australian magpie contributes to that. It often targets the sulphur-crested cockatoo.

BELOW Woodswallows match real swallows by feasting on flying insects, but they deviate dramatically from that by also taking nectar and (shown here) lerp. This dusky woodswallow has found a very dense lerp (*Glycaspis* species) infestation.

Lyrebirds (ABOVE) are outstanding survivors from the world's first flourishing of songbirds. Australia had them 20 million years ago, before Europe had any of its current songbird groups. Australia was the first continent with song, and it has always had the 'best' song, if we accept opinions by international experts that lyrebirds are the world's best songsters.

Bowerbirds have also lasted through deep time. The western bowerbird (BELOW) shows they have kept up with the changing climate, by occupying regions of central and western Australia that became very arid millions of years ago.

Most Australians do not realise that four bird of paradise species occupy Australian territory, including the magnificent riflebird (ABOVE), and the very similar paradise riflebird, found in rainforest 200 kilometres from Sydney.

Some birds are bound to one grass – spinifex – and the endangered mallee emu-wren (BELOW LEFT) is one of these. The splendid fairy-wren (BELOW RIGHT) is one of the world's most promiscuous birds. The cheek feathers of this male are raised in display.

ABOVE When a Baudin's black-cockatoo scoops seeds out of a marri gum nut the operation is, for two reasons, very efficient: this intelligent parrot has sophisticated control over its jaws and feet, and is so well adapted to marri seeds, its main food, that it has an elongated mandible just to reach them. Specialisation like this is extremely rare.

BELOW Instead of woodpeckers drumming neatly on trees for grubs, Australia has parrots tearing them open destructively. This yellow-tailed black-cockatoo has ruined a wattle tree to extract a witchetty grub.

Cassowaries can rise more than 2 metres high to pluck fruit.

ABOVE Owing to its unique anatomy, the magpie goose is the one waterfowl on earth to be placed in its own family. It should not, strictly speaking, be called a goose.

BELOW A male musk duck takes courtship to extremes, offering grunts, whistles and a strong musk scent as part of its bold display. It belongs to an ancient lineage.

ABOVE Australia is globally outstanding for all its pigeons behaving like pheasants and partridges, including the partridge pigeon, which lives and breeds on the ground, in and around grass. It usually walks rather than flies to water.

BELOW LEFT The Tasmanian native-hen reveals something important about Australia by violating the global rule that flightless rails keep to islands without predatory mammals.

BELOW RIGHT It belongs to the world's oldest landfowl group, but the brush-turkey thrives in cities, undeterred by dogs, cats and traffic.

ABOVE LEFT Pied imperial-pigeons are adept at dangling to reach fruit.

ABOVE RIGHT The extinct Norfolk Island kaka belonged to the world's oldest surviving parrot lineage, three members of which survive in New Zealand.

BELOW Owlet-nightjars have as their closest relatives hummingbirds and swifts. Early hummingbirds resembled owlet-nightjars more than anything else (see fossil page 163).

The rose-crowned fruit-dove wears more bright colours than all the pigeons and doves in North and South America combined. Across most of the world pigeons and doves are grey or brown, but in fruit-doves, probably because they are attracted to colourful fruits, mate selection has produced flamboyant birds. Common in northern Australia, this small rainforest dove sometimes shows up as far south as Tasmania.

ABOVE Metallic starlings assemble messy enclosed nests that sometimes fall out of trees. These handsome birds evolved recently from starlings that built nests within the security of crevices, so their building style represents something new that may improve over time. Like many Australian birds they sometimes drink nectar.

BELOW This honeyeater, the gibberbird, inhabits one of the harshest of all habitats – the stony deserts called gibber plains. Rather than consuming nectar it has the lifestyle of a desert lark. Larks reached Australia only recently, which explains why something else has taken the place of one, but not why that something should be a honeyeater.

ABOVE The channel-billed cuckoo is the world's largest cuckoo to insert its eggs into the nests of other birds. In the nineteenth century it was sometimes called a hornbill or toucan because, like these birds, it has a long bill to reach fruits suspended on flimsy stems.

BELOW The wandering albatross is so at home at sea that individuals may go years without touching land. They are assumed to sleep on the wing when the sea is rough, which it often is in the latitudes they prefer.

era when it was kept as a pet, mentioned an intelligence commanding respect, along with this: 'It repays kindness with gratitude, and is as affectionate as a dog, and as playful as a kitten.' When Diamond and Bond made their global list of birds that play they came up with only twenty-five species, including every living member of family Strigopidae.[65]

When did their forebears become intelligent? This family began to separate from other parrots sometime between 45 and 72 million years, according to a 2011 genetic study. Apes split from monkeys about 30 million years ago, which was followed by a separation of gibbons (lesser apes) from great apes just over 20 million years ago. The New Zealand parrots show more aptitude than monkeys, and even gibbons, while belonging to a much older lineage. The dates make it likely that parrots had a long reign as the most sentient animals on the planet until large apes came along.[66]

Miocene Australia had a massive perching bird, the size of a large crow, which may have been as smart as one, and a rival to those early parrots. In perching birds, although not as consistently as in parrots, aptitude appears on deep branches. Lyrebirds have not been tested but their unsurpassed mimicry counts for something. An early superintendent of London Zoo, Abraham Bartlett, whose knowledge of the world's birds impressed Darwin, found 'a great amount of inquisitiveness and intelligence' in a lyrebird he kept, and young ones taken from their nests a century ago to become free-ranging pets became 'as full of tricks as a monkey'. They are the world's largest songbirds after northern and thick-billed ravens, with even bigger brains than their size would suggest. (Diamond and Bond overlooked them by not consulting books from the nineteenth century, when they were kept as pets.) Bowerbirds are placed on the next branch of the songbird tree, just above lyrebirds, and as well as their bower artistry, they show remarkable mimicry and spatial memory, linked to particularly large brains. Satin bowerbirds paint their bowers with a plant-saliva mixture. Female lyrebirds and bowerbirds reward intelligence by accepting advances from the most skilled males. Intelligence seems to be the norm for large songbirds as well as parrots, and bower-birds, like lyrebirds, are much larger than most songbirds.[67]

The notion of some birds matching primates for intelligence is very new, since until recently no one thought to test birds. They were simply dismissed as 'bird brains'. Mammals and birds are not easy to compare anyway as their accomplishments vary so much between tasks. Birds are better with words and mammals with mirrors. Parrots seldom use tools, but they have hinged beaks instead.

In 2006 a leading champion of bird intelligence, neurologist Nathan Emery, made poor use of outdated evidence to argue that songbirds, parrots and primates evolved their brainpower recently in parallel responses to unstable climates from the late Miocene onwards. Mammal intelligence does seem recent, but there is no reason to think this of birds. They may have approached a peak early on then not increased much because flight limits brain size. In capuchins a diet of hard nuts in place of fruit is thought to have driven tool use, and it may have driven parrot intelligence as well, but far earlier on. Big brains are costly to feed, but the parrot toolkit provided access to the richest of all plant foods – oily and protein-rich seeds and nuts.[68]

I began by talking about bird noise, which led to sugar and starch, then to songbirds and parrots and their Australian origins. Australia was made very different by the bird groups it inherited. That point holds true about other birds as well, and the most singular example is the next one.

Six

The Last of the
Forest Giants

The Cassowary Coast in far north Queensland is home to the world's largest and most dangerous garden bird. Accost one and you risk being kicked to death. Residents there have told me about taps on windows and thumps on doors, of fish heads and rats swallowed whole, of black feet emerging from garden sheds, of raids on pet-food bowls, and mishaps in swimming pools.

Cassowaries are the largest birds alive today after ostriches. They are shorter than emus but weigh more. They became a sensation when they reached Europe on early spice ships and are important today as rare evidence of the first division in living birds, since evolutionary distances are nowhere greater than between cassowaries plus emus and everything else. Cassowaries are vital to rainforests for all the seeds they spread. No other bird is so obviously a keystone species – one whose removal weakens a system, just as a church arch becomes unsteady without the support of its keystone.

Birds that Kick

My route to understanding these tropical birds was curiosity about the dangers they pose to people. In 2006 I went out with Queensland government officers who were responding to a routine complaint. In a comfortable lounge room a woman gave voice to her fears. There were ornaments on the sideboard, lush carpet on the floor, a poodle on her lap. Except for the talk about giant birds lurking outside, this might have been any suburban home.

To defend themselves, cassowaries kick. Their main weapon is a spike-like claw on their inner toe, up to 12 centimetres long. They also peck, push, headbutt and jump on people.[1]

In 1990, out walking on a country road with his wife, Barry Tuite noticed they had company. The *Courier-Mail* reported that he 'narrowly escaped death after being gouged' when the bird kicked his chest, knocking him unconscious. A surgeon's probe found a 12-centimetre hole close to his heart. Six years later, in 1996, a man's leg was broken and a woman's face slashed and thigh punctured in two separate incidents.[2]

In New Guinea, where they are hunted and caged as livestock, cassowaries regularly kill. Four deaths were reported in the Southern Highlands in one sixteen-month period. I suspect that birds kill many more people in New Guinea than tigers do in most countries in Asia, but statistics are not available. Cassowaries rated as a danger to allied troops during World War II.[3]

Only one death has been reported on mainland Australia, a number kept low by the lack of data about Aboriginal deaths in the distant past. In 1926 a bird fought back when confronted by two boys with dogs. Fleeing from the fracas, sixteen-year-old Phillip McLean received a slash in the neck when he stumbled and fell. So much blood escaped that when he tried to rise he collapsed and died.[4]

If these birds were allowed to keep to rainforest, conflicts would seldom occur, but encounters with humans keep rising as development booms. The Cassowary Coast – a 180-kilometre stretch of land running south from Cairns – is crowded with farms, roads, tourist resorts, and housing estates old and new. 'When taking a drive through the Cassowary Coast,' I read on the Cairns Visitor's

Information website some years ago, 'you'll pass through small townships that are surrounded by banana and sugarcane farms.' The rainforest these birds rely on survives as strips and pockets and in small national parks. Sea-changers moving north for beaches and rainforest find birds in their gardens that can weigh more and stand taller (2 metres when erect) than they do. Cassowaries have become much tamer over time, upping the risks of something going wrong.

Early fears usually yield to acceptance and delight. Locals are proud to have giant birds strolling across their lawns to raid tomato patches and pet-food bowls. Fences go unbuilt to encourage visits, one resident told me, while something shaggy marched past us. Food is often provided, even though this is illegal. A cassowary can swallow an orange or banana whole. So rare are attacks that most people make light of the risk.

The woman with the poodle lived in fear of four big birds her neighbour was feeding, as their approaches were ambiguous. Even behind her fence she felt unsafe. Cassowaries do look very intimidating to the uninitiated. French naturalist Comte de Buffon wrote of them having warrior heads and eyes like lions. No one ever said that about emus or ostriches.

Environment officer Scott Sullivan tried to divine if the birds in the woman's garden had aggression or food in mind.

'That woofing noise they make, do you ever get a hiss out of them?'

'Do they fluff their feathers out on their bum when they approach you?'

It was soon agreed that, while they were not showing aggression, her poodle should be kept out of harm's away. 'They do some pretty nasty damage to little dogs,' Scott told me later. 'We've had a few go down to the vet in Tully with big holes in them.'

The man feeding the birds two doors down regaled us with tales about hungry faces peering through windows, of thumps on the door with both feet, and of one bird that launched itself against a pawpaw tree to dislodge a fruit he had planned to eat the next day. He could have been fined for feeding dangerous wildlife, but Scott took a softer line, asking him to place food further away from the house.

The laws against feeding become particularly contentious whenever a cyclone reduces natural food supplies. Cassowaries became world news after Severe Tropical Cyclone Larry struck north Queensland in March 2006. Deprived of fruit, they were said to be invading towns. It became one of those stories foreigners expect of Australia, like tales about kangaroos hopping down main streets. *The Australian* ran a spectacular image of resident Jan Shang hanging out washing while a cassowary stood by, above a story that opened with this: 'They have borne Cyclone Larry and weeks of torrential rain, but now the luckless residents of Innisfail face a new dilemma – a posse of hungry marauding cassowaries. The critically endangered and famously testy flightless bird, famous for its ability to disembowel humans with its razor-sharp claws, is running amok through the backyards and suburban streets of north Queensland in search of food.'[5]

Reading this in Brisbane, I persuaded *Australian Geographic* magazine to fly me north to write about the cyclone's animal victims. From Cairns I drove straight to Innisfail in search of invading cassowaries.[6]

I discovered that Jan Shang can be said to reside in Innisfail only because the town boundaries take in sizeable areas of bushland and farms. To reach her home I had to drive 3 kilometres beyond the suburban edge, past paddocks and a small national park to a cluster of houses ringed by rainforest.

Approaching her house I spied a figure poised in the shrubbery near the door, and backed away. Jan soon appeared and I heard her story, shorn of spin, while Faith the cassowary looked on. Jan had brought up three children here among these birds, training them from early on to keep safe distances. One daughter, now living in England, had seen a photo of Faith in full colour in the London *Telegraph*. Not everyone had been raised safely, though; one pet dog was torn open.

As Jan talked, Faith drank from a trough while a second bird looked on. In the load of faeces dropped on the drive I could see plenty of chopped fruit. Faith was like anything tame awaiting food, but she was nearly as heavy as me, with a blue and pink wrinkled neck, a thick cloak of lustrous black feathers, and a horny casque like a helmet on her head. Cassowaries are spectacular birds, however often you see them.

The media stories proved untrue. There was no flock 'striking fear into the hearts of Innisfail residents', as wire service AAP alleged, nor any birds 'menacing' the town, as the *Sydney Morning Herald* claimed. The real story was that of a long-term bond between people and birds brought to world attention by extreme weather.[7]

I did find something authentic, an article written in the *Innisfail Advocate* in 2005 for local readers, about a cassowary that entered the commercial centre of town a year before the cyclone struck. It strutted past the clothing store, pharmacy and Catholic church. 'It went for a walk on the pedestrian footpath amongst everyone,' a National Parks ranger was quoted as saying, 'dashed across the main intersection, slipped on the tiles outside the Commonwealth Bank and started galloping off towards the taxi rank.'[8]

About a metre tall, the cassowary was a youngster that hadn't strayed far. The small town of Innisfail straddles a river bordered by regenerating rainforest. Every 'urban' cassowary I heard about was at most a street or two from forest, usually less. Not prime rainforest, but thick enough to deter dogs. Environment officers place signs on footpaths when a cassowary is about.

In Mission Beach this bird has iconic status, as shown by all the cassowary road signs ringing the coastal township, by all the stickers on cars. I doubt that any other place on earth has as many signs to aid birds. Oncoming cars will flash their headlights to alert you not to traffic police, but to the bird up ahead. One cassowary road death I read about rated more space in a local newspaper than the obituary of a dignitary. 'They're my pets,' said one resident I met. 'I don't have any cats or dogs.' While this man shovels dirt or uses an electric drill, the cassowaries watch. Big dogs are tied up because they might kill the

Around Mission Beach it is impossible to forget that giant birds are part of the landscape because of all the road signs warning drivers not to collide with them.

birds, and small dogs are kept indoors because they might get killed. The woman who feared them insisted she was not against them. 'We've chosen to live here, and we've got to live with them,' she said.

Cassowaries bring tourist dollars into eager communities. The Queensland government acknowledged this by merging two shires in 2008 to create the Cassowary Coast Regional Council. Local stores are crowded with cassowary kitsch: statues, models, murals, stickers, posters, toys, wind chimes.

The level of risk they pose is difficult to quantify. By sifting through decades of Queensland newspapers and reports, National Parks officer Chris Kofron found evidence of 150 attacks. Twenty-eight people had been kicked, eighty-five chased, nine pushed, eight pecked, four jumped upon, two headbutted and two robbed of food. Many slashes and slits were stitched up in hospitals. Most attacks came from birds that were being fed by people, and many of the remainder were probably motivated by hunger. Three of the attacking cassowaries lived in tiny pockets of rainforest where they survived by begging.

The man whose leg was broken was protecting a cassowary from two dogs. Two of the worst attacks remain unexplained. Apart from the attacks on people there were thirty-five on dogs, three on horses and one on a cow. Various cars were chased and windows broken. One popular rainforest trail was closed for a time to prevent injury.[9]

Despite seven serious attacks – one ending in death – Kofron decided that cassowaries 'should not be considered dangerous', although he did advise that children be kept at safe distances.[10]

Another bird may have taken more human lives in the past. In western Victoria, pioneer James Dawson was told by Aborigines that wedge-tailed eagles were hated for their readiness to attack young children. One baby crawling outside the family hut near today's Caramut was 'carried off'. Wedge-tails can lift weights of 5 kilograms. Indigenous children were disciplined with warnings about eagles, supernatural and real. A child was reportedly eaten by a New Guinea eagle, and children are sometimes taken by African crowned eagles. Wedge-tails are so common today that a baby could easily vanish from a picnic rug, but this would be most unlikely unless an epic drought took out their usual prey. Many wedge-tails today live on roadkill.[11]

Cassowaries have a relationship to people which is full of unusual dimensions. These birds reached Europe in the same spice ships that carried bird of paradise skins and cockatoos, long before Captain Cook 'discovered' Australia. Cassowaries occur in New Guinea and in Seram, one of the Spice Islands. The Dutch brought a live one to Europe in 1597, and by 1611 London had one, courtesy of the East India Company. So taken was philosopher John Locke by a pair kept in St. James's Park that he wove them into his seminal *Essay Concerning Human Understanding* (1690): 'the shape of the horse or cassiowary will be but rudely and imperfectly imprinted on the mind by words: the sight of the animals doth it a thousand times better'. As birds that never lose the power to provide an emotional charge they were an ideal choice for Locke's essay. They inspired others to powerful language too. Here is Oliver Goldsmith in *A History of the Earth, and Animated Nature* (1774):

> . . . formed for a life of hostility, for terrifying others, and for its own defence, it might be expected that the cassowary was one of the most fierce and terrible animals of the creation. But nothing is so opposite to its natural character, nothing so different from the life it is contented to lead. It never attacks others; and, instead of the bill, when attacked, it rather makes use of its legs, and kicks like a horse, or runs against its pursuer, beats him down, and treads him to the ground.[12]

The taste of cassowary meat was mentioned in a Dutch book in 1726. Emus came to Europe's notice much later, as 'New Holland Cassowaries'. Not until 1854 did survivors of Edmund Kennedy's fateful expedition stagger out of the jungle with news that Queensland had real cassowaries. Like 'cockatoo', the name comes not from any Australian tongue, but from Malay. 'Emu' is corrupted Portuguese. She-oaks gained their proper name – *Casuarina* – from foliage that droops like cassowary plumes.[13]

These birds provide another reason to include New Guinea in Australia. The island's three cassowary species give it the world's highest concentration of big birds – a distinction lost by New Zealand after the extinction of its moas. New Guinea shares one cassowary species with north Queensland, and another stands only 1.1 metres tall. The dwarf cassowary is nonetheless dangerous enough for a former cassowary hunter, Daniel Wakra, to warn me never to run downhill from one. Three people were killed during his childhood in a Central Highlands village, which has around 1200 residents today. I have never heard of tigers killing at that rate in Asia, except in the Sunderbans. The northern cassowary of northern New Guinea also kills people.[14]

Ostriches, being much larger than any cassowary, have a kick that can crush a head, but as birds of open places they prefer to flee than fight. With a sprint speed reaching 70 kilometres an hour, they show that over short distances two legs outdo four. I have seen a group asleep at night in South Africa's Karoo, their heads perched above the sward they were seated inside like a series of periscopes. The only animal I have had to run from in Africa was a tame ostrich in Swaziland that wanted my lunch. Its looming eyes, larger and higher than mine, were intimidating. Emus flee when they can, although a toddler near Canberra was trampled by one half-tame bird.[15]

Tropical jungle conceals enemies and impedes quick escapes, explaining why cassowaries are primed to attack. This raises a question: why should Australia, including New Guinea, stand alone in having giant forest birds? South American rheas, African ostriches and Australian emus are birds of the savanna; they rely more on vigilance and speed than kicking feet and rarely come near people or houses. Until humans arrived, both New Zealand and Madagascar had enormous birds in their forests (moas and elephant birds), but no giant birds inhabit the rainforests of Africa, Asia or the Americas, despite those continents having birds in every other niche. Big birds may be incompatible with big cats: leopards, tigers, jaguars. Australia once had marsupial 'lions' and giant goannas, but presumably not in wet rainforest.

Gigantism

The fact that Australia is the only continent with a bird as the largest animal prowling through its rainforests adds weight to the notion of a land whose birds are ecologically powerful. Any large rainforest marsupials living in the past succumbed to either climate shifts in the Pleistocene or to arriving humans. Cassowaries are the last of the rainforest megafauna. They are several times the mass of the largest rainforest marsupials, namely tree kangaroos.

Big mammals – including us – so dominate the world today that giant birds stand out as oddities. They were more successful before the rise and spread of large carnivores and humans. South America until the last glacial peak had 'terror birds' (Phorusrhacids) as top order predators, including one with a 71-centimetre skull. Europe and North America, further back in time, had their own giant birds with huge heads (*Diatryma*, *Gastornis*), which may have been the main predators of their day.

Australia had its mihirungs or thunderbirds (Dromornithids). The massive bills on the larger ones suggest they were savage predators. *Bullockornis* was described by one expert as having had 'military-issue jaw muscles' on a horse-sized head. *Dromornis stirtoni* stood 3 metres tall, and if, as some say, it weighed 500 kilograms – far more than your average horse – it was the heaviest bird ever. Some outcrops at Riversleigh are 'riddled' with thunderbird bones, arguing against a diet of meat for their owners, since large predators cannot be this plentiful.[16]

Mihirungs were long thought to be relatives of ducks and geese, but newer evidence suggests that, like cassowaries, which they out-numbered for most of their history, they were ratites, a group I will return to later. Mihirungs towered over the marsupials of their time, until a few million years ago, these latter increased in size, culminating in the mighty diprotodons. In other words, Australian birds peaked in size well before mammals did. Grinding teeth and a foregut for fermentation are more efficient at processing coarse vegetation than are the beaks and stone-filled crops of giant birds, which may explain why mihirungs declined in size and diversity as large mammals pros-pered. Only one species remained, of reduced size, by the time people arrived, although it was still three times the weight of a cassowary.[17]

An Arnhem Land cave painting of something much stouter than an emu with a shorter bill may well depict this last mihirung, called *Genyornis newtoni*. So might stories that reached Victorian pioneer James Dawson, about past birds that were larger than emus and feared for their courage, speed and lethal kicks. Only the bravest warri-ors dared spear them, and only from the safety of trees. In his book about Victoria's Aborigines, Dawson featured them not in his section on superstitions, but in his chapter about animals. He expected their bones to be found, and *Genyornis* remains have turned up in many locations. The word 'mihirung' came from the Djargurd Wurrung people befriended by Dawson.[18]

Extinction is the fate of most species, and especially of giant birds. The complete loss of flightless birds from Europe in the Eocene is blamed on climate change in concert with new mammals, including carnivores, arriving on land bridges. Extinction of most terror birds 2 million years ago is also blamed on predatory mammals, which stole south when North and South America met. The world was further depleted of big birds by early humans, who are implicated in the demise of Australia's last mihirung, elephant birds, New Zealand moas and adzebills, *Sylviornis* in New Caledonia, and a giant Fijian megapode.[19]

Pieces of mihirung eggshells have been found across inland Australia, but none are younger than about 45 000 years old, matching human arrival in Australia rather than any serious climate change. (*Genyornis* remains have been found in every Australian state, implying very wide climatic tolerances.) Some of the eggshell pieces are burnt, and because they show temperatures too high for bushfires they are taken to be litter from Aboriginal feasts. The picture is one of giant birds prospering among mammals, often dominating and preying on them, then succumbing to other mammals – typically humans and carnivores. But the question of whether humans eliminated the megafauna, including *Genyornis*, remains fiercely contested, and the topic needs further analysis.[20]

New Zealand's remarkable fossil site at St Bathans shows that moas, some 16–19 million years ago, were living alongside mouse-sized mammals. But when Polynesians arrived there they found a land of giant birds from which all mammals apart from bats had vanished. It is commonly thought that while dinosaurs existed they suppressed mammals, although we now know that birds are one line of dinosaurs that survived. (Crocodiles and turtles are closer to birds than to lizards, having shared an ancestor with them more recently.) New Zealand may be a place where dinosaurs (as birds) kept suppressing mammals until people arrived. The disappearance of New Zealand's mammals is otherwise very difficult to explain.[21]

Examples are useful to explain principles, and cassowaries have often been enlisted as evidence for theories that were later discredited. Writing in *The Development of the Continents and Their Living World* in 1938, the German geographer Theodor Arldt told of cassowary ancestors journeying from North to South America then swimming across the Pacific. In 1949 Perth geographer Joseph Gentilli, an immigrant from Europe, devoted six pages to whether cassowaries and emus wandered down from Asia or crossed over from the Americas via Antarctica and possibly New Zealand. Since emus occur more widely, he thought they must have arrived first. One of

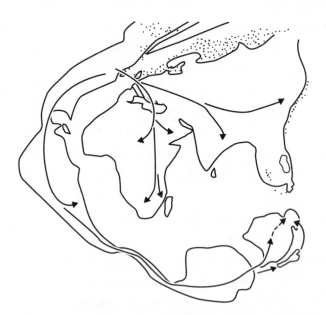

The eminent geographer Joseph Gentilli never doubted that cassowaries, emus and moas originated in the Northern Hemisphere. One map he produced (redrawn here) showed them journeying from North America through South America and Antarctica to reach the Antipodes.

his maps showed cassowaries and emus originating in the vicinity of New York. When continental drift gained credence in the 1970s cassowaries found their way into textbooks as evidence for Gondwana being a centre of origin. All the big flightless birds, known as ratites – the ostrich, emu, cassowaries, rheas, moas, elephant birds, kiwis – occupy southerly lands. They seemed to vindicate the concept of vicariance – of a common ancestor evolving into different forms on different lands as continental plates parted.[22]

The molecular evidence obtained in order to prove this disproved it instead. New Zealand's kiwis turn out to be closer to emus and cassowaries than to New Zealand's lost moas. There were bigger upsets than this, but first, some background.

Archaeopteryx gets the billing as the world's first bird, but there was another ancestor, living far more recently, that was the *last* common ancestor of all living birds. It was the unknown bird that on the DNA tree would be at the point where the trunk first branched. It left two descendants, each of which has many descendants alive today. One

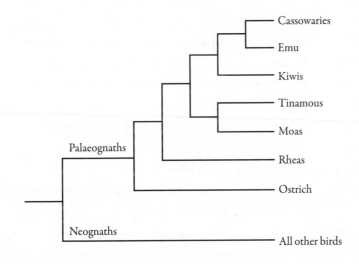

All the world's large flightless birds are found on southern continents, but assumptions about what this means were ruined by the finding that South America's tinamous (which can fly) are closer to emus and cassowaries than to ostriches. The exact position of rheas on the tree is uncertain.

gave the world the Neognaths, which constitute 99.9 per cent of all living birds, and the other the Palaeognaths.

Until a few years ago the Palaeognaths were thought to fall into two groups – ratites, and the forty-five or so tinamous of South America. Tinamous fly, though not well, and they vary in size from a turkey to quail.

But recent molecular studies place tinamous among the ratites. Cassowaries and emus are closer to tinamous than to ostriches. This means that the first Palaeognath produced the ostriches on one hand, and the rest of the ratites plus tinamous on the other. Ratites, as a group united by common descent, do not exist separately from tinamous, just as mammals do not exist separately from bats. They are an artificial assemblage – a concocted category that fails to reflect genealogy, so we should focus our thinking instead at the level of the Palaeognaths – of which some are small and some can fly. Palaeognaths have left many ancient remains in the Northern Hemisphere, mainly of birds that could fly.[23]

The new evidence shows that ratites are eclectic survivors that lasted through deep time by becoming big and flightless. Their smaller relatives, with the exception of kiwis and tinamous, were winnowed out. Thomas Huxley thought ratites had descended from dinosaurs without ever acquiring flight. In truth they lost it multiple times. Cassowaries being closer to tinamous than to ostriches, the common ancestor of cassowaries and tinamous must have had proper wings, since tinamous have them, which means a cassowary ancestor lost them long after it diverged from an ostrich ancestor, which in turn lost them after separating from that last common ancestor. The different ratite groups are far from closely related. Seven thousand species of perching birds have been accepted into one order, while the thirteen or so surviving ratites are commonly divided among four. They differ in such important matters as the number of toes (ostriches have just two) and the presence (or not) of rudimentary wings.[24]

Given all the Palaeognath fossils found in the north, an origin for them in the south has often been questioned. They could have originated anywhere and spread partly by flying. They may be southern today because mammals in the south were more accommodating to large birds (a point I will return to in chapter eleven). Swift legs allow ostriches to survive among dangerous mammals of northern origin, but flightlessness could not evolve in Africa today. Before the truth about perching birds became known, ratites were put up as the best reason to doubt a northern origin for all birds, when as evidence goes they are not convincing at all.

Bound as they are to rainforest, Australia's main cover when it was younger and wetter, cassowaries have probably changed less over time than emus.

Ecosystem Services

Young plants need water, nutrients and light, and inside rainforests light is often the scarcest of these. Seedlings in the gloom grow slowly if at all, unless falling trees or limbs let light in. Some trees help their young get by without the sun by producing giant seeds full of starch or fats to provide energy to their seedlings. But big seeds need something big to move them from the shade of the parent tree.

Other continents have elephants, apes, bears and tapirs moving large seeds. After leaving the trail in a Thai rainforest I once found some trampled ground – an elephant campsite – in the midst of which was a great dung heap with a stout seedling rising from a very large seed in the middle. Without knowing it, elephants had rewarded the tree that fed them by clearing some ground and 'planting' and fertilising a seed. Divested of mihirungs and large mammals, Australia only has cassowaries to swallow the largest seeds. With no teeth or gizzard stones their bodies provide gentle passage, fruit flesh often coming out looking intact. 'My poo grows trees', says one government sticker for children. Flying foxes, rats and musky rat kangaroos carry large fruits as well, but only over short distances.

By roaming with full stomachs through farms and woodlands cassowaries can move seeds between gullies kilometres apart. Six kilograms of seed and pulp can be dropped at one toilet stop, sometimes releasing thousands of seeds, although few can grow unless they are then scattered by rats or torrential rain. Cassowaries swallow more than 230 kinds of fruit, although many are so small that pigeons and bowerbirds disperse them as well. The larger fruits include the cassowary plum, cassowary gum and cassowary pine, each of which is as shiny-blue as a cassowary's neck.[26]

In northern New South Wales, all the big fruits show that until recently some rainforests harboured cassowaries. On one property I saw black walnut fruits (*Endiandra globosa*) bigger than golf balls, and durobbies (*Syzygium moorei*) that can reach apple size. Because nothing eats them they fall and rot. A thousand kilometres to the north cassowaries eat closely related fruits. Botanist Lui Weber has identified fourteen rainforest trees in New South Wales that 'fit the cassowary fruits syndrome', all with fruits as large as their distributions are small. They are parked in perpetuity inside their ice-age refugia because their transport service is defunct – a victim of glacial drought or of deftly thrown spears. It was probably a dwarf cassowary, one that is known about from a fossil leg bone.[27]

Signs of lost birds are especially marked in New Zealand, where giant moas, the main browsers, shaped their world in extraordinary ways. New Zealand has some of the strangest plants found anywhere. Wire plants are shrubs, saplings and creepers whose long thin stems bear small sparse leaves that speckle the understorey like confetti fixed in space. If you pull some leaves towards you there's next to no resistance, whole plants bending your way and springing back when released. Evolved to survive clamping beaks on long necks, the wiry stems are amazingly tough. I could not break off tufts of leaves with one hand without taking steps backwards. Emus that were offered these plants in experiments could not peck enough to sustain themselves, while goats consumed them at four times the rate. Moas would

have confined their interest in such plants to passing pecks. Spread
across sixteen plant families, wire plants include a conifer, a southern
beech and a raspberry. Upon rising above moa height these divaricating plants, as they are also called, often morph into large-leaved trees.
Some experts have read their form as a response to cold weather, but
New Zealand is not especially cold.[28]

Many wire plants have close relatives in Australia (including
Coprosma and *Pittosporum* species), which form ordinary shrubs and
trees. The outback has trees with divaricating saplings, such as leop-
ardwood (*Flindersia dissosperma*), but the stems are stiff, woody and
thick because diprotodons would have been deterred by a high ratio
of fibre to food. Because it had elephant birds, Madagascar had some
wire plants with tensile strength, while Africa's divaricating plants are
more like those in Australia, but often sport stout spines as an extra
deterrent. Ostriches and emus don't eat enough leaves to have left a
clear signature on these plants.[29]

Thanks to moas, New Zealand also has a nettle so fierce it killed
someone, unusual prickly plants, and native flax so tough the British
Navy imported the leaf fibre to make rope. Honeyeaters pollinate this
flax, which has ended up with leaves that deter birds and flowers that
attract them. There is also a pungent spicy tree, horopito (*Pseudowintera
colorata*), with red warning marks on its pale leaves, and a non-spicy
plant (*Alseuosmia pusilla*) that mimics it with uncanny accuracy. Other
plants are so good at looking dead or denuded that even botanists walk
past them. Plant mimicry is rare outside New Zealand, for mammals
use smell as much as sight to find food – although Australia does have
mistletoes that deceive possums by mimicking the trees they infest. In
The Life of Birds David Attenborough paid homage to New Zealand
as the world we would have if birds instead of mammals ruled, but he
never even got to elastic stems and disguised leaves. In *The Lord of the
Rings*, filmed in New Zealand, Frodo hides under moa plants in some
scenes and plods past them in others.[30]

Wire plants reflect millions of years of evolution but bird–plant relationships can form quickly. Cassowaries now dote on the pond apple, a Neotropical tree whose fragrant green fruits evolved to attract mammals, and which has become one of Australia's worst weeds. Cassowaries spread the seeds in extreme numbers: up to 2500 per dropping. The situation is blackly ironic: an endangered bird threatening rainforest by helping a major weed.[31]

Emus, while nearly as large as cassowaries, avoid big fruits, and the gizzard stones they imbibe (and the occasional bolts, nuts, even car keys) grind down seeds, save those with hard coats. After they swallow quinine tree fruit (*Petalostigma pubescens*), the pits explode in their dung to scatter the seeds, which are taken further by ants enticed by an attached morsel of food, in a three-stage dispersal. Nitre fruits (*Nitraria schoberi*) give emus diarrhoea, and instead of voiding the seeds in piles they drop them individually, to the plant's great advantage. *Genyornis* must have spread the seeds of some of today's idiosyncratic fruits, including lady apples (*Syzygium suborbiculare*), which ended up with a new transport service – Indigenous people, who by scattering the giant seeds turn some of their campsites into fruit orchards.[32]

Conserving Difficult Birds

All the plants they aid are one reason why cassowaries are considered ecologically crucial. The rainforests in northern New South Wales are today less rich at a fine scale than they would be if cassowaries were still strewing the seeds of durobbies and black walnuts; instead, these trees keep to the tiny refuges they occupied during bleak glacial times.

The cassowary therefore rates as a conservation priority. Rainforest Rescue, which I serve as a patron, has purchased twenty-two rainforest allotments on the Daintree coast to save its habitat. Further south at Mission Beach another group, the Community for Coastal and Cassowary Conservation, lobbies hard for the bird's protection. Far more government dollars go to cassowaries than to most birds, especially when food stations are established in rainforest after cyclones.

The conservation of ratites is not like that of other birds. Cars are not usually a problem for birds but they take out many cassowaries. I have seen one right by the side of Australia's national highway. The kill rate is made worse by people who feed them beside roads after cyclones, encouraging them to identify roadsides with food. 'A fed bird is a dead bird', warns one government slogan.

Dog attacks are another problem particular to these birds. During an autopsy on one hapless young cassowary, I saw the grey holes rent by canines in its abdomen. In a world so dominated by humans and their companion mammals, space for giant birds keeps contracting. Evolution will never again turn a flighted bird into anything like a cassowary, so they are irreplaceable in more ways than one.

Often mist nets are used to capture birds for research, but cassowaries require a different approach; the government unit I went out with, to save a youngster starving after Cyclone Larry, wore leather chaps like those used to protect against chainsaws.

The prospect of these birds killing someone is always kept in mind, since it would tilt the conservation equation. One government official told me about a large cassowary that approached a lady with a pram. 'What would happen if that bird had killed that kid in that pram?' Tolerance prevailed some years ago when a bird was 'terrorising' children leaving the school bus, but all the vitriol aimed at flying foxes shows that rural people are losing tolerance for inconvenient animals, including ecologically important ones. Cassowaries were killed in retaliation many years ago when a boy on a horse was attacked. A fatality today should not be allowed to erode the goodwill that now prevails.

Seven

Australia as a
Centre of Origin

For most of the twentieth century biologists were held back in their thinking about birds by all the dogma about northern origins. Australia is easier to understand when you know it had songbirds, parrots and giant flightless birds as part of the cast early on. Identifying other early players can help us understand the continent better, and also the impact it had on the wider world, which went beyond the birds I have mentioned so far. We have three kinds of evidence about origins to draw upon – fossils, current distributions and DNA trees, but they don't always follow the same script.

The debunking of the northern dogma was a liberating experience for many scientists, and not just in Australia. In northern museums and universities new ideas brewed, including a grand claim that, instead of the north, modern birds came out of Gondwana. Joel Cracraft put the case for this in 2001, attracting support, at least initially, from Walter Boles.[1]

The giant Chicxulub meteorite, which unleashed a global cataclysm when it smashed into Mexico 65 million years ago – taking out the dinosaurs and virtually all birds, marking the end of the Mesozoic – may have been kinder to birds in the south, Cracraft proposed. Most planetary life succumbed to the fires, acid rain, dust-induced darkness and cold, or whatever it was that ensued after impact, but in West Antarctica, 10000 kilometres south of the impact site, fossil layers show higher than usual plant and animal survival into the next time period, the Paleogene, suggesting this site was somewhat protected. Sedimentary sites in New Zealand and Patagonia also show very high plant survival. Tasmania had the only seed ferns we know about to survive the dramatic transition from the Cretaceous to the Paleogene (known as KPB, or K–T). The south may well have saved those few birds that spawned the rich variety we see today – although it was no haven. In New Zealand, sediments that span KPB show a forest rich in beeches and Proteaceae suddenly replaced by fungus (in a layer of spores and fragments 4 millimetres thick), after which some ferns and then trees returned, from the sheltered locations they must have survived in.[2]

In 2012 the Swede Per Ericson came out with a different theory about southern origins, which, like Cracraft's scenario, is somewhat speculative. Ericson interpreted various DNA trees as evidence that Africa gave the world a large array of forest birds, including kingfishers, woodpeckers, eagles and owls, which he called the Afroaves, while Australia and South America between them produced the perching birds, parrots and falcons (Australavis). The Australavis part of this theory is plausible and relevant, as we have seen, but there are fossils to complicate the matter, which I will come to.[3]

Whenever the data is piecemeal and stakes are high – meaning that many scientists would love to be first to answer a major question – theories flourish. If Cracraft is correct, Australia was a major centre of origin, with other groups beside songbirds and parrots that have an old and special relationship to the continent. 'Modern' birds emerged during a period (Late Cretaceous to Paleogene) when Australia, South America and Antarctica were all that remained of Gondwana, and most of Antarctica, at least in winter, was off limits to birds.

Biogeography as Embarrassing Science

Before I say any more about southern (and Australian) origins I should offer a comment about the field I am operating in – biogeography, the science of origins. Reconstructions of the past are acts of imagination, born of regions of the brain that don't always obey the dictates of science. Darwin built a cogent picture of the past by giving it twenty years of careful thought, but many who came after him were bewitched by notions of industrial progress, wandering tribes and the sinking of Atlantis.

Biogeography has had to struggle to free itself from Old Testament notions of creation. Faith in a grand narrative made this discipline a fantastic endeavour, an art as much as a science. Animals were considered to have poured out of centres of origin on odysseys round the globe. In his 1949 *Emu* article Gentilli was 'practically certain' that most Australian birds came from Central Asia. Why Central Asia? Because it has the big temperature swings needed to drive the mutations that enable evolution.[4]

Six years earlier George Gaylord Simpson had remarked that the psychology of scientists can hinder the solution of scientific problems. But he said this only to denigrate those who questioned northern origins, not to be honest about himself. The language of biogeography is sometimes snide because the divides run so deep: one international meeting in 1998 decided that 'Biogeography is a mess.' For an embarrassing science, it's hard to go past biogeography.[5]

One reason origins are difficult to think about is that we're so apt to see the past through our own backgrounds. There is 'the vanity of the present', to quote Richard Dawkins – the presumption that the past had a purpose, which was to deliver this particular present. Fossil experts discuss 'experiments in evolution', as if birds had an adolescent phase during which they tried out jaws of teeth before settling on the design they have today. Gene trees imply a destiny by showing a trunk from which 'primitive' species veer off to the side as stunted limbs. The language misleads. Birds keep evolving and invading and arriving, suggesting a resolve that does not exist, yet these verbs are nearly impossible to avoid. Other words popular fifteen years ago – *primitive, archaic, prehistoric* – are far worse. The avian fossil record has gaps that can be filled only with leaps of imagination, and bones so fragmentary you can see in them what you will. A Cretaceous parrot beak that was announced to the world in *Science* in 1998 was soon denounced as a possible dinosaur fragment, and a 'parrot' leg from Denmark was later found to be from an ibis.[6]

The southern-versus-northern genesis feeds into something larger: professional credibility. The question becomes one of who speaks for the past with more authority – those who pore over fossils or those who sequence genes. The fossil evidence suggests northern roots for most birds, but most fossil hunting goes on in Europe and North America, skewing the database. Australia, being largely flat with few eroding cliffs, is not very forthcoming about the birds it once had, while its important neighbour Antarctica hides under ice. Genetic studies regularly suggest southern origins, but DNA trees are merely lines drawn between survivors and can't capture all the complexity of the past. Many bird groups centred today in the south have left fossils in the north, raising doubts about where they arose.[7]

The disputes also pit geneticists against traditional museum biologists, who, after a century of effort, showed that all too often they could not discern true relationships. Parallel evolution can emulate real kinship, and spurts of evolution can disguise close ties. Classification is meant to indicate blood relationships, not surface similarities. In the decades since Sibley's boiled DNA showed the way, geneticists have moved in on classification in a big way. I thought the kioea I examined at Honolulu's museum in 2004 was a giant honeyeater, and had the genetics not been done four years later, linking this extinct bird to northern waxwings, I would have written bunkum about honeyeaters reaching Hawaii. But molecular studies can also go wrong; for example, when the same mutation in two species produces matching gene sequences, suggesting a relationship that does not exist. Geneticists deserve reproach when they ignore inconvenient fossils, as often happens.[8]

I mention all this to show why, for many birds that could possibly have a southern past, the evidence is confused and unsatisfying. Biogeography is a better science than it once was, but not yet as good as it could be.

Waterfowl and Landfowl

In Europe, going back to Roman times, a black swan was a symbol of the impossible, a creature with less credibility than the unicorn. When a near-mythical land in the south was found to have some, imaginings stirred. Australia certainly has strange waterfowl, although black swans barely count as such today, having nothing more than dark plumage to invite attention. The real oddity is the magpie goose.

All the world's ducks, swans and geese are close enough to fit into one giant family (the Anatidae), while the magpie goose, with its knobbed head, hooked bill, looped trachea and half-webbed feet, rates a family all its own (the Anseranatidae). It can tell us more about the past than any other bird with a bill because it is made so differently – a goose it is not. In the Northern Territory, on swampy plains that formed just 6000 years ago, flocks can reach a quarter of a million, showing that an 'old' bird can prosper in young country. During the last ice age they probably used Lake Carpentaria, the vast wetland that forms when falling seas expose the plain between New Guinea and Queensland.[9]

Ducks and geese occupy a special place on the bird tree of life. The first branch of this, as we've seen, separates the sixty or so species of Palaeognaths, just 0.1 per cent of the world's bird species, from all the rest, the Neognaths. These fall into two groupings, one made up of the waterfowl (ducks, geese, swans) plus landfowl (pheasants, quail, partridges and megapodes, which are also called gamebirds), and the other of everything else. Waterfowl fall into three families, the

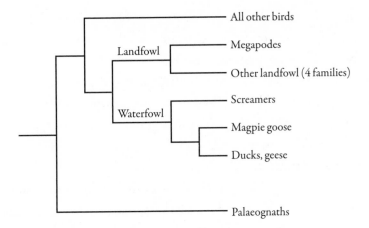

The bird tree of life shows the Palaeognaths branching off from the majority of birds first, followed by the waterfowl and landfowl. Palaeognaths, megapodes, screamers and magpie geese are all confined to southern continents.

third comprising the screamers (Anhimidae), three South American birds of swamps with chicken-like beaks. Screamers were given little thought in Australia until an Eocene bone of one was found in southern Queensland – at the site that yielded that early perching bird.[10]

Screamers are the earliest surviving waterfowl group, which is something they show by having slender beaks. Noting that the Queensland screamer is the oldest one yet found, Walter Boles and a Polish colleague suggested in 2012 that screamers might have evolved in Australia. At the time this bird was alive Australia remained joined to a winter-dark Antarctica, which retained a link to South America.[11]

This screamer and the magpie goose provide reasons to wonder if Australia was a centre of origin for waterfowl. A German DNA study of the world's ducks, swans and geese published in 2009 provides further reason.[12]

Because it is mostly dry, Australia doesn't have many waterfowl – a mere twenty-two species in the whole continent, against twenty-nine in much smaller Britain and forty-two in Alaska. But a large proportion of the Australian species are without close relatives, which can be taken as evidence that they are lonely survivors from old radiations. Six species have a genus all to themselves. (Black swans are close to other swans and aren't among these.)[13]

For Gould it was the musk duck (*Biziura*) that helped convince him Australia has odd waterfowl: 'Like many other of these antipodean forms, it must be regarded as an anomaly. It is, in fact, a *Biziura*, and nothing more, for it stands alone.'

The first of these ducks collected for science was found to have 'a bag like that of a lizard hanging under its throat, which smelt so intolerably of musk that it scented nearly the whole ship'. This 'bag' helps the males attract mates, and very unusually for a bird the perfume does too.[14]

The German study placed the musk duck as a sister to the branch producing the world's swans and geese. Another Australian oddity, the pink-eared duck, sat on a higher branch, just below those for swans and geese. Swans and geese are effectively big ducks with long necks, which means the pink-ear may be revealing about their last short-necked ancestor, which (like all direct ancestors, as far as we know) no longer exists.[15]

But Australia's expert on defunct ducks, Trevor Worthy, questions this German study for comparing just two gene sequences and for leaving out some species. He agrees that Australia gave the world shelducks (*Tadorna*), since Australia had them 10 million years before Europe, and because Australia's radjah shelduck is the sister of all other shelducks.[16]

Australia could have been a cradle for other waterfowl groups, but stronger evidence is needed. Fossils show that Europe and North America had had long-extinct birds that were evidently related to the magpie goose. Wyoming, during the Cretaceous, had a modern bird (a Neoave) that predated the divide into waterfowl and landfowl. Early waterfowl may have spread around the world early on rather than enriching one continent first, as songbirds did. Waterbirds tend to be good travellers because wetlands in one place often dry out, only

The radjah shelduck provides one of several reasons to suspect that Australia served as a significant centre for waterfowl evolution. It is so different from the shelducks on other continents that it can be taken to represent the earliest lineage. Central Australia has yielded the world's oldest shelduck fossils, going back 25 million years.

to fill in another. Australia has ducks that seem quirky but they don't stand out ecologically – musk ducks don't eat unusual foods. Perhaps it was different when Australia had vast inland wetlands, but water-fowl are not one of the groups that help define Australia today.[17]

Landfowl include the bird the world turns on – a small rainforest pheasant bred to become the chicken, *Gallus gallus*. Britain's 155 million domesticated fowls are estimated to outweigh by thirteen times all of its wild birds. The chicken's forebear, the darkly iridescent red junglefowl, whose cock-a-doodle calls have entertained me during searches for elephants in the Thai rainforest, has typical pheasant habits. Interbreeding with village chickens has left few pure ones behind.[18]

Because landfowl are bigger, plumper and easier to keep than most birds, their value to humans goes far beyond chickens to include turkeys, partridges, grouse, quail, peacocks and other pheasants. DNA work places the lower landfowl branches in the Southern Hemisphere, with the megapodes (mound builders) of Australia and nearby islands in the lowest position as the oldest family, a finding which has anatomical support. That Cretaceous bird in Wyoming had a jaw closer to a megapode's than to anything else.[19]

Megapodes are often called 'primitive', although they did not inherit from reptiles their habit of laying eggs in the mounds of mulch and soil they rake together. They operate far outside the norm for birds, by neither sitting on their eggs nor giving care to their young.

Their eggs are massive, to accommodate chicks that can fly the day they are born, and which fend for themselves from birth. The 3-metre-high mounds built by the orange-footed scrubfowl are said to be the world's largest nest made by one bird. In the outback, mounds of sand 20 metres wide are regarded as the work of extinct giant malleefowl, known from old bones. A solitary childhood could explain why megapodes seem more obtuse than most birds. Their brains are small. This has not stopped brush-turkeys thriving in gardens on the grubs and vegetable snacks they claw up from beds and lawns, exacerbating urban erosion (and frustrating me). They helped persuade Gould that Australia abounds in anomalies.[20]

Like lyrebirds, megapodes form the oldest surviving branch of a large group, but this doesn't mean that Australia gave the world landfowl. The continent has no other early landfowl branches, and few other landfowl, only quail, including one (the king quail) that is shared with Asia and probably came from there recently. Southern Africa has seventeen landfowl compared to six in mainland Australia, a land more than twice its size. The fossil record shows that Europe and North America had very early landfowl, of stem lineages that predate megapodes. The current distributions of landfowl point south, but the fossils point north. An origin in Europe seems plausible to me.[21]

A poorly known fact about birds is that very few species – only 3 per cent – sport a penis. Most practise the 'cloacal kiss', a transfer of sperm when their cloacas (lower openings) briefly meet. Penises are made interesting by how well they fit the phylogenetic tree, in that nearly all birds with a phallus are landfowl, waterfowl or Palaeognaths. Penises were inherited from dinosaurs, then disposed of by the line of birds that gave rise to most species alive today.[22]

Hygiene probably explains why penises were lost. Birds face more disease risks than mammals since they use the same opening for defecation and sex. Reptiles have a single opening as well, but they have lower temperatures to reduce the risks of infection. In birds, to limit the contact between warm damp skin, evolution may have favoured smaller and smaller penises until none remained. Penises are to birds what platypus eggs are to mammals – a 'primitive' feature. Teeth are something else that in birds but not in mammals count as 'primitive'.[23]

Some bird penises are exceptional. An aroused Argentine lake drake finds himself in command of a 40-centimetre implement. The sharp spines along the base probably facilitate rape. Australian blue-billed ducks may be similarly endowed, since they are closely related, but no one has produced a tape measure at the right moment. To thwart rape, the females of some ducks possess multiple false vaginas. Ducks probably face little disease risk because they have so much water washing over their cloacas. Landfowl are by comparison modestly endowed.[24]

Dodos and Hummingbirds

After waterfowl and landfowl we can't be sure what emerged next, because DNA sequencing (for reasons that are too complicated to explain here) can fail to properly group birds that radiated quickly a long time ago. When different gene sequences are compared, different trees often emerge, although some branches always recur in the same or similar forms. One plausible tree that appears divides the Neoaves – all those birds that are not Palaeognaths, waterfowl or landfowl – into two nearly coherent groups, called Metaves and Coronaves. A southern past for Metaves has been proposed, but because this division only shows up when one particular gene is compared, it might not show the directions taken by evolution, in which case these groups do not really exist. Not knowing the whole tree does nothing to stop us from allocating some groups to Gondwana, as the example of pigeons shows.[25]

This group too has been assigned to the south, on evidence that is unusually strong. Australia has produced all the earliest pigeon remains: a 25-million-year-old shoulder bone at Riversleigh, announced to the world in 2012, and two limb bones found in the Lake Eyre catchment. Europe's rich strata have yielded no pigeons older than 5 million years. Three large genetic studies fit in with this by rooting pigeons in two Gondwanan lands, Australia and South America.[26]

Small, drab pigeons in South America are the sisters to all other pigeons, but Australia and the islands around it became the stage on which pigeons showed what they could become by sponsoring a radiation unrivalled on earth. Australia is as much a land of pigeons as of parrots. The numbers show it: Australia has ten times as many pigeon genera as Europe, and twice as many as North and South America combined (they have ten altogether to Australia's twenty).[27]

Australian pigeons succeeded in two ways that, for pigeons, are unusual: by turning to fruit (a subject discussed in the next chapter) and by operating like landfowl. There are savanna and spinifex pigeons which, like partridges, nest on the ground and run to elude enemies, while New Guinea offers pigeons in place of pheasants. At 2.6 kilograms, the world's largest, the Victoria crowned pigeon towers above domestic fowl. As for the pheasant pigeon I followed down one jungle trail, nothing about that lanky iridescent bird could persuade me it was a pigeon apart from the face; my eyes kept registering a pheasant whenever they left its head. All the pigeons assuming the places of landfowl are evidence that real landfowl have done poorly in Australia, and so are the two lyrebirds, which used to be called 'native pheasants'.[28]

It is unclear how many of the world's pigeons can be traced back to Australia (rather than to South America) but those with definite Australian origins have done especially well on islands, spreading east through Polynesia as far as Pitcairn, to become the main land animals in remote Oceania. On many Indonesian islands, Wallace found parrots and pigeons to be 'by far the most prominent and characteristic of the living creatures that inhabit them'. Only seven forest birds inhabit Christmas Island, in the Indian Ocean, and two are pigeons. Further west, Mauritius had a pigeon that was especially large, a metre-high bird that gave up on flight – the dodo.[29]

That dodos were close to pigeons was recognised long ago, and in 2002 Oxford biologists went further by extracting DNA from a museum specimen to find that dodos were genuine pigeons rather than early offshoots. The article announcing this, 'Flight of the Dodo', linked them with New Guinea's crowned, pheasant and thick-billed pigeons. The dodo is another bird with Australian roots.[30]

Hummingbirds are something else with a connection to Australia, although the nature of the link is ambiguous and perhaps tenuous. They have swifts as their closest relatives, followed by Australia's owlet-nightjars. The evidence for this is so strong – genetic, anatomical and fossil – that these birds are grouped together in their own order (Apodiformes). As shadowy inhabitants of the Australian night, owlet-nightjars verge on the obscure, but they have attracted interest from far away. When the great nineteenth-century anatomist Thomas Huxley saw their jaws he was reminded of swifts. Europe's leading fossil expert today, Gerald Mayr (no relation to Ernst Mayr), found their bones to 'compare well' with those of long-extinct swifts in Germany and France. On one German hummingbird (*Parargornis*) that lived 47 million years ago, Mayr found a beak befitting a swift and feathering which 'resembles that of owlet-nightjars' (see below). This rainforest bird looked far more like an owlet-nightjar than a hummingbird or swift. Wallace remarked that one treeswift in the Asian rainforest

The earliest hummingbirds looked more like Australian owlet-nightjars (see colour section) than anything else, telling of an ancient relationship. This is *Parargornis*, found in Germany.

'resembles the larger hummers in everything but the bill', and New Guinea has a closely related bird.[31]

Gerald Mayr has suggested that the southern continents had hummingbirds at the same time Europe did. South America, the centre for hummingbirds today, was connected via Antarctica to Australia at the time when *Parargornis* was alive, but for reasons I will come to I doubt that Australia ever had hummingbirds. The early forms probably fluttered about leaves and flowers snatching insects, before nectar became the main attraction. Most owlet-nightjars hunt insects inside New Guinea rainforest, but there is one species that takes much of its insect food on the ground in woodland, including large numbers of ants – an odd food for a bird of the night. I suspect that owlet-nightjars are relicts from some old, wide-ranging group.[32]

South or North?

In the years before Charles Darwin, men who studied fossils thought the world had come through an Age of Fishes and an Age of Reptiles (Mesozoic) before reaching the Age of Mammals (Cenozoic), which doubles as the age of modern birds. Their thinking was coloured by the Book of Genesis, and by human history, which showed one civilisation or tribe inevitably supplanting another. A version of this thinking pushed in the 1990s had all early birds dying out along with the dinosaurs, except for a couple of lucky survivors, which filled the void with the explosion of birds we see today. Mammals supposedly went through their own bottleneck and explosion.[33]

But like Noah and his ark, the idea of a couple of survivors is so melodramatic it has lost support. Recent fossil finds suggest that several birds came though the massive KPB extinction event, while molecular clocks, which use rates of genetic change linked to fossil dates to infer ages of groups, imply that many birds did so.[34]

A 2007 molecular study on waders proclaimed that fourteen species must have lived through KPB to produce the fourteen very different groups alive today, despite there being few fossils to back this up. Many bird fossils from before KPB have been found, but nearly all of them are from long-extinct groups or are too fragmentary for any confidence about what they are. The most dramatic claim yet made is that sixty lines came through Chicxulub's infernos, including birds of paradise, honeyeaters, and even pardalotes, which allegedly escaped the flames and radiation by hiding in their holes. This flight of fancy, which arose from a loose reading of the evidence, appeared in the *Geological Society of America Bulletin* in 2004.[35]

Joel Cracraft remarked that the rock and clock dates can be reconciled if modern birds emerged from Gondwana, since its fossil record is so poor we can't expect to have the hard evidence showing that modern birds lived among dinosaurs. Keith Barker said this about perching birds in his famous 2004 paper. Gondwana has yielded two of the very earliest and most important fossils of modern birds: that of *Vegavis*, believed to be an early waterfowl that lived among dinosaurs in Antarctica, and a stem penguin (*Waimanu*) that was sporting in New Zealand's waters perhaps only a few million years after KPB, near the place where those forests gave way to fungus. With big, solid bones for diving, and roosting where sediments settle, penguins are good at leaving fossils, and we can suspect there were many bird species alive back then, since penguins are not close to waterfowl.[36]

But the clash between the clocks and rocks can also be explained another way, as evidence that geneticists use dubious fossils and geological events to calibrate their clocks. That Danish 'parrot' leg that really came from an ibis was used to date cockatoo origins in a major paper in *Nature* in 2012 called 'The Global Diversity of Birds in Space and Time'. Gerald Mayr complained that four of the eleven calibrations in that paper were flawed. The mistakes usually push roots back in time, often coming about because someone is too hopeful they have found the earliest evidence of some important group.[37]

Europeans have good reasons to doubt Gondwanan theories, namely all the sensational fossils showing that Europe, when it lay closer to the equator, had jungles lively with hummingbirds, frog-mouths, trogons, broadbills, mousebirds, and other tropical and 'southern' specialities. Most of the world's hummingbird fossils, including exquisite skeletons, have come from Europe.[38]

Because they lived and breathed some 10–20 million years after KPB, these birds had time to come from somewhere else, and we don't know where they started out or how widely spread they were. Europe's fossils leave us with more questions than answers, but they do reinforce my point about ratites, that southern distributions today can't be taken as proof of a southern genesis. Marsupials show how a group with northern ancestors (in Asia) can entrench itself in the south. Cracraft thought that megapodes and Australian frogmouths had southern roots but fossils in the north trump them by displaying more basal features.[39]

Fossil beds everywhere produce surprises, and in Australia what stand out are all the flamingo bones. For some 20 million years these pink birds graced inland lakes, before bowing out maybe a million years ago. The Lake Eyre region at one time had three species, more than Africa has today. One of the last to vanish from Australia was the greater flamingo – the main flamingo in Africa. By living where silt accumulates and tending to be large, waterbirds leave behind more fossils than do forest birds, and nothing shows this more clearly than the fact that flamingo fossils in Australian museums outnumber those of parrots and pigeons combined. Only two Australian parrots of any age are known about from fossils: a Miocene cockatoo and Pliocene budgerigar.[40]

Flamingoes aside, Australia has not yielded as many fossil surprises as Europe or North America. Australia had mihirungs, screamers, and storks (*Ciconia*), like the European birds said to deliver babies, but Walter Boles has noted an 'apparent conservatism' over time that set Australia apart from the rest of the world. We can measure this by comparing the survival rates of families. If we take a key time period, the late Oligocene to mid-Miocene, Australia had seventeen families of forest birds that we know about, and except for the mihirungs, which succumbed *after* humans arrived, all of them survive today – including megapodes, owlet-nightjars, hawks, swifts, kingfishers, pigeons, lyrebirds, logrunners, orioles and honeyeaters.[41]

In Europe, a sizeable proportion of forest families from that period are extinct there, and more than a few are extinct globally. Australia is partly responsible for that, because whole families vanished soon after its songbirds arrived, including birds related to swifts and woodpeckers. Most of Europe's small forest birds could not, it seems, compete with songbirds, unless they had unusual habits such as pecking on wood or catching insects above the canopy. Australia, from what we know, did not suffer extinctions in the same way.[42]

Some of the fossils found in Australia are relatively old. They include the world's earliest-known 'modern' rail (*Australlus disneyi*), darter, crown swift, and possibly the oldest typical eagle (*Aquila bullockensis*). Australia had some of the earliest grebes, cormorants, stone-curlews and kingfishers ever recorded. But none of these fossils are old enough to reveal much about origins. These bird groups, like those in Europe's rainforests, were probably widespread in their time, and indeed, slightly younger fossils of most of them have been found on other continents. But two hold special interest for different reasons: the rails, because of flightless species that I will mention later, and the kingfishers.[43]

Australia is unusually well endowed with giant kingfishers, which include the world's largest – the laughing kookaburra – and the shovel-billed kingfisher of New Guinea, an earthworm specialist that operates like a mammal by ploughing up rainforest soil with an exceptionally stubby bill, caking it up with mud inside and out.[44]

Like songbirds, parrots and pigeons, kingfishers have done especially well in Australia, but the kingfisher tree does not show deep roots in Australia and kookaburras are not an especially old line. Why Australia has so many large kingfishers is unclear to me, unless it has to do with all the lizard prey. Often hunting near water, kingfishers show a willingness to cross wide rivers that would sometimes lead to much longer journeys over water. If Asia was their land of origin, they show how well a northern group can do in the south, but kingfishers have not achieved anything like the success in Australia of songbirds, parrots and pigeons.[45]

As for Europe's fossil record, it is rich enough to suggest that much of what is missing really wasn't there in the first place. Europe early on had no flamingoes, pelicans or storks – or none that we know about. There were no kingfishers, falcons, cuckoos or bee-eaters, and of course no pigeons or songbirds. Connecting land gave North America and Europe similar birds, but Asia remained separate until a collision with Europe some 30 million years ago. Some of these groups would have started out in Asia, or in Africa, which left Gondwana early on to become a separate crucible for evolution, or in Madagascar. The fossil records of these lands are very poor.[46]

Knowing where a group began is not important if it spread around the world early in its past, as swifts, eagles and many waterbirds surely did. Evolution would often have been a decentralised affair, with an evolutionary innovation on one continent promoting proliferation and spread, followed by further shifts on others. A centre of origin doesn't mean much if this happened. Gondwana could well have been left with the only meteorite survivors, but if they quickly reclaimed the north, any evidence of southern origins might not survive. Songbirds, parrots and pigeons stand out because they proliferated in Australia over deep time, achieving an influence there that remains singular today.

Palaeognaths, with far fewer species, are important too, but the other groups linked to Australia contribute far less to its distinctive ecosystems. Australia would not change much if owlet-nightjars went extinct and pheasants took the place of megapodes. Songbirds, parrots and pigeons achieved great success outside Australia, but their influence in Australia remains distinctive.

I have no more birds to mention but I can't leave the topic of fossils without returning to parrots, since their fossils and DNA could be saying such different things. Europe, North America and India may have had 'parrots' 55 million years ago. England in the balmy Eocene had three species flitting past possible magpie goose relatives. Wyoming in the Eocene was equally blessed, but we would not, if we saw these birds today, know what to call them, for they were 'stem' parrots, lacking the full toolkit of living species, while still carrying enough features to suggest they were closer to parrots than to anything else. Some had stout beaks without the curved tips we expect today. Those that did have hooked beaks had raptor-like feet, suggesting they were predators.[47]

Can something without a parrot's beak be called a parrot? What rules should language follow? To capture the variation of the past, including ancestral forms and extinct sister lineages, palaeontologists stretch words in strange ways. Birds that wore the garb of owlet-nightjars get called 'hummingbirds', while 'whales' galloped about on hoofs, because palaeontologists define the latter to include every extinct twig on the branch that ends with living whales.[48]

Gerald Mayr has written in the past about 'unambiguous' parrot remains in Europe, but in 2014 he conceded that these early birds might not have been parrots. If they were, they probably lived all over the world. In Australia, or perhaps New Zealand or Antarctica, one line came into that powerful hinged beak, to crack open seeds of rainforest trees or possibly banksias, thereby becoming the first crown species – meaning it had everything that defines parrots today. Certain plants, and an absence of certain mammals and birds, could explain why the south had the only stem parrot to cheat extinction by leaving descendants. From Australia one of these crown parrots reached South America, probably crossing the Antarctic coast one summer. Others reached Europe 20 million years ago, including a small German species whose fossilised bones recall those of Australian king parrots and red-rumps. But Europe became hostile to parrots and none remain today. Even if these birds were stem parrots we can say that Australia gave the world 'parrots' if we keep to the usual meaning of the word.[49]

Australia's Sister Lands

Ornithology has moved on since Cracraft tendered his theory about Gondwana, and I've not seen it referred to in recent years – except by critics. But it raises one important issue that I haven't mentioned yet, and that is the nature of Gondwana itself. By the time the first modern birds arose, Africa, India and Zealandia had long departed, leaving Antarctica with slim connections to Australia and South America. Antarctica became Gondwana's hub, but during the crucial time for early bird evolution – the Late Cretaceous to early Eocene – it was a land that is difficult to imagine, accustomed as we are to its images of ice.[50]

Botanists have an easier time of it because Antarctica's last forests have living analogues. In the Antarctic twilight, I have perched on a pink outcrop near the abandoned Wilkes Station while calving ice struck the sea, sounding like cannon fire, and tried to visualise beeches and conifers growing there, like those in New Guinea, New Zealand and Tasmania. Pollen dredged from the nearby seabed by Douglas Mawson's polar expedition confirms that rainforest once dominated the Antarctic coast. The scrubbier rainforests further west, around Prydz Bay, must have had small birds pollinating some of the twenty-eight Proteaceae species identified from ancient pollen. The birds could have migrated in winter across the narrow seaway that, by 84 million years ago, separated Australia from Gondwana, or travelled along the lingering land connection that ran through Tasmania to the mainland. Antarctica's only small birds today are Wilson's storm-petrels, one of which hovered near me like a dark swallow, investigating gaps between rocks.[51]

Antarctica's living past aroused interest in the nineteenth century when, after a long voyage that showed up similarities on cold southern lands, Darwin's young colleague Joseph Hooker proposed that plants had somehow spread between Tasmania, Antarctica and other southern lands. Most experts who followed accounted for southern plant distributions by concocting immense land bridges that connected distant places before subsiding below the sea. Speculation was so respectable that Frederick Hutton felt his dignity safe when he admitted in 1896 to having 'abandoned my former idea of a Mesozoic Antarctic Continent, and substituted for it a Mesozoic Pacific Continent, stretching, more or less completely, from Melanesia to Chili'. The Paris Academy of Sciences offered 'Proofs of the subsidence of a Southern Continent during recent Geological Epochs'. None of these theories invoked land that moved as it really had, by drifting; it always sank instead, Atlantis style. Faith in grand bridges survived into the 1960s, when Dutch botanist van Steenis maintained that the evidence for a bridge between Australia and New Zealand was to most Australian botanists 'obvious'.[52]

Antarctica early on was not, most of the time, especially cold, but from the Cretaceous onwards winters were dark. Plants could survive them in a dormant state. Incredible as it may seem, buttercups and stunted beeches were flowering 500 kilometres from the South Pole during a warm phase 3.8 million years ago (if, as it seems to be, the pollen dating is correct).[53]

Australian trees in experiments have endured ten weeks in darkness to little disadvantage, showing that the coast of Antarctica was suitable for today's plants, given that Australia's Davis station (at 68°S) only has a month of winter dark. At Toolik Field Station in the Alaskan tundra, at the same latitude north, I was told that owls, ravens and ptarmigans survive winter on a few hours of dull light in the middle of the day, showing that Antarctica's coast would have suited large forest birds. Toolik has rich herbfields and various mammals surviving far colder winters than Antarctica once had.[54]

But the coast near South America veers close to the pole, and if it was similar in the past it must have posed a barrier to birds. Polar dark fascinates plant and dinosaur experts, but none of the biologists who propose bird spread across Gondwana mentions it. Richard Dawkins in *The Ancestor's Tale* claimed that 'Ancestral ratites wandered freely over the whole continent of Gondwana', when winter dark made this impossible. So did ancient ice sheets, if claims about them appearing briefly in the interior (from the Late Cretaceous to the Eocene) are valid; either way the Antarctic coast remained 'relatively warm'.

By where they grow today, some plants tell of ancient journeys around the Antarctic coast. Australia's hoop pines (*Araucaria* species) share a name with the Arauco people of South America, because *Araucaria* trees grace both places. There are other shared plants (*Caldcluvia, Eucryphia, Lomatia, Nothofagus, Orites*), as well as large stag beetles (*Sphaenognathus*). Other journeys are indicated by fossils, including remains found at the very site in Queensland that yielded those early perching birds. The tiny fossils Walter Boles showed me predate Australia's final separation from Antarctica by about 10–15 million years.[55]

The teeth of a smallish marsupial (*Chulpasia jimthorselli*) found in Queensland look so much like teeth found in Peru (belonging to *C. mattaueri*) that they have been identified as coming from a close relative, although not everyone is convinced. Two snakes at this Queensland site had South American relatives as well, including one whose name (*Patagoniophis*) means 'Patagonian snake'. Patagonia once had platypuses, and eucalypts, generously illustrated in an article available online: 'Oldest Known Eucalyptus Macrofossils Are from South America'. The nicely preserved leaves, flowers and gum nuts are roughly the same age as the Queensland birds, from a time when warmth made travel easier. Eucalypts may have grown on stony slopes right around the coast of Eocene Antarctica, within sight of giant penguins sporting in the water. The Antarctic beeches (subgenus *Brassopora*) in New Guinea are something else South America once had. The plants, for the most part, spread from Australia to South America, while the marsupials and snakes went the other way.[56]

But the giant terror birds (*Phorusrhacids*) of South America reached the tip of Antarctica (Seymour Island) in the company of sloths and hoofed mammals, without finding a way to Australia. Unable to hibernate, most birds were probably blocked by winter dark. DNA trees suggest that modern parrots, pigeons and perching birds made one journey along the Antarctic coast, although these travels are far from proven. Gerald Mayr doubts that perching birds crossed Gondwana at all, and suggests instead that South America obtained suboscines from Australia via Asia and North America.[57]

Other journeys were probably made by waders and waterfowl. The closest relatives of Cape Barren geese and plains-wanderers (coscoroba swans and seedsnipes) live far away in South America, implying long journeys made by ancestors in the past. The screamer fossil found in Queensland tells of a similar journey. If the Arctic is anything to go by, Antarctica had waterfowl and waders nesting in tundra around pools filled by summer snow melt. I envisage flocks in the interior withdrawing for winter to the coast, or sweeping on to Australia and New Zealand. Accidental trips across Antarctica were likely for these strong-flying birds. Antarctica would have offered pastures free from grazing mammals, a resource that only one bird in Australia feeds on today: the Cape Barren goose. It does well on cool southern islands with no mammals.[58]

The unearthing of fossils would trump speculation about all this, but thick ice conceals most of Antarctica's sediments, or has gouged them from the surface and dumped them in the sea.

New Zealand is another key piece of the Gondwanan puzzle, hence the dictum, 'Explain New Zealand and the world falls into place around it.' As if tuataras and kiwis were not enough, New Zealanders have found they have the world's most 'primitive' parrots and perching birds.[59]

But another comment counts even more here. In a biting essay, 'Goodbye Gondwana' (2005), New Zealand botanist Matt McGlone

demanded a rethink about his land: 'for New Zealanders in particular, abandoning "Time Capsule of the South Seas" for "Fly-paper
of the Pacific" will be a wrench. But it has to happen: goodbye
Gondwana.'[60]

Like Australia but even more so, New Zealand is romanticised as
a land adrift, a lonely 'moa's ark' that drifted away from Gondwana
far earlier than Australia did – some 82–90 million years ago – with
an assortment of oddities. The fragment it was part of, Zealandia,
also called Tasmantis, was a vast land before most of it subsided.[61]

But the romance has been thoroughly spoiled by genetic evidence
showing that various plants and animals living in New Zealand are
too close to their Australian kin to have lived through 82 million
years of isolation. Their ancestors must have arrived much later over
the sea. Even if New Zealand kept a land bridge to Australia until
54 million years ago, as some say it did, travel over water becomes
necessary to explain New Zealand's honeyeaters, robins and other
birds, and hosts of plants and insects.[62]

This paradigm-smashing has a bearing on New Zealand's special 'wrens' and parrots. Geneticists have assumed that these diverged
from other parrots and perching birds when Zealandia broke away
from Australia and Antarctica (simultaneously). The date of 82–85
million years has become the main one used to calibrate the molecular
clocks that time songbird and parrot evolution. It has justified claims
that lyrebirds diverged shortly after KPB, that cockatoos split from
other parrots before the time of *Tyrannosaurus rex*. In a recent study of
fish, thirty-six fossils were used to set a clock, but bird geneticists, with
few fossils to turn to, have placed immense faith in a single speculative
date, one that ruled out the possibility of birds flying to New Zealand
long after it drifted off.[63]

A mistake often made in the north is to think of parrots as inherently tropical. Saturated colours suggest equatorial rainforest, the birds
of which, from a dislike of open places, often shun the sea. But parrots
often cross water. The sealers who chanced upon remote Macquarie

Island found small green parrots living among the penguins. To reach this cold slice of treeless land halfway to Antarctica, those red-fronted parakeets had to cross at least 640 kilometres of icy ocean. The sealers and their cats found them tasty and they did not last long. Parrots have reached small cold islands so often that five species rate a mention in *The Complete Guide to Antarctic Wildlife.*[64]

Norfolk Island at one time had two parrots, including a kaka whose ancestors flew 750 kilometres from New Zealand. Gould watched a pet one skip along a floor, but by the time he published his remarks about their fondness for butter and cream, none survived. Parrots have reached many tropical islands. The collared lories I've seen in Fiji were certainly not passengers on moving land, and New Zealand's birds might not have been either. The KPB event, at which time Zealandia was low and shrinking, could have removed them all apart from early moas, providing a clean slate for wanderers from Australia and Antarctica.[65]

A commendable lack of nationalism was shown by two New Zealanders who in 2010 dubbed their birds 'a subset of Australia's'. Steven Trewick and Gillian Gibb went on to dismiss the use of 82 million years in clocks. In an article that leaves many scientists looking careless, Gerald Mayr did so as well. The latest DNA papers are more circumspect about this date, and in one 2011 study about parrots it was spurned altogether. So strong are the winds from Australia that even spiders cross in large numbers, as hatchlings floating on threads. Red desert sand ends up on New Zealand's alpine snow. Cattle egrets and double-banded plovers journey back and forth each year, probably bearing occasional seeds. Some birds arriving recently from Australia have stayed to profit from farmland in ways the resident rainforest birds cannot. Australia the continent (including New Guinea) dominates the Pacific biologically as Australia the nation does politically, and it overtook Gondwana long ago as a source of birds. Experts who assume that early parrots did not cross any sea have to accept that later ones did, for there is no other way to account for the parrots found in Africa, Asia and Oceania.[66]

Fossils show that New Zealand's unusual bats (Mystacinids) lived in Australia early on before going extinct there, and New Zealand's parrots and wrens probably did this as well. The wrens have feeble wings, but that doesn't mean their ancestors did. New Zealand still counts as special, but it is not a Gondwanan ark so much as a sanatorium for old Australian groups.[67]

Australia as a Centre

Uncertainties about Gondwana do not damage the conclusion that Australia was a land that kept giving. For its size, no other land has mattered so much, for the simple reason that it provided the world's most successful groups of birds.

Songbirds and parrots each left Australia more than a dozen times, when once might have been enough to fill the world with their colours and calls. Two Danes, Knud Jønsson and Jon Fjeldså, have remarked that in effect, Australia was 'pumping corvoid oscines across to Asia, from where some continued onwards to Africa and even to the Americas'. The nightingales and larks I saw in Europe bear witness to what was probably the first successful songbird invasion, and the woodswallows in Asia may mark the most recent. Africa was a surprisingly regular recipient of birds crossing the Indian Ocean or spreading along a forested Asian corridor, by receiving, one after another, the Passerida, parrots, bush-shrikes, orioles, and two or three arrivals of cuckoo-shrikes. We will never know how many invasions sputtered out without leaving descendants alive today, like those long-extinct songbirds that showed up in Europe but left no successors.[68]

The northern bias in biology is something I keep lamenting, but it wasn't always so strong. Alfred Russel Wallace saw New Guinea as the home of the first pigeons, but his kind of thinking had died by the time Ernst Mayr became active. One man did more than anyone to destroy it. A bizarre essay penned by American palaeontologist

William Diller Matthew in 1915 – the very year Alfred Wegener drew scorn by proposing continental drift – was embraced in its day as a classic, leading a generation of scientists astray. Central Asia, declared Matthew, was the centre of origin for all life, and in map after map he showed animals flowing out to conquer the world, like soldiers galloping from a fort. His take on Darwin's teaching was heretical: 'Whatever be the causes of evolution, we must expect them to act with maximum force in some one region; and so long as the evolution is progressing steadily in one direction, we should expect them to continue to act with maximum force in that region. This point will then be the center of dispersal of the race.'[69]

Zoogeographers in 1915 who could have chosen Wegener's continental drift embraced nonsense instead, stunting their science for decades to come. Matthew's article was reprinted again and again, as recently as 1974. Modern critics note that his very notion of a centre of origin 'derives from the creation myth of the Bible – the Garden of Eden'.[70]

Matthew did not cite the Bible but he did invoke the Levant, arguing that the old civilisations in the Middle East, India and China showed that *Homo sapiens* evolved near their borders on the great plateau of Central Asia. Mongols, Caucasians and Negroes flowed as spreading arrows across his map. Europe was always invaded from the east, he explained, China from the west. After accounting for humans he provided text and maps showing primates, carnivores, tapirs, birds and everything else pushed outwards by periodic ice ages. Many researchers wanted an oracle like Darwin to lead them through the biogeographical morass, and Matthew's supreme confidence attracted legions of Matthewians.[71]

His portrayal of humans as one species among many was thoroughly Darwinian, but his final logic must have left Darwin writhing in his grave. The poverty of evidence for his theory he explained away by the 'fact' that new races appear during dry epochs when bones seldom fossilise. Joseph Gentilli became his Australian disciple. When continental drift became credible enough to spawn vicariance biogeography, its adherents were so hostile to Matthew that many rejected any spread of species over water, perpetuating extremist thinking.[72]

The great palaeontologist Louis Agassiz, writing in 1850 in the *Christian Examiner and Religious Miscellany*, urged scientists to reject the idea of one centre of origin, insisting that nothing in the Book of Genesis justified this, but even by offering an alternative scenario he could not hold back the tide. Matthew was clear that 'The distribution of modern birds is universally interpreted in terms of Northern derivation', by which he meant the cold lands of today's Kazakhstan, Mongolia and Tibet.[73]

No one invokes centres of origin today, but when we look at songbirds, parrots and pigeons, we see that Australia almost behaved like one, 'pumping' out birds to Asia, Africa and the Pacific. One songbird that wandered out tens of millions of years ago did extremely well, producing today's 3500 species of Passerida, which include everything from thrushes to sunbirds to swallows, making up 35 per cent of the world's birds. But they did not fill enough niches to stop other songbirds from achieving success outside Australia. One invasion gave the world intelligent crows, jays and shrikes, another produced Asia's colourful minivets, another gave vireos to the Americas, bush-shrikes to Africa, and so on. Australia played a similar though more modest role for parrots and pigeons, providing, for example, the ancestors of the grey parrots I saw commuting high above Kenyan rainforest. Tropical Asia is far richer in birds today than Australia, and might always have been that way, but Australian birds did better in Asia than the other way around. Why?[74]

With good brains and good feet for gripping, songbirds had so much going for them that they could succeed in many ways, and the first species to leave, and the second and third, did not carry with them all the innovations produced by early proliferation in isolation. Australia early on suited parrots and songbirds by not having certain birds and mammals, such as rodents. Perhaps this advantage applied as well to the ancestors of songbirds and parrots.[75]

If it is true that perching birds, parrots, falcons and seriemas shared an ancestor, where did it live? Current distributions point straight to Gondwana, since it has the only living seriemas and the oldest lineages of the other groups, as Per Ericson stressed in 2012. But the 'parrot' fossils found in the Northern Hemisphere, if they do come from early parrots, suggest other possibilities. Stem parrots and perching birds could have originated in the north then spread south, where one species from each group, evolving in Australia into something better, spawned all the parrots and perching birds we see on earth today. It is also possible that Gondwana provided the Northern Hemisphere with stem parrots, if that's what they were. Any theory will remain just that until the Southern Hemisphere disgorges more fossils. Unfortunately, information from key ages is completely missing. Australia has so far produced only a handful of fossils from the 30 million years after KPB.[76]

Before I leave this topic I should mention the zone of complicated islands borne of Australia's collision with the Asian plate. The islands east and west of New Guinea offered more opportunities for birds to prosper away from certain birds and mammals, later in time. We know from their DNA that Australia's grey and rufous fantails, and black-faced monarch, had a set of ancestors that spread from Australia to Melanesia and back again. Ernst Mayr, like Darwin before him, thought birds only went from big to small lands, but the DNA has forced a rethink about even very small islands, voiced in articles such as 'Are Islands the End of the Colonization Road?' and 'Is a New Paradigm Emerging for Oceanic Island Biogeography?' Biogeography is maturing, and Australia keeps contributing to that.[77]

My interest in birds turns on their origins and place in the environment. I have talked about the strong bird–plant relationships in Australia, of enhanced polination and cassowaries that assist seeds. I have little more to say about bird origins but there are other bird–plant relationships that make Australia distinctive, and the next two chapters look at these, starting with a visit to a famous region that was fashioned by the journeys of strong-winged birds.

Eight

The Forest Makers

The tourist industry on the Daintree coast of north Queensland is proud to offer customers 'the oldest living rainforest on earth' – a habitat that dates back, some say, 100 million years. On show is 'the ancestral stock of all native fauna and flora in Australia', gushed a tour guide waving at trees in one documentary I saw. 'This is the evolutionary cradle for much of our wildlife,' chimed in the narrator. 'A fragment of Gondwana miraculously alive.'[1]

These claims are far from realistic. Forests of eucalypts met early explorers on many flats where lush rainforest now grows. From the Daintree road you can see grey eucalypt crowns shimmering above dark jungle. Charcoal in the soil records the fires that burned when these trees dominated. The rainforest expanded quickly when burning by Aborigines stopped.[2]

Local tour guides like to point out the lumpy seeds of an extremely 'primitive' tree, ribbonwood (*Idiospermum australiense*). Ancient rainforests accumulate relict trees with no close relatives, and ribbonwood counts as one of these. But the Daintree is not rich in 'primitive' plants. In the Wet Tropics region, of which the Daintree forms a part, they are concentrated in mountains and plateaus further west. From a local biologist, Hugh Spencer, I learned that ribbonwood grows near streams because lowland rainforest during the last ice age clung mainly to stream edges. Three biologists who modelled Wet Tropics rainforest over time concluded that 'lowland rain forests were largely extirpated from the region during the last glacial maximum, with only small, marginally suitable fragments persisting in two areas'.[3]

The Wet Tropics region, which includes the largest tracts of rainforest on the mainland, is famed for its unique rainforest birds. But to see a tooth-billed bowerbird, something that chews leaves with a serrated bill, you must go above an altitude of 600 metres. To see a golden bowerbird, a being of sublime beauty, you must go still higher. For mammals it's the same; when I spotlight in the lowlands at night I seldom see anything better than giant native rats. Diversity is crowded into the mountains because they attracted rain that kept the rainforest alive during long ice-age droughts, and they boast a suite of species missing from the lowlands. A guide to Wet Tropics birds describes the main tract of lowland rainforest, with good reason, as 'rather poor for birdlife'.[4]

Hugh showed me native nutmegs (*Myristica insipida*), one of the main Daintree trees. Their aromatic seeds are spread by pigeons, but nothing about nutmegs suggests a long residence on the Daintree coast. The nutmeg family boasts 440-odd species, of which Queensland lays claim to just four. The fossil record of nutmegs in Australia goes back only tens of thousands of years. Africa, possibly Asia, but certainly not Australia, was the cradle of nutmeg evolution.[5]

The Daintree abounds in plants which, on similar evidence, look like latecomers from Asia – including weeping figs, gingers, bananas, raspberries, and the Daintree's iconic fan palms (*Licuala ramsayi*). Most plants in tropical rainforest depend on birds to spread their seeds across the landscape, and this is the key to understanding how the Daintree rainforests came to be.[6]

Asian Immigrants

Charles Darwin's close friend Joseph Hooker was the first to notice that the vegetation of northern Australia has a strong Asian flavour. He declared that India and the flatlands of the Malay Islands shared the same flora, 'whence it is continued in great force over the whole of tropical Australia'. The essay he wrote this in was praised by Darwin as the most interesting he had ever read.[7]

Australia's northern flora, and especially the rainforest, has since been dubbed Indo-Malayan, Oriental and Malaysian, on the assumption that most of the plants 'invaded' from Asia in the past. What Mayr said about birds was assumed about rainforest from very early on. The zenith of this thinking was a defining article about Australia's flora penned by Nancy Burbidge in 1960. The tropics she depicted as 'rich in Malaysian elements' while also harbouring old Australian endemics. North Queensland had been the main portal for entry and exit of plants, in traffic that was uneven: 'Northward migration of Australian elements has apparently been less successful than southward migration of Malaysian elements.'[8]

Her paper and others like it drew fire in the years that followed, as surging nationalism inspired scientists to reclaim the rainforests for Australia. One concern was that paranoia about the Yellow Peril pouring in from the north would undermine efforts to protect Queensland's forests. To explain how Asia and Australia shared related plants, and often the same plants, learned fingers pointed to India. The new thinking about continental drift marked it out as a piece of Gondwana that had sailed north with a cargo of plants before docking in Asia. But DNA evidence gathered in recent years suggests that, unlike Mayr, Burbidge was right about a northern 'invasion'. A large majority of the shared rainforest plants have their relatives centred in Asia rather than Australia. Very few have fossils of any age in Australia. Botanists have concluded from this that immigration exceeded emigration 'more than nine-fold'.[9]

Burbidge did not envisage an empty vessel that filled from Asia, but a large Gondwanan legacy that was enriched over time by immigration. Australia's northern vegetation is a mix of new and old, Asian and Gondwanan, with the old element strong on mountains and the new doing best at their feet.[10]

As for the Gondwanan inheritance, north Queensland has no monopoly over that. The uplands of south-eastern Australia and New Guinea are also richly endowed. They have relict plants such as beeches and celery-top pines that north Queensland has lost. Every corner of Australia boasts something that qualifies as a Gondwanan relict. All that hype about the Daintree as a miraculous fragment should stop.[11]

A former university lecturer of mine, Ray Specht, was one to invoke a role for India, back in 1988. But timing was a problem. India broke free from Gondwana during the Cretaceous. If we leave the rainforests for a moment and turn to woodland grasses, kangaroo grass (*Themeda triandra*) and black spear grass (*Heteropogon contortus*) grow in Asia as well as Australia, and Specht was persuaded by this that they 'may have remained relatively unaltered' for 125 million years. This would take kangaroo grass further back than *Tyrannosaurus rex*. No one thinks this way today.[12]

So how did these grasses become so widespread? The very different mammals in Asia and Australia show that no land bridge ever existed. Grass seeds sink within days when placed in water, ruling out simple oceanic transport. (Most seeds, with a few exceptions, mainly seashore plants, are designed to imbibe water to germinate when they become wet, and salt soon kills them.)[13]

A particular bird comes to mind. Sleek and brown with slick tapered wings, oriental pratincoles breed on the grassy plains of India, Pakistan, Burma, Thailand, China and Vietnam. The northern winters bring them to Australia. In 2004 close to 3 million were counted on plains and beaches south of Broome, corralled there by quirky weather.[14]

These birds must carry an occasional seed in mud on their legs or enmeshed in plumage. They commute without touching the sea on the way. Often they roost in grass beside muddy pools. Seeds need not travel often to account for the plants we see today: one seed on every hundred million birds would suffice. Kangaroo grass fed the hungry flocks of Australia's ambitious squatters. Did pratincoles make that possible? This seems very plausible but there are other birds that might have brought in this grass.

Hooker was not the only botanist intrigued by all the 'foreign' plants growing naturally in Australia. When Robert Brown sailed around Australia with Flinders in 1802–03 he encountered scores of small plants he knew from England, including purslane, loosestrife and self-heal, to name a few. His list – 'Natives both of Terra Australis and of Europe' – was deemed important enough to be published with Flinders' journal of discovery. But the DNA tools that shook up thinking about birds have done the same for plants as well, albeit in a different way. Usually, species on different continents have been found to be too closely related to have separated when the continents parted. Seeds cross oceans far more often than anyone had guessed, although seldom by floating. A central role is implied for birds. For just as many plants rely on birds to move their pollen, so do many Australian plants need birds to spread their seeds.[15]

Almost Perfect Dispersal

Seeds are the main food of pigeons. Unlike parrots, which husk and crush seeds, pigeons simply swallow them, and they have much smaller heads for that reason. All their food processing takes place in a muscular horny gizzard containing small stones. When I kept pigeons as a teenager I knew to give them grit as a digestive aid. Nicobar pigeons will swallow crystals a centimetre wide.[16]

But one group of pigeons has deviated from the default pigeon diet by not digesting seeds. Fruit pigeons strip flesh quickly from fruits and berries, leaving the seeds to pass unharmed though a soft loose gizzard that never sees grit. Rainforest expert Richard Corlett calls these pigeons 'almost perfect dispersal agents'. Asia's green imperial-pigeon has 'a gape and gullet enormously extensible, ridiculously large nutmegs being swallowed entire, two or three being accommodated in the crop at one time – a seemingly impossible feat'.[17]

The first vertebrates to crawl onto land hundreds of million years ago would from time to time have spread seeds and spores. Any seed that evolved a shell to resist desiccation might have survived simple jaws and a gut if ingested with leaves. The next step would be a coat of food around the seed to invite consumption. The world's oldest evidence for this is the fossil of a small dinosaur, *Minmi*, found in central Queensland with three flesh-covered seeds in her stomach area. Australia's marsupials are oddly unimportant today as consumers of fruit, because birds either got in first or were better at surviving long-term climate change. The one fruit specialist is the musky rat-kangaroo of Wet Tropics rainforest, which keeps to the ground. Many possums eat some fruit, but not in large amounts. Stomachs made for slow extraction of nutrients from leaves do not work well on acidic pulp. Flying foxes love soft fruits, but they only swallow the smallest of seeds.[18]

As for birds, Palaeognaths were probably spreading seeds during the Mesozoic, just as they do today. Among songbirds, bowerbirds are the world's oldest fruit-eating lineage – based on their place on DNA trees, between lyrebirds and everything else – although we can't be sure that their earliest ancestors ate fruit. Parrots usually chew seeds in fruits rather than helping them spread. The key seed couriers in Australia today, judged by their value to plants, are cassowaries and pigeons, with flying foxes coming third.

That pigeons can move seeds was impressed upon me when I stayed in a cabin near Mt Warning, in northern New South Wales, topknot pigeon country. Topknots are steely-grey with long plumes swept back into a brown bun – a topknot. They are darker and sleeker than the crested pigeons in city parks that sometimes go by the same

In Australia's subtropical rainforests the topknot is the foremost 'forest maker', responsible for distributing large numbers of seeds.

name. Some mornings, in a succession of flocks, more than a thousand would pass by. From hillside roosts they would hurtle down the narrow valley towards fruit trees below. Some flocks, hugging the canopy, looked like the shadows of clouds racing over the land. Their wings brought an unsettling roar that sounded as if the roof were lifting off. Groups flying higher recalled the sound of fast-fluttering flags. Distant flocks came to my ears like a wave on its long roll to the beach, or an old car on a gravel road. Groups that loafed in trees, spraying seeds down on quickly discolouring leaves, brought to mind intermittent rain. The largest flock I saw held 120 birds.

I could only wonder about the sounds made in times past, when flocks of 3000 went past. Their reign ended when pioneers brought down the Big Scrub, Australia's largest subtropical rainforest, a feat that was mourned by one Harry Frith, whose grandfather had participated in the destruction by carving his farm from the Scrub near the present-day hippy town of Nimbin: 'Until 1842 no white man had penetrated it and, until 1862 no farmer had dug its soil. But by 1900

the forest was gone and its ashes, washed into the deep red soil had
left not even a black stain on the surface. There have been few more
rapid and complete ecological disasters in Australia's long history of
thoughtless destruction of its resources.'[19]

The remnants today make up less than one per cent of what once
covered 75 000 hectares of fertile basalt soil. I am a patron of a group
working to protect them, Rainforest Rescue.

The grandson of a pioneer, Harry Frith went on to become an
ecological pioneer, one of the first to study Australia's birds in depth.
Upon being made chief of the CSIRO Division of Wildlife Research
in the 1960s, he installed a pigeon aviary in his Canberra office in
order to study them. His academic work cited his grandfather's and
father's recollections as well as his own: 'In 1900 it was possible for the
son of one selector, in one day, to shoot for his family's larder over 300
Topknot Pigeons from a tall fig tree, using a clumsy and inefficient
muzzle-loader.'[20]

Topknots obviously thrive on fruit flesh, for they are not known
to eat anything else – although grubs in fruits may provide essential
protein. There are another seven pigeons in the subtropical rainforests, some of which digest the seeds in fruits and some of which don't,
but none attained such high numbers.

Those large flocks of topknots in the past must have moved millions of seeds. When wetter times returned after ice ages, topknots
would have helped rainforest expand from sheltered refuges, benefiting all rainforest animals.

The services provided to plants by fruit pigeons are exemplary.
In his classic text about seeds written in 1930, Henry Ridley singled
pigeons out for their ability to 'fly very long distances, not only
over land, but far over the sea, and they move very fast, so that they
can disperse seeds swallowed at a very great distance from the spot
from which they took them'. Many frugivores (fruit-eaters) seldom
cross water at all. Bowerbirds and birds of paradise are confined to
the Australian region for that reason. The largest seeds borne across

the seas rely on pigeon wings. One reason for that distinctive pigeon profile of small head and wide chest are the flight muscles, which make up 31–44 per cent of body weight. The wings are large.[21]

Pigeon power won respect during wars thousands of years ago. Domestic pigeons were enlisted as message bearers, historians say, by the Assyrians, Egyptians, Persians, Phoenicians, Romans, the crusaders, and by Britain in two world wars. The Germans tied cameras on them for aerial reconnaissance during World War I. More recently one racing pigeon made the news by flying from Britain to Brazil, while another, from Tasmania, was found among penguins on Macquarie Island, 1500 kilometres away. While they don't spread seeds internally these pigeons show what others can do.[22]

On mainland Australia's northernmost beach, on Cape York, I have stood at day's end watching squads of pied imperial-pigeons, ten or twenty together, heading out to islands low and fast as if on urgent summons. They move over water with the confidence of seabirds. In 1606, some way to the north of me, the expedition of Luis Váez de Torres became the first from Europe to note an Australian bird, when these same white and black pigeons were spied on an isle that had 'plums' with meagre flesh and big stones.[23]

Around the coast to my south lay the ruined homestead of pioneer Frank Jardine, who saw these pigeons repair to islands to sleep in what was 'an unbroken, continuous stream, thousands upon thousands', that took two and a half hours to pass. Dunk Island beachcomber E.J. Banfield wrote in 1908 of a 'never-ending procession from the mainland to the favoured islands – a great almost uncountable host'. Trees swayed under their feathered burden. Explorer-policeman Robert Arthur Johnstone wrote of a deafening flapping of wings, of flocks that could be heard coming from miles away to congregate at night in millions. Early anthropologist Donald Thomson was reminded of fluttering snowflakes, roaring frogs and myriad bees. Their cooing at dusk, to quote Jardine again, was 'one deep, unbroken, monotonous boom'.[24]

On small islands their flimsy twig nests once crowded every tree, and when these were full they laid eggs on bare ground, inside rings of fallen rainforest seeds. These had been voided by the flocks returning from mainland feeding grounds. An early name for these birds – nutmeg pigeon – denotes a liking for that one fruit. The bird and nutmeg tree have nearly the same Australian distribution. Although the bright red flesh (mace) around the nutmeg seed looks insubstantial, it is super-rich in oil (32 per cent), which fuels long journeys.[25]

These amazing pigeons fell victim to shooting parties in boats, sailors restocking larders, and farmers felling rainforest, but the small flocks operating today still shift seeds. On Green Island near Cairns they brought the rainforest back after bêche-de-mer fishermen cut down and burnt it in their fuel-greedy smokehouses, leaving an 'ugly tangle of shoulder high burrs'. Returned by pigeons were nutmegs, palms, figs, lawyer vines and laurels, driving succession that continues to this day.[26]

On every Barrier Reef island with rainforest, pied imperial-pigeons are credited with getting it there after sea levels stabilised. Journeys by these and other imperial-pigeons with strong wings, over time, can explain a long-standing puzzle, namely the 'hybrid' nature of New Guinea. 'The rain forests of New Guinea are a paradox, with a basically Asian flora, but a very un-Asian fauna,' claimed Richard Primack and Richard Corlett in their book *Tropical Rain Forests* (2005), exaggerating slightly. Marsupials and Australian birds occupy the lowlands, which are rich in trees derived from Asia. The Daintree has a subset of these plants. The mountain flora, as we've seen, is more Gondwanan, although some experts overlook that. One theory is that Australian plants proved ill-adapted for the tropics when the continent drifted north, giving an edge to trees with an equatorial past. This might help explain why, when birds were spreading seeds in both directions, Australian trees achieved limited success in Asia.[27]

Pittosporum trees from Australia have perhaps done the best, getting as far as China, Africa and Hawaii, evolving as they went. Glue-like pulp on the tiny seeds buys travel outside as well as inside birds, by adhering to feathers. Other members of family Pittosporaceae lack sticky fruits and never left Australia (appleberries, bursarias, and the like). Doughwoods (*Melicope*) are another group to have reached Asia and Hawaii, but forest loss has set them back. At the herbarium in Honolulu I saw a few dried leaves, all that remains of *Melicope balloui*, last seen in 1927. Of *Melicope obovata* only a photo of dry leaves remains. The leaves themselves went when Berlin was bombed, American botanists having taken photos of the only specimens in the 1930s after anticipating the worst. Five of Hawaii's *Melicope* are feared extinct.[28]

Most Australian trees stayed home, and so did major groups of Asian trees. The leviathans that dominate Asia's rainforests – the dipterocarps – target local breezes rather than birds by producing long wings on nut-like seeds, and only a few reached New Guinea, and none Queensland. Trees with fruits suited to large mammals also stayed behind. Thanks to pigeons carrying the seeds of whatever fed them, New Guinea probably gained bird fruits in higher densities than Asia. When American biologist David Pearson compared rainforest birds around the world he found more fruit-eating birds in New Guinea than anywhere else. Birds of paradise are one manifestation of that and so are all the multi-hued pigeons.[29]

Pigeons big and small and drab and colourful impressed Wallace when he worked his way to the eastern reaches of the Malay Archipelago: 'The two most remarkable . . . groups of fruit-eating birds—the Parrots and the Pigeons—attain their maximum development as regards beauty, variety, and number of species, in the same limited district, of which the great island of New Guinea forms the centre, and which I have proposed to call the Austro-Malayan subregion.'[30]

Pigeons were in his purview when he drew his mighty line – Wallace's Line – marking out Asia proper from Wallacea, a transition zone of low diversity between Australia and Asia. Having worked each side of the line, I can see why it is drawn where it is. In ten days' field-work in lowland Borneo I encountered pigeons just three times, while in Sulawesi I could not escape them cooing and chuckling and clapping overhead. Often enough I was listing fruit pigeons in my notebook in the first hour of each morning. The black-naped fruit-doves were my favourites – little green, white, yellow and red things that scampered over foliage like frisky squirrels. Borneo has fourteen pigeon species, New Guinea, of similar size, has forty and much smaller Sulawesi twenty-one.[31]

Why such differences? One reason of course is the southern origin of these birds. I mentioned the dodo in the previous chapter, but the pigeons that move seeds fall into a different group, the fruit pigeons. It was a world-changing event when somewhere in Australia, or per-haps on a large island nearby, one line of pigeons switched from a diet of seeds plus fruit pulp to one of pulp alone. The pigeons that made that shift left unusually large numbers of descendants. The largest pigeon genus on earth is that of the fruit-doves (*Ptilinopus*), and imperial-pigeons (*Ducula*) have also done well. Fruits provide a poor food for growing birds and the new diet was only possible because pigeons, like flamingos and some penguins, feed their chicks 'milk' – a thick secretion containing more fat and protein than does human milk. It is disgorged from the crop (part of the digestive tract) of both parents.[32]

In a region stretching from India to Polynesia the consequences of that dietary shift were profound. Pigeons spreading out of Australia became the forest makers on remote islands. As the largest and often the only frugivores on hundreds of islands, they provide the only explanation for trees such as rainforest palms and nutmegs with large, non-floating seeds. Had that first seed-friendly pigeon not evolved, forests would be poorer today, not only on Pacific islands but in Indonesia, New Guinea and Queensland. The nutmeg of commerce is one tree that exists only because imperial-pigeons in Maluku (the Moluccas) dote on the oily mace surrounding the seeds. Some islands have lost their large pigeons to hunting, raising questions about the future of rainforest.[33]

In the Daintree the pied imperial-pigeon seems pivotal. Wallace 'discovered' this bird by obtaining the first specimens for science on the Aru Islands, below western New Guinea, where traders kept them as pets. Its range is idiosyncratic, extending from Queensland to coastal New Guinea and its islands, across to Maluku, and taking in islands offshore from Java, Borneo, Malaya and Burma. On mainland Australia it uses most of the lowland rainforests, but in Asia keeps to tiny islands.[34]

On first impressions this makes no sense, but there is a way to understand it. With Melanesia in mind, Ernst Mayr and Jared Diamond have called this pigeon a 'supertramp', a bird excluded from species-rich or large islands by competition. Its habit of nesting on small islands suggests an origin as an island bird. It is said to drink seawater.[35]

The monumental flocks in Queensland are a sign of ecosystem poverty. Earlier birds went extinct when the lowland rainforests shrank during ice ages, thus creating a sudden opening, when higher rainfall brought the rainforest back, for a small-island opportunist to move in. These pigeons avoid the mountains, which retained some rainforest and some fruit-eating birds. They inhabit reconstituted landscapes, breeding on islands which, like the lowland rainforests, reformed after the Last Glacial Maximum ended 18000 years ago. Imperial-pigeons are not the only lowland frugivores, but Queensland has surprisingly few. Winters are spent by them in southern New Guinea, away from major rainforests, in savannas, riverine forests and other habitats with few fruit-eating birds. Millions of imperial-pigeons flying down from New Guinea each year provide a comfortable explanation for Queensland's 'Asian invasion'.[36]

North-western Australia has these pigeons as well, but instead of spending winters in New Guinea they circulate between small 'islands' of rainforest in the region, as they do in Asia, although the 'sea' they cross is one of savanna. There are more rainforest patches today than during ice ages, and many are expanding. Those near the coast have large numbers of succulent fruits, biologists have noted, implicating birds and bats in their birth and maintenance.[37]

The Northern Territory shares an amazing 40 per cent of its rainforest trees with Asia, and 14 per cent with Africa. Each ice age would have deleted some Gondwanan plants, leaving openings for northern imports when the monsoons returned in force. The region lacks mountains large enough to protect ancient rainforests by guaranteeing rain. One of the main trees, black lillypilly (*Syzygium nervosum*), has the genetic signature of a new Australian. Its native range includes India and South-East Asia but not (yet) Queensland. Some of the rainforest patches it dominates are only a few thousand years old, on silt deposited after sea levels stabilised. Its fruit is so tasty I am not surprised it has reached many places, including remote Christmas Island. Its success in the Northern Territory is attributed to imperial-pigeons and bats circulating its seeds between rainforest islands. Flying foxes (fruit bats) only swallow tiny seeds, putting them well below pigeons as 'friends' of trees.[38]

More Fruit Couriers

I have talked up pigeons for good reasons, but there's another fruit-loving bird in Queensland that advertises rainforest poverty. Noisy and energetic, metallic starlings have bulging, blood-red eyes, shiny black coats, and a habit of scampering over leaves to feed. Any tree they roost in becomes like a crowded country hall in which everyone chatters and no one listens. They do everything in groups, including flying down in summer from New Guinea to breed in boisterous colonies in tall trees. Aboriginal people had a calendar tree whose ripening fruit heralded their flocks: the nutmeg. Below starlings' nests, nutmegs accumulate, although these small birds do not carry them far. Under one nesting tree I found a carpet of rainforest seedlings so dense they looked like alfalfa sprouts. A month later they were dead, every one, withered by a dry spell.[39]

Only a few Australian songbirds deserve English-sounding names, and this is one. It is a true member of the starling family and one of a small number of passerid songbirds that 'returned' to Australia. Earlier on I emphasised clever songbird nests, but most starlings, because they build in protected holes and crevices, get by with rude bowls. The nests of introduced starlings, sited under eaves or in walls, are only ever described as 'untidy' or 'roughly constructed'. Evolution dictates that useless skills are lost.[40] The metallic starling is one of only a few starlings to build a globular nest in the limbs of trees, but being poorly built and crowded together, the nests often falls out. One naturalist observing them in 1929 noted that 'down came babies, cradles and all'. Reptiles loiter below for the hapless 'babies'. No other bird I know of does so badly from sudden gusts and snapping limbs. I have watched these birds tug tendrils from vines to serve as spiral clasps on their nests, and these are poor substitutes for good weaving skills (see colour section). African weavers also crowd vast numbers of globular nests into trees, but the contrast in quality is so striking it is almost embarrassing. Why should this be?[41]

Evolution operates at a great variety of speeds, and metallic starlings mark the fast end of the spectrum. They turn out to be part of a species 'swarm', an explosive radiation in the genus *Aplonis* which, DNA researchers say, is characterised by very low genetic divergence. The metallic starling evolved recently from other black starlings, which look nearly the same but nest in crevices rather than on branches. It is one of Australia's younger bird species, and its change in nesting style represents an extension of what starlings do. Weavers make better nests because their skills go back millions of years.[42]

Why adopt a risky new nesting style? Why forgo the security of nooks and crannies for the vagaries of branches? The obvious reason would be to overcome a shortage of cavities. If birds were multiplying beyond the number of tree holes, the drive to breed would force experimentation. As many as 400 metallic starling nests have been found in one tree – an impossible density were crevices to be used. These birds can produce three broods of up to four young a season, a phenomenal output for a frugivore and one that more than compensates for flaws in craftsmanship. They roam in much larger flocks, of up to 5000 birds, than most *Aplonis* starlings, which go about 'in pairs or small groups'.[43]

One thing could drive this success: superabundant fruit. The range of this bird is centred on New Guinea – the land of abundant fruit. Starlings probably spread south when each recent ice age ended and rainforest in Queensland expanded. The point here is that the two main fruit consumers in the Wet Tropics each suggest a young, reconstituted ecosystem, not 'a fragment of Gondwana miraculously alive'.[44]

Something else about these birds is unusual – their habit of migrating. Birds that live entirely on fruit are rare among the ranks of seasonal migrants. Africa has no fruit migrants while Australia has five.[45]

One night on Cape York I shone a spotlight into the northernmost trees on mainland Australia to find branches laden with sleeping channel-billed cuckoos. These big grey birds live largely on figs, and this flock was en route to New Guinea for winter. A drooping wing marked one out as unfit to complete the journey. The world's largest parasitic cuckoos, these mysterious birds slip their eggs into the nests of magpies, currawongs, choughs and butcherbirds. Another night, after hearing its piercing calls, I spied one on a crow's nest near my house. The agitated crows were lucid enough to abandon the nest. Cuckoos only lay eggs in the nests of songbirds, so it is apt that a land of large songbirds should have the largest cuckoo cheat. Australia, the home of songbirds, is richer in cuckoo parasites than its size would suggest, and could be the place where the deception first began. Channel-billed cuckoo eggs mimic those of currawongs rather than crows, suggesting that currawongs became their victims first, giving evolution time to fool these birds by producing look-alike eggs.[46]

The other seed couriers that migrate between Queensland and New Guinea are the koel (another fig cuckoo) and superb fruit-dove. The five fruit migrants can be found as well on the archipelagos just north of Australia, which, by offering fruit in an island setting, may have encouraged these birds to travel widely, with ensuing great benefits for plants.

Seed Highways

Birds move plants in every habitat, not just rainforest, including plants with no food around their seeds. Some seeds become sticky when wet, or have bristles or hooks, to suit them for journeys on the outside of animals. Other tiny seeds travel when mud sticks to birds, or when they're pecked up but not digested. Plants with such seeds can turn up anywhere, including places where birds seem unlikely couriers.

Some of the plants on the Australian Alps are shared with mountains far across the sea. Large numbers are shared with New Zealand and New Guinea. One sedge (*Carex jackiana*) goes further, gracing peaks in Java, Sumatra, Malaysia and India as well. Star sedge (*C. echinata*) takes in mountains as far off as Europe and North America. Tufted hairgrass (*Deschampsia cespitosa*) turns up in cold places all over the Northern Hemisphere. I have seen it in northern Sweden. Bristle grass (*Trisetum spicatum*) is the most widespread example, with a presence in South America as well.[47]

A book about the Kosciusko alpine flora lists forty-seven small plants with an international distribution – nearly one-third of all the plant species, a proportion demanding an explanation since it is so much higher than the average for Australia. The situation is stranger than it seems because many alpine plants that *are* confined to Australia (cresses, carraways, eyebrights) have near relatives on northern lands, implying older but equally improbable journeys.[48]

In times past, alpine plants were seized on as evidence of simultaneous creation. Designing plants perfect for one summit, the Creator also bestowed them on others. While rejecting this, Darwin's supporters could offer nothing better. Look at how Hooker struggled with the problem in the essay I mentioned before:

> When I take a comprehensive view of the vegetation of the Old World, I am struck with the appearance it presents of there being a continuous current of vegetation (if I may so fancifully express myself) from Scandinavia to Tasmania; along, in short, the whole extent of that arc of the terrestrial sphere which presents the greatest continuity of land. In the first place, Scandinavian genera, and even species, reappear everywhere from Lapland and Iceland to the tops of the Tasmanian alps, in rapidly diminishing numbers it is true, but in vigorous development throughout. They abound on the Alps and Pyrenees, pass on to the Caucasus and Himalaya, thence they extend along the Khasia mountains, and thence to the peninsulas of India to those of Ceylon and the Malayan archipelago (Java and Borneo), and after a hiatus of 30°, they appear on the alps of New South Wales, Victoria, and Tasmania, and beyond these again those of New Zealand and the Antarctic Islands, many of the species remaining unchanged throughout! It matters not what the vegetation of the bases and flanks of these mountains may be; the northern species may be associated with alpine forms of Germanic, Siberian, Oriental, Chinese, American, Malayan, and finally Australian and Antarctic types; but whereas these are all, more or less, local assemblages, the Scandinavian asserts its prerogative from Britain to beyond its antipodes.[49]

A continuous current, a Scandinavian prerogative, what did Hooker mean? The problem runs deeper than it seems. Outside the Australian Alps, vast numbers of southern Australian species, usually of small size, have near relatives growing far north of the equator. Many a eucalypt casts thin shade over native mints, carrots or raspberries.[50]

In Hawaii the question of plant origins weighs heavily, and birds are often invoked as part of the answer. That these islands rose naked from the sea is not contested, since they are still rising. Near the Kilauea Caldera I have walked over the hot slabs of recent lava flows above a buried highway demarcated with blistered and bent 'No Parking' signs, to sit at night with hundreds of people watching glowing lava strike the sea.

An estimated 75 per cent of Hawaii's plant lineages came with birds. (The others had seeds that could float or blow across the sea.) Wondering where Hawaii's woody violets came from, two biologists were surprised to find the genetic link was to a little violet in Siberia and Alaska. Twenty-one species of bird, mostly waders, commute each year between Hawaii and the Arctic. Violet seeds readily survive digestive acids, although they might also travel on plumage. Many waders that visit Hawaii from Siberia and Alaska, including tattlers, turnstones, godwits and golden plovers, also reach Australia.[51]

Much about waders recommends them as vectors for small seeds. In their millions they cross between hemispheres, staying aloft for days at a time, never landing on the sea. One radio-tracked godwit caught in headwinds flew nonstop for nine days across the Pacific. Waders in one study were regurgitating viable seeds up to 340 hours after ingestion. A wader flying for that long could reach Victoria from Siberia. Although they feed largely on animals, waders have been found with seeds in their guts and faeces, eggs and seeds on their bodies, and bivalves clamped to their toes that were heavy enough to break them off.[52]

During his reign at CSIRO Harry Frith found time to study Latham's snipe in the Australian Alps. One photo he published of their habitat showed a scene dominated by that plant I saw in Sweden, tufted hairgrass. Snipe are waders that feed largely on earthworms, but, in an operation that would be forbidden today, Frith's team dissected 500 of these birds and found seeds inside most of them, mainly those of sedges, rushes, grasses and buttercups.[53]

Latham's snipe breed on the other side of the world – in meadows and grasslands in Japan, including on Mt Fuji. They fly directly to Australia without stopping in South-East Asia. In New Guinea, one was found in the Snow Mountains at 3550 metres. The related Swinhoe's snipe proceeds from Mongolia to the high peaks of Indonesia and New Guinea, and sometimes comes further.[54]

Snipe could account for the strange distributions of some sedges and rushes. They could also explain the overlap between England, Australia and Japan: the latter has nearly every plant on Robert Brown's list of shared species, or a close relative. In *On the Origin of Species* Darwin remarked on Australia and Japan sharing plants. Ducks and other birds moving locally between wetlands would have spread plants between Europe and Asia, leaving snipe to bring them south. One snipe I flushed in southern Queensland was resting beside a native raspberry, *Rubus parvifolius*, a species shared by Japan, China and temperate Australia.[55]

Something like thirty-five species of wader visit Australia to escape winter in China, Mongolia, Siberia and Alaska. Travelling south through Asia, some of them spend time at freshwater wetlands. Birds that might have acquired seeds from such places include the wood sandpipers, marsh sandpiper, greenshank, and long-toed stint I saw one spring at a lagoon near Darwin. The sandpipers were strutting like carnival stilt-walkers, but jabbing quickly as if frantic to replace fat lost during travel. On the far side of the lagoon Aboriginal women were doing their own jabbing, with sticks. One held a long-necked turtle by the neck, while men in the reeds sat watching. These waders had paused on arrival in Australia, but others probably keep going before they land.[56]

Seeds travel north as well as south, and Asia has an interesting 'Australian' contingent to show this (which I list in the source notes). Fringed lilies are popular Australian wildflowers, with one (*Thysanotus chinensis*) that has reached Asia. Like nearly all the plants mentioned here, its seeds lack wings or plumes for airborne travel. Earlier travellers include Europe's sundews, which can be traced by their DNA back to Australia. Tasmania and New Zealand share 200 plant species, of mostly Australian origins, and a wader that commutes between them – the double-banded plover – helps explain that. It often feeds on grasslands, sometimes on mountains.[57]

Freshwater plants have some of the widest distributions in the world, a situation that can only be explained by invoking birds. Ducks and herons moving between wetlands can, when considered over a large enough scale, account for otters in Britain and platypuses in Australia swimming among some of the same plants. Waterbirds can explain why lagoons in tropical Australia and Asia share most of the same algae, and why Norfolk Island has a freshwater snail (*Posticobia norfolkensis*) whose nearest relatives are 1400 kilometres across the sea in Australia. Bird travel tens of millions of years ago has been proposed to explain how one freshwater plant (*Aglaoreidia cyclops*) could leave its pollen in Eocene freshwater sediments in Europe, North America and Australia. A few years ago, at a sewage farm in Darwin, I saw dark legs rising above yellow feet, marking out the owners as little egrets born in Asia, telling of one journey that might have aided plants. Waterbird services are studied the world over, often because seeds, snails, leeches, and the like are readily found on ducks shot by hunters. Some crustacean eggs are like seeds in sporting spines, or becoming sticky when wet, or having shells that readily survive stomach acids.[58]

By what they have carried, birds have left their mark everywhere. On northern lands they helped oaks, beeches and chestnuts march north after the last ice age. In their winter larders blue jays store thousands of seeds, carrying them miles at times before hiding them in shallow soil, then failing to retrieve them all. In the Swiss Alps, above the tree-line, I have seen young pines sprouting in groups from the forgotten caches of nutcrackers. Far to the north, geese are credited with re-greening Greenland after glacial retreat.[59]

People also benefit from the work birds do. Black lillypillies and lotus lilies are just two of many Aboriginal foods whose spread can be

attributed to birds. Beneath unwelcoming eucalypts, European settlers found plants they could recognise: violets, buttercups, cresses,
mints, flax, to name a few. Into medicine chests went native cen-
taury, linseed, speedwell, selfheal, brooklime and pennyroyal, while
raspberries and elderberries were served in homestead kitchens. These
Australian plants were trusted because they were known.[60]

The contribution of birds remains undervalued since it has to be
inferred rather than observed. Until recently the focus was elsewhere.
When a few decades ago continental drift took centre stage, moving
lands were invoked to explain most distributions. This seemed plausible until DNA testing savaged the paradigm. Two biologists summing
up a symposium in 2004 reached the 'surprising conclusion' that
'long-distance dispersal may be more pervasive than all but its most
ardent advocates (e.g., Iltis, 1967) had previously proposed'.[61]

Dispersal is inconvenient. Botanist Robert Hill explained why.
'No doubt many biogeographers wish that long-distance dispersal had
not occurred – it is likely to be capricious and thus difficult, if not
impossible to reconstruct in what can be regarded as a scientifically
stringent way.' But the international flavour of the Australian Alps
shows that seed movement need not be capricious; bird flyways are
seed highways.[62]

Understanding lags because those collecting DNA data don't
always know what to make of it. A paper about sedges (*Caustis*) shared
by continents invoked 'a relatively recent, long-distance dispersal
mechanism, such as mountain hopping'. But mountain hopping is
not a mechanism. Plants do not hop, not even between mountains.
Given the direction of rotation of the planet and the winds that flow
over it, I can find no science to suggest that alpine seeds could blow
from one hemisphere to another, or sail on volcanic vapours, or proceed in any other way, although they probably do so on exceptional
occasions, even though most of them lack wings or plumes. Mountain
hopping is like Hooker's 'current of vegetation' in lacking a meaning
and a valid place in science.[63]

For lowland plants such as kangaroo grass, possibilities are provided by floating islands (rafts of vegetation and soil that dislodge from tropical river banks) and by large fallen trees with soil bound to their roots. Often observed in the eighteenth and nineteenth centuries, floating islands are seldom seen today, presumably because there is less forest along rivers to dislodge. Writing in 1930, Henry Ridley told of a big tree that arrived on the Cocos-Keeling Islands with dirt on its roots along with a clump of sugarcane that was subsequently planted. When debris runs aground, any seeds upon it, to escape ruin from salt water, have to get past the beach to proper soil. For small seeds at least, birds provide millions more opportunities than do floating trees.[64]

Since birds are the only way to explain all the improbable alpine and freshwater distributions, this argues for a role in other habitats as well, where the evidence is less clear. Birds receive their due in studies of small islands, rainforest and freshwater ecosystems (including studies of snails and crustaceans), but no audit has ever been done of all the services they provide. What is clear is that much of the world's greenery has birds written all over it, and that Australia and other lands are richer in plants because of this.

Unwelcome Forest Making

Bird dispersal will attract more interest as climates change. For just as we want birds to aid plants by moving pollen, so will birds carrying seeds help plants track their climatic space. But dispersal has become dysfunctional as unwelcome seeds are being spread.

Harry Frith saw his biggest topknot flocks around piccabeen palms. 'The beating of the wings and the rattle of the displaced berries falling is audible for hundreds of yards,' he remarked in 1952. Having trunks that neither thicken nor branch with age, palms attain a finite height then allocate all their output to reproduction, resulting in regular generous crops.[65]

The flocks I watched were not visiting palms. The seeds raining down were those of camphor laurels, one of Australia's worst tree weeds.

'In 1955 there was a great failure of the fruit crop in the rainforest,' wrote Frith, 'and great numbers of Topknot Pigeons came out into the open dairy country, searching widely for food. They began feeding on the ripening berries of camphor laurel, which was widely planted as a shade tree after the destruction of the native rainforest. This has now become an annual event . . .'[66]

Where dairy farmers kept the land tidy, these Asian trees are now advancing over hills and down streams. Forest is returning, but forest comprised of one tree. Fruit birds such as topknots have rebounded from the 1970s, when Frith seldom saw more than fifty together, but insect-feeders gain nothing because the camphor in the leaves is a potent insecticide, the ingredient in mothballs. Except when there is ripe fruit, the groves are lifeless. Free of insect enemies, the trees bear reliably and bountifully. The topknots I saw flew straight past the meagre crops of native laurels to reach them.

In Orara Valley, near Coffs Harbour, one biologist traced tens of thousands of wild camphor trees to just three that had been planted at the school in 1902. Who is more culpable: humans for planting three seeds or birds for 'planting' thousands?[67]

Birds are complicit in many other weed problems. Pioneer James Fenton confessed to being one of the 'miscreants' who, by planting three sprigs, inflicted blackberries on Tasmania:

> I put out my valued cuttings in my garden on a rich plot of chocolate soil. I watered them! I watched them! I weeded them! with all the tender care of a nurseryman, and had the unspeakable pleasure of seeing them put forth buds in due time . . . Two or three years after that date the blackberry quite astonished the settlers (including myself). Round the stumps and logs in every direction young blackberry bushes grew up luxuriantly and spread

out violently. At first it seemed mysterious how they managed to spring up so far away from the parent bushes, but it soon was apparent that the birds were fond of blackberries, and that they carried the seed far and near.[68]

Blackberries won a place on Australia's list of its twenty worst weeds in 1999. Six of the weeds on that list (Weeds of National Significance) owe a debt to birds, including the pasture grass hymenachne. At a meeting in Darwin in 2009 I was stunned to hear that the fight to remove this rank water grass from Kakadu National Park had been lost, that it keeps finding its way to remote lagoons, carried in, it seems, as fine seeds on birds. On an adjoining station I saw where this grass had been planted to feed cows through the dry. Those iconic wetlands with their waterlilies and magpie geese, potent symbols of wilderness, will require herbicides in perpetuity to suppress a smothering weed. The Northern Territory's magpie-goose management plan identifies hymenachne as a 'significant long-term threat' to the bird, a problem partly of its own making.[69]

The main mutualisms between birds and plants – pollination and seed dispersal – are unusually strong in Australia, as shown by nectar serving as a major food and pigeons and cassowaries providing exemplary dispersal. Weeds provide the most powerful evidence of plants benefiting from dispersal, but it is evidence we could do without.

Nine

Of Grass and Fire

By a very large margin, grasses are the world's most successful plants. A third of all global vegetation is grass-dominated, a figure that rises to one half in Australia. Grasses provide six of the world's top seven crops, feed our herds and flocks, and surface lawns and parks. Songbirds are the world's most successful birds, humans the dominant mammal, and grasses are their plant equivalents.[1]

Ask why, out of more than 600 plant families, one should so dominate global production, and an odd idea comes up: grasses thrive on their own destruction. They dominate pastures and wild places by sustaining heavy consumption by animals and fire. They specialise not in protecting themselves against damage but in quickly replacing themselves afterwards. They can do this because they are exceptionally productive, fixing carbon more efficiently than most plants, in leaves that are short-lived and cheap to replenish from growing points kept near the ground. They are the only plants to suit lawns for this reason.[2]

Fire was a partner to their success. Many grasses are helped by and help fire. The idea of flammability as an adaptation was first put by American Robert Mutch in 1970, and it has advanced in journal papers such as 'Kill Thy Neighbour' and 'Are Some Plants Born to Burn?' Some ecologists insist flammability cannot evolve in plants, but claims that adaptations to fire are really adaptations to grazing and drought miss the point that many plants benefit greatly from fire.[3]

In most fires grasses are the main fuel, providing the continuity flames need. Grasses and eucalypts benefit alike from regular fires destroying their competitors, but trees need occasional fire-free years for their seedlings. A succession of hot grass fires will kill even adult trees, allowing grasses at times to displace rainforests, eucalypts and acacia woodlands. Many grasses thrive on yearly fires, and some can survive two a year.[4]

When fires are stopped, trees often thicken up and suppress grass. In national parks along the east coast, missing out on Indigenous burning and on lightning fires that don't travel far because roads and farms serve as firebreaks, rainforests are advancing via seeds dropped under eucalypts by birds. Australia would carry far more rainforest (supporting rainforest birds) had grasses never evolved. The area of forest on earth would double were fire switched off, according to modelling by leading biologists, with Africa and South America gaining the most.[5]

Although dinosaurs are never portrayed eating grass, in 2005 *Science* carried an article about fossil coprolites (dung) showing that late-Cretaceous dinosaurs did just that. All the oldest grass fossils come from Gondwanan lands (although not from Australia), suggesting a southern start for another dominating group. Three rush-like plants in Western Australia – making up one of the least successful of all plant families today (the Ecdeiocoleaceae) – are their closest relatives.[6]

Grasses came to success slowly, advancing in steps. Fossils reveal surges in pollen or charcoal as the planet cooled, dried, and became more seasonal, acquiring monsoon climates in which a wet growing season alternates with a dry burning one. The electrical storms that mark the beginning of monsoons often spark fires over vast areas a week or more before the first rain arrives.[7]

Domination by grasses came at different times on different continents, with Australia, going by fossil pollen and fossil teeth, lagging

well behind. Her marsupials were late to evolve the high crowned
teeth needed for abrasive grasses and few did so, grassland kangaroos
and wombats only appearing a few million years ago. The biggest kangaroo ever, the extinct *Procoptodon goliah,* ate saltbushes rather than
grass, judging by the wear marks found on its teeth. The timing meant
that grass birds from Asia could move into grasslands before many
indigenous birds had evolved to fill the niches.[8]

Grass-assisted Invasion

One article by Ernst Mayr, 'Timor and the Colonisation of Australia
by Birds' (1944), stands up better than his other Australian work
because it mentions some birds that really did enter from Asia:

> It is a striking fact that most of the recent arrivals in Australia
> from Timor are grassland birds. This would seem to indicate
> that grasslands were the prevailing type of habitat on each side
> of Timor Strait during the period(s) of the greatest proximity of
> Timor and Australia, and this in turn would lead to the conclu-
> sion that the climate at that time must have been as dry or even
> more arid than it is to-day.[9]

About all of this Mayr was correct. Fossil pollen has been obtained
by Australia's Ocean Drilling Program from 5200 metres under the sea
on the Argo Abyssal Plain, and it shows that north-western Australia
had grassy woodland at least 7 million years ago, which became
pure grassland – the oldest in Australia we know about – a couple of
million years ago. If birds came from Timor via small islands, they
only had 90 kilometres of sea to cross at a time. Cyclones can carry
birds much further than this. Timor's honeyeaters, lorikeets and
pigeons show that birds also went the other way, although the island
is too young – only a few million years old – to have helped early
songbirds reach mainland Asia.[10]

Talk of an 'explosion' may go too far, but the word has been applied to the way in which one songbird spawned the Passerida, that giant, rapidly evolving group that makes up a third of the world's birds. Most of their success occurred outside Australia, but some of them flowed back from Asia, and they are prominent on Mayr's list of grassland birds: pipits, bushlarks, reed-warblers, cisticolas. Any bird you spy in grass in the north of Australia could well have a recent Asian past, which is not something you could say about any other vegetation. Grasses abetted an invasion, lending their success to birds from Asia. At a remote dam in the Northern Territory, in a vast plain of treeless Mitchell grass, I watched hundreds and hundreds of tittering Horsfield's bushlarks come to drink one morning, and they outnumbered other birds a hundred to one. These chirpy characters enliven grasslands as far away as China and Burma. Java comes up in their scientific name: *Mirafra javanica*. Recent DNA work puts the age of the species at under 1.2 million years. Their time in Australia could be less than half that. Other birds invited in by grasses include finches, owls, quail and birds of prey.[11]

Some of the grasses are themselves new. Australia was made more inviting by pratincoles or other travellers bringing in, for example, cockatoo grass, a pivotal finch food. The Asian influence shows up strongly in Australia's tropical rainforest trees, and in the grass birds and smaller plants of the savanna, although 'Asian' can be too simplistic a term. The bird with the quirkiest name, the zitting cisticola, spread through Asia from Africa. I have seen it in Queensland, Zimbabwe, and in Bali 'zitting' above fields of rice, a grass it likes.[12]

Travel attaches to certain lifestyles. The sickle wings that steer swallows to insects high above fields helped them become one passerid success. The world's most widespread songbird, found on every continent bar Antarctica, is the barn swallow. But wings matter less than minds in determining who goes where. A willingness to explore helps some birds find islands of grass inside forest, and when the forest reclaims these, to go somewhere new. New Guinea's mannikin finches and pied chats soon find the grasses that inherit gardens carved from rainforest, and not by coincidence can pied chats be found as far away as Iran.

Birds that live on grain maximise food-finding and safety by usually roaming in groups, a habit that suits colonisation. (Arrive alone and you can expect to die alone.) Mannikins are the ultimate colonisers. DNA trees suggest four sea crossings in their past: a departure from Australia by the first passerids, the return of finches millions of years ago, a spread back to Africa (via Asia) where mannikins evolved, then expansion by some manakins (presumably again through Asia) to Australia. The first finches reaching Australia stood to benefit from a complete void of songbird seed specialists.[13]

Australia was a global centre for the world's three main seed-eating groups – parrots, pigeons and songbirds – but songbirds only embraced seed after they left Australia and the competition imposed

Finches represent one of Australia's newer groups and recent success stories. Australia has very old bird lineages living alongside young ones. These are chestnut-breasted mannikins, which live among rank grasses, including sugar cane.

by parrots and pigeons. That's what the evidence implies. On every continent outside Australia, songbirds are the dominant seed birds, in which role, noted three Swedish biologists, they are 'among the most successful within the entire class of birds'. The winners include sparrows, 'finches' (a word applied to several groups), and Africa's red-billed quelea, the world's most plentiful bird. Flocks swirl along valleys like swarms of impatient bees. To protect grain crops, up to 100 million queleas are killed each year in South Africa alone, while elsewhere in Africa dangerous poisons and giant fires are set upon them. Australia does have some 'old' songbirds with mixed seed-insect diets, including grasswrens and white-faces with beaks approaching those of finches, but these have not achieved much success.[14]

Some passerids are so made for seeds they can crack open cherry stones and pine cones, but Australia was not reached by the stronger-jawed passerids, only by the grassfinches (Estrildidae), which entered multiple times. We now have a contest over grain between two groups with superior beaks, one old, one new, the multifunctional parrots and the fast-evolving finches.[15]

Darwin was alerted to natural selection by the beaks of finches that had changed so fast he could see it had happened. The Galapagos finches display exceptional plasticity, evolving quickly to exploit different foods. Researchers on one island detected changes in just twenty years, when seed size changed after a drought. American house finches, because of differences in diet, have bills that vary between adjoining urban and rural flocks. A single gene – Bmp4 – profoundly influences bill shape. Those vivid red 'i'iwis I saw in Hawaii resemble honeyeaters, but are recently descended from finches (family Fringillidae) that gained tubular beaks after chancing upon Myrtaceae nectar. Finches evolve fast by breeding and dying much younger than most Australian birds. Zebra finches can nest when two months old and they only live twelve years, compared to a budgerigar's twenty-one.[16]

For a very long time before it had finches or grasslands, Australia had parrots. We can be confident they predated Australia itself if we define it as an island rather than a piece of Gondwana. Their beaks evolved inside rainforests to breach kernels and nuts, or perhaps inside woodlands bearing banksia cones and gum nuts, tens of millions of years before Australia had finches husking grain.[17]

Evolution retrofitted some parrots for grass seeds, but there were limits to what could be achieved. For pecking small seed from rough ground, pointed finch (and pigeon) beaks are much better than curved mandibles mounted on large heads. Finches eat many of the same seeds that feed much larger parrots and pigeons, so that during times of scarcity, which sometimes become times of extinction, they can not only peck faster, but live from fewer pecks. After finches began teeming into Australia they may have replaced small parrots during ice age droughts. Small parrots remain plentiful in the arid zone, where an ability to travel further between water and food could explain why budgerigars thrive alongside smaller zebra finches.[18]

The concept of competition favouring one whole group over another is not one I am drawn to, but it does suit the evidence. Experts have noticed that the northern Australian savannas are as rich in finches as they are poor in small parrots, while in southern Australia it's the other way round. Competition has been invoked to explain why, during a time of scarcity, hooded parrots – one of the few small parrots in the Northern Territory – gave up eating fallen spinifex seeds before local finches did. The south now has other finches – introduced sparrows, goldfinches and greenfinches – posing a threat to Australia's most endangered bird, the orange-bellied parrot, because they feed in winter in the same locations, in flocks that can outnumber the entire parrot population. The biologist tasked with saving the species told me that to ease competition the finches should be culled. Introduced finches are a concern as well for endangered parakeets in New Zealand.[19]

The only bird lost from the mainland since European arrival was a close relative of the hooded parrot, one that overlapped in range with eight finch species. Explaining the demise of the paradise parrot has become a popular guessing game, with a catalogue of possibilities that include sheep, cows, foxes, cats, fire, drought, ringbarking, tree thickening, prickly pear invasion, poison, trapping, termite mound removal, and combinations thereof. Penny Olsen's colourful book about this

The paradise parrot achieved far more fame by going extinct than it ever could have acquired by remaining alive. The only photos were taken when naturalist Cyril Jerrard discovered the last known individuals. This image he took in 1922 has become one of Australia's most widely published wildlife photos.

bird, *Glimpses of Paradise*, runs through them all. The last problem was unusual, but paradise parrots nested in termite mounds, which, crushed and blended with water, made durable homestead floors and country roads, and sturdy tennis courts, like the one on which champion Pat Rafter practised as a child. The impact of humans on this parrot may have been compounded by competition from finches during lean times.[20]

The Volatility of Savanna

One reason for my dwelling so much on competition is that nature is portrayed poorly whenever harmony is implied. Leading ecologist Charles Elton dismissed the notion of a balance of nature back in 1930, and no biologist since then has risen to its defence, except for some who flirt unwisely with visions of Gaia. Harmony and balance are Greek aesthetic ideals with no rightful place in biology. In most habitats there is ebb and flow and occasional catastrophe, even over short time scales, with some species increasing while others founder. Because of its glacial cycles, the last few million years have been especially unstable. New birds can drive change, but climate has been the main force. Savannas are particularly volatile because grasses are ultra-responsive to shifts in rainfall, amplifying the impacts of these by spreading – or not – fire. Heavy outback rains can raise so much grass that shrubs and saplings over vast areas are killed by the next inferno, which becomes a forest-delete button.[21]

Grasses are highly responsive to rain because they are super-efficient chemically and physically, using low-cost tissues to fix carbon in large amounts. They produce a lot of flimsy leaves that can dry out and burn or disappear inside animals, to different ends. Cows, by removing grassy fuel that might kill seedling trees, help trees thicken.[22]

This has become an issue for the other close relative of the hooded parrot, the golden-shouldered parrot, an endangered bird bound to the open savannas of Cape York Peninsula. Two biologists who spent three years getting to know it, Stephen Garnett and Gabriel Crowley, found it enduring in two small areas, and at Stephen's advice I went to one, the cattle station of Sue Shepherd and husband Tom, who hosted the biologists during their time on the Cape.

Sue, who also became their field assistant, showed me the steep mounds the birds nest inside, while regaling me with tales about parrot-parenting ups and downs. Meat ants had claimed one old hole, while another had become a tomb for three chicks that died inside, signalled by a tiny feather winking yellow in a cobweb outside. With all the banter about lost broods, the grey mounds soon took on the air of tall headstones, but the young birds skittering about a nearby dam, reckless in the face of their rarity, gave hope that this parrot is not on an unstoppable slide.[23]

Lodged in the walls of most nest chambers, like rounds of bullets fired at close range, were the pupal shells of special moths. The bird collector William McLennan in 1922 recorded the moths' strange diet: 'The young birds were frequently voiding excreta and at times [it] would get all over their feet and tail feathers; instantly caterpillars would swarm out and devour it, eating up every scrap even off the feet & feathers of the young, thus the young birds were kept scrupulously clean.'[24]

The grassy patches this parrot needs are disappearing under tea-trees at the rate of 5 per cent a decade, cattle having replaced fire as the main consumers of grass. The image of habitat loss in Australia is commonly one of bulldozed trees, but this bird faces too many trees. Tree suckers and seedlings once suppressed by fire are now thickening into dense copses. In Lakefield National Park, set up partly for this bird, a no-fire policy doomed the local population as trees reaped the benefit.[25]

Sue's parrots are faring well as her property, unlike most, is managed with fire. Early in the dry season, when there are night dews to stop them, small burns are lit to encourage sweet new shoots for cows. Just before the wet season larger fires are lit, in grasses bulked up enough to immolate the suckers of paperbarks that rise among them. Lit at the same time as any lightning fires would start, these 'storm burns' help parrots and pasture alike by keeping trees out. On many properties fires are feared, and torched grass seems wasteful, so pastures are converting to trees, as I saw just north of Sue's land.[26]

Fire

Questions about their parrot drove Stephen and Gabriel to investigate Aboriginal burning. The journals of explorers tell of fires lit all through the dry season, and especially towards its end. Robert Logan Jack encountered 'pretty good feed – the grass having been burned three weeks before'. Another traveller found 'beautiful feed for our horses and sheep'. Gabriel and Stephen believe that Indigenous fires maintained the golden-shouldered parrot's grassy habitat. To produce kangaroo meat, the land was managed in a way that aided certain parrots.[27]

Fire gives a unique flavour to conservation discourse in Australia. Nowhere else are threatened species talked about like this: 'Aboriginal people formerly burnt in a way that produced a spatial diversity of vegetation structures. These appear to be required by the Partridge Pigeon.'[28]

The Gouldian finch recovery plan compares the 'patchy mosaic of smaller burns' of Aboriginal people in the past with today's intense fires, which removed the Carpentarian grasswren from seven of its eight Northern Territory sites. At one of these, near Borroloola, I scrambled in vain over rugged hills, hoping to find it. Australia has fourteen endangered land birds, and because mosaic burning has largely lapsed, fire, or lack of it, is thought to threaten all but one. Birds Australia has described fire shifts as the main threat to birds after habitat loss and degradation. Some forty-seven threatened bird species face problems from fire. In the northern savannas ecologist Don Franklin has identified 'a faunal assemblage in crisis in the absence of extensive vegetation clearance', because fire and cows are removing seeds. Rather than dying during fires, birds succumb later to lack of food or cover.[29]

Fire has a drama to it that invites exaggeration, so I should fill in the picture just drawn by adding more detail. Fire is not the main problem for many imperilled birds in Australia, just one among many. Some birds at risk come from places with little grass, telling of another side to the story which I don't have space to mention. For many birds the issue is too little fire, not too much. More birds face threats from changes to fire in Australia than anywhere else, but the number at grave risk remains small. Some birds benefit from excess burning, particularly those that feed on bare ground, while others benefit from complete fire exclusion. The spread north of glossy black-cockatoos into tropical Queensland is attributed by Stephen Garnett to she-oaks multiplying since Aboriginal burning ceased.[30]

Fire in the northern savannas is a compelling issue. Three-quarters of the burning in Australia happens in this region. Rather than too little fire, as in eastern Queensland, a large part of the north has too much. The fires are too big and too frequent. In Arnhem Land they may travel for months over tens of thousands of square kilometres, until the first wet season falls quell them. Large though lightning fires can be, they never reach such sizes, as electrical storms arrive just a few weeks before the wet, albeit when grass is extremely dry. Small mammals suffer more from over-burning than do birds, one reason being that cats easily find them after fire has removed their cover. Leading ecologists warn that 'A new wave of extinctions is now threatening Australian mammals, this time in northern Australia.' Although it remains unproven, fire and feral cats offer the best explanation for the disappearance from vast tracts of the likes of phascogales and rabbit-rats.[31]

Fire is exceptionally important to Australia because it's a highly flammable land that until recently was occupied by fire-using peoples. All too often, although not everywhere, explorers saw grey plumes and fields of ash. Outside Australia burning was important to hunter-gatherers and some farmers and herders, but not, during the eighteenth century, over a whole continent. In Australia it was missing only from waterless tracts that were avoided and from certain habitats, including rainforest, mangroves, saltmarshes and saltbush plains, which were unprofitable or impractical to burn.[32]

Fire was the most powerful force at Aboriginal disposal and grasses provided most of its power. They reward fire more generously than other plants by sending up new shoots that attract kangaroo prey – although fires were lit for other reasons as well. Major Mitchell's famous assertion that 'fire, grass, kangaroos and human inhabitants seem all dependent on each other for existence in Australia' was the first of many extravagant claims made in Australia.[33]

Aboriginal fires have more recently been blamed for devastating the rainforest, extinguishing the megafauna, changing the climate, and stimulating eucalypt evolution and expansion. While this list displays imagination, these theories have failed to win much support. The ascendancy of eucalypts over rainforest is real enough but mainly linked to 40 million years of drying. Mt Etna's rainforest animals disappeared well before people arrived, and the same is true of southern beeches and conifers across much of Australia.[34]

Thanks to lightning, which strikes the planet 8 million times a day, fire has been a force for 400 million years. In Queensland, while I was writing this chapter, half a dozen large and damaging fires were caused by lightning in one day. The fossil banksia cones I mentioned earlier tell of dependable fires 40 million years ago. Charcoal layers in south-western Australia suggest that fires were burning every 6-10 years, 3 million years before people reached Australia.[35]

Aboriginal fire management has licensed some writers to adopt charged language by portraying Australian forests as 'artefacts' (Marcia Langton, Tim Flannery) or 'estates' (Bill Gammage). In his beguiling book *The Biggest Estate on Earth*, Gammage claimed that 'Australia in 1788 was made, not natural.'[36]

But philosopher Val Plumwood warned that 'Counting something (e.g. a place) as purely human (or "cultural") when it involves the labor of nature jointly with human labor hides or denies the work of ecological systems and human dependency relations on it.' I have written elsewhere that Aborigines 'appeared to wield enormous power only because Australia carries so many flammable plants'. Flammability preceded people by tens of millions of years. Humans, grasses and eucalypts served each other's interests, and by starting fires people entered into mutual relationships with these plants. People often pulled the trigger, but the vegetation served as the gun. Before humans arrived there had been lightning to set it off. Aborigines increased the burning in many places but did not initiate something fundamentally new.[37]

Gammage invoked an Aboriginal 'estate' as a sincere gesture to highlight Indigenous skills, but the agency of plants should receive its due. The term 'ecosystem engineer', used for plants and animals that change or maintain habitats, offers a different slant on what happens when someone responds to a drying sward by setting it alight. Humans and elephants operate as ecosystem engineers when they push over trees and create paths through forests, and flammable grasses are as ecologically powerful in their own way. Words such as 'artefact' and 'estate' deny that ecological power.[38]

Although fires preceded humans, we should not presume that

birds become *adapted* to human fire. Benefit and need are not the same thing. Partridge pigeons like patchworks of burnt and unburnt ground for the simplest of reasons: open ground to feed on and thick grass to nest inside, positioned side by side because, like real partridges, they are made for walking, and need them close together. Lightning fires leave edges that suit them, and so does ground that morphs from gravel to soil. Mosaic burning helps them just as land management in the east is helping another edge bird, the noisy miner.[39]

In truth, what we know about fire is not easy to mould into a coherent story. Next to nothing is known about the *history* of Aboriginal burning, and land managers have difficulty replicating it today. How closely it should be copied is unclear, since it was not done to maximise biodiversity. In a recent paper in *Emu* about northern savanna birds John Woinarski and Sarah Legge warn that paradigms of fire management remain 'sketchy and contested'.[40]

Fires deep in time have left clues in the landscape – charcoal traces in lake and sea beds – the reading of which has changed over the years. When Tim Flannery invoked fire in *The Future Eaters* the limited evidence then available suggested that burning had multiplied after people reached Australia. Writing twelve years later, ecologist Chris Johnson noted 'very little evidence for an increase in burning as a direct and immediate consequence of human arrival'. In 2011 nineteen experts spoke of an 'urgent need to re-assess our understanding of the late Quaternary history of Australia'. Their review of all the studies on charcoal deposits came up with no evidence of a continental-scale change in fire at the time of Aboriginal colonisation. There could have been some changes to the sizes and timings of fires, but not at a scale that increased the charcoal left behind. The evidence shows that climate was the main influence on fire. Burning increased during warm wet periods and waned when Australia was cold and dry. (The paper explaining this can be found online by its title: 'Late Quaternary Fire Regimes in Australia'.) *Flammable Australia*, a CSIRO book produced by Australia's community of fire experts, has had two editions a decade apart, 2002 and 2012, which record the switch in thinking.[41]

One of the authors of the 2011 paper was Peter Kershaw, whose earlier work convinced many Australians, including me, that Australia was set alight by humans. Findings from a few renowned sites had been extrapolated unwisely. Massive charcoal spikes were used to argue for impossibly early human arrivals, more than 100000 years ago, by experts convinced that fires need hands to light them. Lynch's Crater in north Queensland is one famous site showing more charcoal (and grass) after human arrival, but the sediment core shows that rainforest eventually came back, achieving more cover than it had just before humans arrived.[42]

A new site to attract interest is a spring in the Kimberley, in the heart of partridge pigeon country, where a 6000-year-old core records a 'mega-drought' that lasted 1200 years. Climatologist Hamish McGowan and his colleagues linked this drought to a dramatic change in rock art, which went from the famous Bradshaw (Gwion Gwion) style, with its finely rendered figures, to the broad-stroke Wandjina art of more recent times. The dates don't fit neatly, but the core has been taken as evidence that Kimberley society collapsed when El Niño–Southern Oscillation (ENSO) weather increased a few thousand years ago, ending the first art style by ushering in a massive drought. What matters here is the 'dramatic increase in charcoal' recorded after the drought ended about 1400 years ago. There was virtually no charcoal for 4000 years, which means the site had no fire management and very few lightning fires, even before the super-drought began. The Gwion Gwion painters presumably burnt some places, but this spring was evidently not one of them.[43]

While Australia was not sent up in flames tens of thousands of years ago, some experts contend that burning increased just a few thousand years ago, due to intensifying resource use by an Indigenous population that was probably rising. Theories to explain this include the arrival of the dingo as a hunting aid, Indian immigrants (detected in recent genetic studies), more coastal food becoming available as rising seas stabilised, and more unstable ENSO weather requiring better land management, including careful burning. Intensification has also been interpreted as an illusion produced by recent artefacts being easier to find, suggesting a population that was larger and busier than it really was. Not everyone agrees, but there could have been changes to burning a few thousand years ago, something embraced as fact by eight fire ecologists who in 2003 wrote this: 'while burning is likely to have been part of the toolkit of the earliest indigenous Australians, such practices, and the scales over which these were applied, doubtless have changed markedly through time'. The last ice age cycle, on its own, would have seen to that. Some of the fire mosaics that benefit birds may be only a couple of thousand years old.[44]

This has interesting implications for the Gouldian finch. With a purple, yellow, red, pink and green costume like something out of a kindergarten art class, this bird is often praised as Australia's most beautiful. Gould named it for his wife who passed away a year after their return to England. A 2008 study revealed a lack of genetic diversity in this tropical bird, blamed on a population crash, presumably during the last ice age peak, followed by a recent recovery that spawned a large population with the genetic diversity of a small one. (Low genetic diversity in big populations is usually interpreted this way.) Genetic variation declines from east to west, implying that all the birds found in northern Australia can be traced back to a cohort that survived in far north Queensland (or even further north on land now under the sea). The author of this study, Rodrigo Esparza-Salas, conjectured that stable shores converted mangroves around the Gulf of Carpentaria into grasslands that finches could cross, and that increased Indigenous burning expanded finch habitat.[45]

The savage aridity of the Last Glacial Maximum 22000–18000 years ago would explain the previous collapse, but Gouldian finches endured another collapse more recently. They were common when, unforgettably, I saw some in cages as a child visiting Brisbane's Royal

Exhibition, but endangered by the time I saw wild ones. In the Northern Territory in 1926 flocks of finches, including Gouldians, were so dense that 'the disappearance of a thousand dozen out of one flock visiting one water hole would not cause a noticeable diminution in the numbers', remarked finch enthusiast G.A. Heuman. The wild population of Gouldians was in 2011 estimated to fall at times below 2400 adults – although numbers go much higher after wet years. This compares with more than 10000 Gouldian finches in captivity – and that's just in the state of South Australia. Wild populations went into a freefall in the 1970s, disappearing from vast tracts while some other finch species remained plentiful. I have walked through the town of Katherine trying to imagine the 1960s when they drank from fire hydrants in the main street.[46]

Mosaic burning sets grasses into a sequence of growth stages, which can result in year-round seed. Gouldians need this more than most finches as they do not turn to insects or shoots when the first rains of the wet season deplete the supply, by causing the seeds to sprout. But Gouldian numbers collapsed many decades after Aboriginal people were removed from their lands, so the decline of mosaic burning does not provide an obvious explanation for their predicament. On the annual Gouldian finch count, I sat beside a remote pool in the Northern Territory with biologist Carol Palmer, who spoke of recaptures 'in the hundreds' out of 'thousands and thousands' of young birds that had been banded at a major site that has Indigenous burning and no grazing. She can't explain why most young Gouldians are dying. This finch is not performing like a superior passerid. Its plight sits oddly with its fame in the world aviary trade. Every wild Gouldian is probably matched today by more than a hundred behind wire. I can walk a short way from my house and see them in an aviary in a stranger's garden.[47]

The problem with mosaic burning is how badly it can be done. Fire is such a potent force that a modest change in its expression can have big consequences. John Woinarski and Sarah Legge were blunt in their *Emu* review to the point of saying that much current fire management is failing. Kakadu National Park is losing whole mammal populations, along with some of its partridge pigeons and grasswrens, even though its managers, right from the start, committed themselves to 'traditional' burning. I have watched an Indigenous ranger flick flaring matches from a car, although most burning in Kakadu today is done from helicopters, following consultation with traditional owners. In some years half the park burns. A retired ranger has voiced regret on ABC radio about damaging fires he was part of. An academic has blogged about 'the ridiculous burning regime' in Kakadu and Arnhem Land. The reduced burning advocated in Kakadu's fire management plan is proving unattainable because fires in this part of Australia are just so easy to light.[48]

Because fires on Aboriginal lands are seldom used today for their original purposes, the burning is in some places excessive and in others non-existent. Once while passing through Indigenous land I found myself stomping out the remains of a fire in what to my eyes was a punished landscape. True fire expertise represents a well of skill and wisdom, but unfortunately it has become scarce today.[49]

John Woinarski and Sarah Legge recommend fire be excluded altogether from some patches of woodland, rather than rotated between them all, as usually happens. They question the dogma that a patch protected in perpetuity amasses enough fuel to become a firebomb, suggesting instead that shrubs eventually curb fire by smothering grasses. Kakadu's fire plan explains that the native annual sorghums, which are extremely flammable, can be shaded out if they are starved of fire.[50]

Sarah runs a project in the Kimberley called EcoFire, which is having a rare success in helping Gouldian finches. Each year her collective of managers in charge of cattle stations, Aboriginal lands and the Mornington Sanctuary, all of which need fire but for different reasons, directs a helicopter to drop 55 000 incendiaries while travelling twice the distance between Sydney and London. The north is one part of Australia where no one advocates a wilderness approach of leaving it alone, even though it is the world's most intact savanna and one of the world's largest belts of continuous vegetation.[51]

When Grass is Eaten

While fire is the most common problem in grassy habitats today, grazing poses challenges for birds as well, and not just by influencing fire. It probably explains why star finches, dainty red-faced birds with white spots which Gould was pleased to discover in New South Wales, soon disappeared from that state. To see them today you must go 2000 kilometres north to remote Lakefield National Park, where they teem in lively flocks. Gould found this bird by a river among 'seeds congenial to its taste'.[52]

The plants that are best at putting out edible seeds also do best at producing edible leaves, and cows take both. One who knew the paradise parrot well, naturalist Harry Barnard, observed how cattle 'fed down the grasses till they completely killed out the best kinds, thus destroying the food of the flocks of seed-eating parrots and finches, also quail and pigeons'. Heavy grazing during drought helped do this parrot in.[53]

Sarah Legge told me that getting cattle out of Mornington Sanctuary was part of the solution to getting Gouldian finch numbers up. Perennial grasses succumb under high stocking rates, leaving gaps in the seed calendar. Cattle numbers jumped in the 1970s when Brahmins, hardier in the tropics against drought and ticks, replaced English breeds, and dams and salt licks went in. Most national parks in northern Australia have acquired some cows from adjoining properties, either feral mobs or farm herds exploiting missing fences. When I met the Shepherds they were just back from mustering feral cattle in Lakefield, the national park that lost its golden-shouldered parrots.[54]

Almost every paddock in Australia has some birds using it from time to time, so we can't say that cows and birds don't co-exist. The question is whether rare birds can tolerate livestock in serious numbers. The answer is usually no, but the exceptions are instructive. Sue Shepherd has endangered parrots using the same dams as her cows, while south-eastern Australia has a vulnerable quail-like bird, in a family all its own, that seems to need grazing. This is the mysterious plains-wanderer, the wader with a Gondwanan connection, a bird that is near impossible to find unless you venture out at night with a spotlight and scan the low vegetation they sleep in.

On two farms turned into national parks largely on the plains-wanderer's behalf, sheep are grazed to aid it. In one, Terrick Terrick in Victoria, where the ranger took me out at night to find some, we drove past the grazing shed and twinkling eyes of sheep. At the other park, Oolambeyan in New South Wales, I was struck by the size of its homestead, which had framed photos of merino rams from earlier times adorning the walls. The captions told me the farm had produced a champion of the Melbourne Sheep Show, and the Grand Champion Strong Wool Ram at the 1973 Sydney Sheep Show. Rams went to buyers as far away as Brazil, and stocking rates were kept low on behalf of valuable stud sheep. Terrick Terrick homestead, which had never been painted, was in the twilight of decay from termites. Its last owner was so thrifty she wore glasses with a missing arm. Having stocked sheep in the same way as her father – at about half the normal rate – she seldom had to destock during drought. (Of the five best properties for plains-wanderers, two were run conservatively by women and the rest were, or still are, merino studs.)[55]

Sheep are run in both parks to give the birds open ground to feed on while leaving low plants to hide among. Grasses can transform a fall of rain into choking growth. A guide published by the NSW National Parks and Wildlife Service encourages farmers with plains-wanderers to increase grazing for a short period when that happens. Like many grass birds, plains-wanderers depend on a habitat that is inherently unstable. Their strong wings can take them great distances if in one place all the pastures are killed by drought or thickened by downpours. In the nineteenth century they sometimes turned up near Sydney, Melbourne and Adelaide.[56]

How and where plains-wanderers lived before squatters brought their sheep is not properly understood. The grasslands they use are mostly not natural, having lost a past scattering of trees. Kangaroos do not graze grass low enough to suit them. The peculiarity of 'needing' sheep is like 'needing' Aboriginal fire. But no other Australian bird has found itself in this situation, and sheep, most of the time, are detrimental to birds.[57]

Grasses as Shrubs

I have emphasised flimsy grass leaves, but nature abounds in exceptions, and Australia has spectacular grasses that exemplify permanence. With long tough leaves, a spinifex (*Triodia*) clump can be as big and stiff and durable as a shrub. Eucalypts and wattles have tough (sclerophyll), long-lived leaves to conserve scarce nutrients, and spinifex matches them as a sclerophyll grass bound to extremely poor soils. Its evolution was a turning point for Australia, as the dense prickly clumps give exceptional protection to birds and other animals over large areas inland.[58]

Deep in the Kalahari I once watched a secretary bird obtain a meal. This big bird stomped and stomped on a clump of grass until the lizard inside was lifeless. No predator ever stomps on spinifex, because mature plants are nests of needles, defences for the plant that also protect anything hiding inside. Like porcupine quills, the leaf tips can initiate infection if they lodge in skin, and 'porcupine grass' is an old name for these plants. All the specialised lizards, mammals, insects and birds that use their clumps make these grasses unique. The many species of spinifex help explain why Australia has the world's richest lizard fauna.[59]

Bound to these grasses are seven grasswrens, two emu-wrens, the painted finch and the spinifexbird. Other beneficiaries include the spinifex pigeon, night parrot, budgerigar and Gouldian finch, which enjoys the seed. The 'wrens' are 'old' Australians, but the spin-ifexbird, like the finch, is a passerid invader that made a remarkable transition from soft Asian grasses to sturdy needles.[60]

The antics around spinifex are amazing. A spinifexbird will dive into a thicket of spines where entry seems impossible. The first striated grasswren I saw retreated so low and fast I thought it was a lizard or a mouse. An hour passed before I took in enough detail to realise it was a bird darting along the ground. One individual I followed for half an hour from clump to spiny clump finally revealed itself to be an emu-wren. I have watched a brown snake try without success to enter a clump, yet small birds move through them so fast – presumably along known paths – you cannot tell whether the grasswren you just saw is nearby or far away.

So concealing is spinifex that while the hummock grasslands it forms dominate a quarter of Australia – more than any other plant type – its birds remain poorly known. Discovered in 1901, the black grass-wren was not seen again until 1968. As for the endangered night parrot, sniffer dogs have been proposed to find it. One reason for the shyness is that spinifex affords so much protection that some of its inhabitants, especially grasswrens and emu-wrens, have reduced wings. Birdwatchers pay well for special grasswren tours, which cross vast tracts of barren land to deliver fleeting glimpses of small reclusive birds that are rather drab, but very enticing for all that.[61]

A story of decline applies to spinifex. Hummock grasslands stretch from Victoria to Australia's north-western shores, but a large number of spinifex birds, including the night parrot, mallee emu-wren and most grasswrens, are doing poorly. Fire is a problem. Spinifex burns readily, in this respect behaving as a typical grass. Its tussocks, made up of numerous well-aerated blades filled with flammable resins, have been described as 'perfectly designed for combustion'. To flush out a fleet-footed dragon I once put a match to a lone clump, but the flames flared too fast for the lizard to escape. A teenager at the time, I learned from my own hands how dangerous grasses can be, by nearly starting a serious fire near a large town.[62]

Spinifex thrives on its own destruction by immolating and killing mulga and other trees that might otherwise shade it out. But fire has gone from benefiting spinifex to often becoming its enemy, because the land is managed today to different values, or not managed at all. The removal of Indigenous people from their lands was succeeded by some exceptional fires, including one in the Western Desert in 1986 that burned out 32 000 hectares, leaving one giant fire scar in place of the 372 small fire scars detected there twenty years before.[63]

Deserts usually lack the connected fuel that fire needs, but ephemeral grasses that sprout among spinifex after heavy rain provide this, benefiting themselves in future by spreading giant fires that burn the spinifex clumps back to their butts, creating more space, for a few years, for smaller plants. Weak wings hinder the return of small birds when spinifex recovers. As well as this, farmers can feed their cows by torching the softer spinifex species again and again to induce new shoots, producing small clumps with no cover or seed for birds.[64]

In the Western Desert of Western Australia burning has been revived by a small group of the Martu, giving anthropologists a chance to learn about spinifex management. Patches greening up after fire bring in bustards, emus and other game, and put out crops of wild tomatoes.[65]

Uncertainties Rule

The only Australian bird still uncaptured on film is a bird of grass. Although it is endangered the buff-breasted button-quail attracts no tangible aid since no one can reliably find it. Only seven specimens have reached museums, all more than a century ago. Its call has never been taped. It is a north Queensland bird missing from the Queensland Museum's vast bird collection, which manages to have fourteen paradise parrots. 'There are no specific recovery, conservation or threat abatement plans for the Buff-breasted Button-quail,' admits a federal government webpage about its endangered status.[66]

The person who comes closest to knowing this bird is a quietly spoken naturalist who has trudged hundreds of kilometres along hot ridges on an unpaid quest to save it, or at least learn how it lives. Lloyd Nielsen put in a year before he saw his first one. His sightings are fleeting because these birds are shy. He drove me to a site where he once saw some. I took photos of the ground.

Spinifex is home to another bird so rare it was not photographed until 2013: the near-legendary night parrot. Fire could be a big problem for these two birds, but we don't know. The past has given us what we see today, but no one can say how birds, people, grass and fire came together in the past. What were finches, parrots and people doing during the Last Glacial Maximum, when there was little or no monsoon, far less fire, since it was cooler, and a sea of rustling grasses reaching all the way to New Guinea? Gouldian finches may have pecked grain where terns and boobies snatch fish today. He may be wrong, but Chris Johnson has proposed that for something like 20000 years of the last ice age 'landscapes in the north became uniformly arid, and tropical grasslands, woodlands and dry rainforest thickets were largely replaced by a treeless mixed shrub/grass steppe'. The grasslands needed by golden-shouldered parrots are ice age habitats kept going by fires lit by people and lightning.[67]

We know that Gouldians are doing worse than other finches because of their specialised diet and unusual nesting needs, but not why they should have these narrow requirements in the first place. The climate they evolved in was surely different from today's, given that the period we are living in, the Holocene, which began about 12000 years ago, is a warm, wet break from the Pleistocene, a mainly cool period that began more than 2 million years ago.

The extinction of the megafauna is something else that raises more questions than answers. Claims that fires raged when Australia lost its diprotodons, *Genyornis*, and other giant animals jar with some of the evidence. In a global review, Chris Johnson singled out Australia as the land where fires did *not* increase when giant mammals were lost. The last megafauna disappeared when it was colder and drier than today and less conducive to fire. Australia's megafauna, judging by their teeth (and diprotodon stomach remains), ate far more shrubs than grass, and because of this Chris suggested that their demise might have triggered an expansion of shrublands and scrub and 'a reduction of grass'. There may have been more fire and grass in some places and less in others. How birds responded we can't say.[68]

There is much about grasses I have not touched on, such as how rocky slopes help birds by limiting fire; the contests between annual and perennial grasses, and the advantages to birds of their different ripening times; and the role of carbon dioxide in vegetation thickening, but none of this makes savannas easier to understand.[69]

Fiery Future

Climate change will shape the next chapter in the story of grass. As temperatures rise and droughts lengthen, fires will become bigger and hotter, and worse for birds and people. Invasive pasture grasses brought from Africa to feed cows will exacerbate the situation, because they are exceptionally generous to fire. Four-metre-high gamba grass burns at eight times the intensity of native grasses, killing trees in the process, and it's spreading rapidly from cattle stations via seeds borne on the wind. A Queensland government report warned that it may 'transform Australia's eucalypt-dominated tropical woodlands into treefree grasslands'.[70]

Australia's most valuable pasture plant, buffel grass, is killing red gums around Alice Springs and woodlands in Queensland. One bottle-tree rainforest in a national park gave way to grassland when buffel spread among the trees and wicked in a lethal fire. Around Uluru, where it is the main weed, buffel grass has ousted Aboriginal tucker plants.

I once saw star finches husking the seeds, but they are rarely eaten.[71]

Extreme grazing pressure could explain why Africa evolved such bulky grasses. A great many were brought to Australia by agronomists, and others came accidentally as hitchhikers. In a 2008 report for the federal environment department, I nominated gamba and buffel as the invasive species posing the greatest threat to Australia in the face of climate change. They are highlighted as well in reports about fire and climate change.[72]

Another imported grass, giant reed, is loved by biofuel entrepreneurs for the same reason weed experts despise it: unsurpassed growth rates. Growing 8 metres high, it lays down carbon faster than almost any other plant. Trees die when it burns. One company wants investor backing to grow millions of hectares of it in northern Australia.[73]

Decades of stored carbon are released after the flames rising from these grasses take the lives of trees. Introduced grasses show clearly that grasses are superior plants that exploit fire, but they do this in the worst possible way.

Ten

Life in a
Liquid Landscape

The tens of millions of birds that feed far out at sea show how useful over water wings can be. But so extreme is the sea that few groups of birds have mastered its challenges. Those that have are possessed of exceptional bodies and lead extreme lives.[1]

The wandering albatross illustrates this by having the world's longest wings. Upon leaving the nest a young wanderer spends up to a decade at sea, mostly on the wing, avoiding land completely for the first five years, mastering a domain where food is rich but scarce. Nights are spent on the wing if not on moving water. This bird, offered expert Terence Lindsey, 'crowds the outer limits of what it's possible for an animal to do'.[2]

Short-tailed shearwaters (muttonbirds) are commuters between continents, flying from Australia to Antarctica's shelf waters to provision their young. The lone chick in a dark burrow waits up to sixteen days between meals brought by parents it does not see. It is abandoned by them some days before it emerges from the hole, because winter calls them north to Arctic seas well before the chick feels summoned to undertake the same journey, a 25 000-kilometre round trip that claims many young lives. Completely different are the lives of young miners and choughs, brought up in the complicated company of parents, siblings and helpers.[3]

In a book about Australian birds, seabirds shift the focus – from the landmass to the waters that surround it, and to oceans and islands more generally. Because seabirds are so mobile, and because the waters associated with each continent have arbitrary boundaries, there is no real 'Australian seabird fauna' to talk about. There is only a Southern Hemisphere fauna, distinguished by all its petrels and penguins, one that varies in richness from region to region, depending on ocean productivity, temperatures, and the availability of islands for breeding.[4]

Two points about seabirds are of special interest here: a global peak in diversity near Australia, and lives which are so extreme they deepen our understanding of what it means to be a bird. The land birds in Australia were shaped by unusual forces, and so it is was at sea, but far more so.

My focus is on the group that dominates the oceans – the albatrosses, shearwaters, petrels and prions, making up order Procellariiformes. (*Procella* means violent wind or storm.) As perching birds dominate on land, so they do at sea, albeit with far fewer species: about 130 in all. They differ from other birds in having two nostril tubes, for extruding salt and for smelling, hence one of their common names, tubenoses; but I will call them all 'petrels'. (Experts often say 'albatrosses and petrels', but no such division actually exists; storm-petrels, for example, are closer to albatrosses than to other petrels.)[5]

Sleeping on Air

Birds went to sea so early in their history that there were birds in water long before there were birds without teeth. A life at sea was not something I could comprehend until I ventured out in it myself, as a guest on the Australian Antarctic Division's ice-breaker, the *Aurora Australis*. I was shown a new world.

Within three days of our leaving Hobart, 60-knot gusts had

whipped the ocean into a foamy mess. Spray hit our windows high on the bridge, and the bow kept plunging into giant chasms that opened in the dark water before us. From the warmth of the heaving bridge all this was sinister enough, but to step outside onto the exposed decks was to feel fear strike like a slap in the face.

How were the birds I'd been seeing coping with this cyclone that we were now inside? They had no land to retreat to. The shadowy shapes of a few shearwaters glimpsed through wave-splattered glass did not tell me much. But then, like an apparition, something white floated before the bow, moving much too slowly and smoothly for the circumstances. In all that cold confusion it was the one stable thing, a wandering albatross, one wing trailing above a rush of foaming water. My heart jumped for its safety, until I saw how at ease it was. I understood the awe these birds inspire in those who enter their domain.

The weather map on the bridge showed a procession of cyclones round the Southern Ocean, promising turbulence in all directions for hundreds if not thousands of miles. Day after grey day the world stayed wild, though the sea was sometimes less extreme. The log filled with entries about 'pitching and rolling' in what was often a 'heavy confused swell'. I saw why sailors in ships of creaking wood so feared the Roaring Forties.

Quiet stole over our ship as the sick hid in their beds, but the birds never showed any discomfort, neither wavering in flight nor opting to rest on the ship. So safe is the sea for albatrosses that if they survive longline fishing they outlive most animals on earth. One royal albatross was laying eggs and rearing young years after the scientist who banded her fifty-eight years before had gone to the grave.[6]

One sleepless night, I passed the hours of sliding back and forth in my bunk by wondering how the birds were spending theirs. Many seabirds hunt at night, but surely, I thought, not on breaking seas. There was nothing stable for a bird to rest on. Can seabirds sleep on the wing?

Most experts think so. Many birds are known to sleep unihemispherically, with half the brain asleep while the other maintains vigilance through the eye it controls. Chaucer wrote of small fowles that 'slepen al the night with open yë'. Some birds alternate, keeping one eye open, then the other, all through the night. Many marine mammals also sleep in halves. A fur seal paddles with one flipper commanded by its wakeful half to keep its snout above water. A dolphin switches side every forty minutes. Seagulls on land flip more often, and go between full and unihemispheric sleep every few seconds.[7]

Seabirds are believed to sleep like this in the air. One frigatebird fitted with an altimeter spent a day and night aloft. Christmas Island frigatebirds must sleep on the wing when they venture out to sea for up to six days at a time. They are birds of the air more than the sea, with long pincer bills for snatching fish from the water surface and for bullying other birds into disgorging their meals. Lacking waterproof feathers, they become waterlogged and drown if they rest on the sea.[8]

Aerial sleep is so unusual that I will digress to mention its best-known practitioners, European swifts. In screaming parties they climb

Frigatebirds are birds of the air that wander far out over the sea, even though they cannot land upon it. They have the largest wings for their weight of any bird, and the buoyancy these bestow help explain how these birds can (apparently) sleep on the wing during long sojourns away from land. This is a young great frigatebird.

high at dusk and disappear into the dark. Radar and radio-trackers detect them one or two thousand metres up, flying mainly into the wind, soaring with some slow wing beats, apparently asleep. A World War I French airman who cut his engine after reaching a great height had an encounter on his glide behind enemy lines:

> As we came to about 10,000 feet, gliding in close spirals with a light wind against us, and with a full moon, we suddenly found ourselves among a strange flight of birds which seemed to be motionless, or at least showed no noticeable reaction. They were widely scattered and only a few yards below the aircraft, showing up against a white sea of cloud underneath.[9]

Two that were caught were swifts. Many swifts are thought to sleep aerially, including fork-tailed swifts in Australia. So rarely do they touch anything solid that articles appear when they do. Australia's leading reference, the *Handbook of Australian, New Zealand and Antarctic Birds*, mentions a few pausing on a branch after bad weather in 1930, and others alighting on tennis-court wire in 1959. From their breeding places in Asia these birds may visit Australia without touching anything apart from insects in the air and water sipped on dives over pools.[10]

Albatrosses probably sleep on water when they can. One sooty albatross that was monitored spent two nights mainly aloft and seven hours on water on the third. Nineteenth-century naturalist George Bennett declared that albatrosses 'appear as comfortable floating in the breeze, as we should be on a bed of luxurious softness'.[11]

Made to Formula

Seabirds test our ability to divide nature into categories, by provoking disputes about how many species to believe in. The gulf remains wide between those like me who accept between thirteen and fifteen albatross species globally and those who push for up to twenty-four, with opinions varying sharply about whether slightly different populations on different island groups qualify as species or not. The larger numbers create more endangered species to (hopefully) elicit concerns and more entities to study, but risk cheapening the concept of species. The Chinese have a proverb endorsed by the International Ornithologist's Union – 'Wisdom begins with putting the right name on a thing' – but no one has found the wisdom needed to broker a consensus on albatross classification.[12]

At another level, however, these same birds show how well classification can work. Calling a bird a petrel means so much: a low body temperature, a long slow lifestyle, the ability to live away from land, loyalty to birthplace, a strong sense of smell, a clutch of one white egg, and a chick fed with stomach oil. So limiting are water and wind that this script for success was repeated over and over with little variation, except in the size of the bird and the wing shape that goes best with that. Sizes vary so greatly that storm-petrels sometimes end up becoming snacks for albatrosses.[13]

These birds' sense of smell is often exceptional. Odour streams are part of the liquid landscape, announcing anything edible, from krill to carrion to the faeces and vomit of whales. Krill betray themselves with the chemical 3-methyl pyrazine, released when they are crushed by whales or seals. The most petrels I saw behind the *Aurora Australis* – nearly a hundred – was on a day when a vent smelt strongly of pork crackling. Storm-petrels in one test were lured to cod-liver oil from 7 kilometres away.[14]

A chamber in the stomach converts some of what they eat into oil – another defining feature. Like the oil humans use, it fuels mobility. Food for chicks can be obtained far from the nest because the oil they are fed by regurgitation does not decay on the long trip home. One wandering albatross with a chick to feed travelled 15 200 kilometres on a 33-day foraging run. Tasmania has muttonbirds in immense numbers because they can raise their young on Antarctic krill harvested over vast reaches of the Southern Ocean, rather than only near the nest.[15]

So slowly do petrels live that an albatross chick can remain in the nest for more than eight months, following two months or longer inside the egg, from which it can spend six days breaking free. The larger albatrosses are aged nine to twelve years and sometimes as old as sixteen when they first breed, and they can keep on breeding past sixty years of age. Songbirds in the Northern Hemisphere seldom live sixteen years. The lives of smaller petrels, although less extreme, are also slow. Shearwaters spend up to twenty days at a stretch on the egg without feeding, and their egg can survive a week without a bird to lend it warmth. One reason why longline fishing drives down seabird numbers so fast is that slow living is matched by slow breeding rates. Fishing would not threaten albatrosses so much if they bred like sparrows.[16]

Penguins (order Sphenisciformes) provide an even starker picture of the constraints imposed by the sea. Few birds, from any angle, are so obvious. Their unmistakable form is the oldest in birds we know about, going back at least 60 million years to penguin bones found in New Zealand. As their sister group, petrels match them in age, but the bones of penguins, big and solid for deep diving, fossilise better. The two groups evolved not from toothed seabirds, but from a toothless flying bird that lived on the land. Their place on the tree of life above Palaeognaths tells us that.[17]

The Great South Wind

Australia lacks any marine birds that stand out as unusual, for the simple reason that marine birds roam so widely. Wandering albatrosses occur right around the Southern Hemisphere, each individual using an area of sea much larger than the continent of Australia. Australian waters attract seabirds nesting as far away as the Atlantic Ocean, while birds breeding around Australia reach Alaska and Chile. Differences are sharp between hemispheres, but not between Australia, Africa and

South America. Australia does stand out, however, for its sheer number of species. Australia's waters are reportedly visited by more petrel species than anywhere else in the world.[18]

This claim came to light in a test of something that invites wonder – the propensity of life to crowd into the tropics. We often hear about this, about all the species we stand to lose if rainforests go unsaved, about the dizzying diversity on coral reefs; it was something that Wallace pondered on his journeys. Yet petrel species peak not around the equator, but in the Southern Hemisphere.[19]

In what counts as armchair science, four biologists in 2009 analysed the maps in a field guide and found a band of high petrel diversity circling the Southern Hemisphere. It was highest around southern Australia and New Zealand, and highest of all in the sea near Tasmania, where forty-two species feed. A 2005 study of *all* the world's seabird species found a global peak around New Zealand.[20]

The bewildering variety of life in the tropics is best explained by the species-energy theory, which rests on the fact that the main energy source for life on earth, the sun, is strongest at the equator. But wind is another energy source, taking some birds large distances on little food. It effectively shrinks the world, reducing the cost and time of travel. The mid-southern latitudes (45°–50° South) are the world's windiest, with mean wind speeds and seabird densities both reaching their maximum at 50° South. Mariners kept seeing albatrosses before their sails because each favoured the same latitudes.[21]

Heated tropical air drawn towards the poles is deflected by planetary spin to produce the westerlies, the world's dominant winds, the ones that made exploration and the spice trade possible. At 50° South there is least land to retard them – only the tip of South America and a sprinkling of islands. These winds drive the Antarctic Circumpolar Current, the world's strongest, something that Australia and South America made possible by leaving Antarctica and drifting north. The extreme temperature contrast between the equator and Antarctica ensures the current's strength.[22]

The exceptional wings of the wandering albatross (reaching 3.6 metres from tip to tip) are agents of its power. This bird needs external energy, and in still air cannot stay aloft. On glassy seas it is becalmed, forced to sit out the days on the surface. Too little wind limits it more than too much. Wind over cold water is its medium.[23]

Albatrosses are ever negotiating intangible realms of air of varying speed and pressure. Temperate cyclones (intense lows) are their conveyors, taking them furthest, and by tacking across currents through different systems they can choose where to go. Winds ahead of cold fronts take them south, while winds from Antarctica bring them back. By locking wings in glide position they can cover 1000 kilometres a day without much effort. One grey-headed albatross circled the globe in forty-six days. To forage, or travel upwind, they use wind deflected from waves. The albatross that concerned me kept a wing near the water for that reason. Nineteenth-century naturalists were convinced they had inflatable air cells keeping them aloft.[24]

Wind also furnishes food by driving nutrients upwards, in upwellings that attract squid, krill, and fish of all sizes. The productivity of the sea is limited by the lock-up of nutrients in dead beings that sink below the light needed for primary production. By pushing water towards submerged ridges and other water masses, wind brings nutrients and light together in special locations that birds are good at finding.[25]

Tropical seas, unlike tropical lands, are not productive, except near coasts, since warm water holds less oxygen and carbon dioxide to drive production, and forms a surface layer that suppresses upwellings of colder water from the depths. The Pacific Ocean has been called a desert. Tropical seabirds rely largely on tuna and dolphins to herd prey to the surface. To a small fish, the sea surface is a wall against which they can be pinned by large fish, one that seabirds punch through for a share. Flying fish can breach that wall, but at the risk of landing in frigatebird bills. In a boat off Cape York I have watched terns, noddies and a booby converge all at once on a swirl of churning water in which, drawing near, we could see the tight ball of desperate fish turning in on themselves, with the tuna arranged evenly around them, blocking escape.[26]

Fishermen can find tuna by watching for excited birds. Tuna are themselves sometimes encircled by dolphins – hence the problem of dolphin bycatch when tuna are netted. Overfishing of tuna reduces feeding opportunities for birds, a potentially serious problem. In cooler waters, whales and other predators also herd prey, but most birds there can survive without them.

The New Zealand Factor

Strong winds in the Southern Hemisphere account for petrel success there, but the peak around Australia and New Zealand is driven more by breeding needs. Seabird wings, being long and thin for soaring, have reduced value for lift. An albatross on land cannot explode into flight like a startled pigeon; it must leap or lean into turbulence, or lumber along the ground flapping. No albatross or shearwater ever pulled up gracefully in a tree, as their wings have poor braking power and their paddle feet lack a toe for grip. All petrels nest on or under the ground, never up in trees, obliging them to find islands without mammals. When mammals do reach their islands, disasters ensue, as we shall see.[27]

New Zealand, 'the capital of the seabird world', was once the largest cluster of islands anywhere without a mammal fauna. Before people and rats arrived, the only predators were the tuatara (a reptile), a small bat, and a few birds, including the kea and, until 700 or so years ago, the world's largest eagle. If certain experts are right, the two main islands supported billions of seabirds from sixteen species, which sometimes bred tens of kilometres inland. Seabirds during peak breeding times may have outnumbered New Zealand's forest birds.[28]

New Zealand is part of a sunken microcontinent, Zealandia, which protrudes at points far removed from the main islands, allowing for breeding over a vast reach of sea. Included are the Campbell Islands far in the south, and the subtropical Kermadecs, Norfolk, and other islands well to the east. Australia has thousands more islands than Zealandia, but only because seas are high. During ice ages the islands of the Great Barrier Reef, Bass Strait and Houtman Abrolhos, which today teem with seabirds, became knolls rising above plains.[29]

Most islands the world over perch on continental shelves that emerge when seas recede, disrupting seabirds, because their islands turn into hills on continents that predators can reach. In temperate Australia, where the vast majority of Australia's petrels breed, the continental shelf slopes away steeply at 100–130 metres depth, so that during the Last Glacial Maximum, when seas dropped 125 metres, there were no islands left and no new ones created. New Zealand had seabirds using its main islands, and its outermost islands survived the rise and fall of seas because Zealandia is mostly too deep to emerge. It could serve as a stable refuge and important evolutionary centre.[30]

As evidence of this, English biologists recently listed thirty-three seabird species unique to New Zealand, including four penguins and an astonishing eight shags. This is an extraordinary number, given that no other nation has more than five endemic species. 'Seabird' was defined in this case to include coastal birds such as shags and gulls, but these numbers are rubbery, because of the rows over seabird classification. If we decide the world has fourteen albatross species, then half of them breed in New Zealand. If we opt for twenty-four species, then thirteen breed on New Zealand islands, and nine nowhere else. New Zealand comes out even further ahead if the country is defined geographically, by taking in Norfolk and Macquarie islands.[31]

Indonesia, in comparison, has 13400 islands spread over a greater area, but no endemic seabirds, and Australia has at most four species, unless you factor in her remote territories, such as Christmas and Norfolk islands, which are Australian only by historical accident. I am talking about numbers of species (diversity), not numbers of birds (biomass). The rich Humboldt Current gives South America the most individual seabirds, but few species breed there because islands are scarce.[32]

I sampled New Zealand diversity from a boat out of Kaikoura, on the east coast of the South Island. A couple of hundred petrels soon crowded around hoping for handouts. There were four, six or seven albatross species depending on your definition of 'species'. The deep trench in these waters has nutrient-rich upwellings to sustain a vibrant food chain, of which discards from fishing boats are now an important part. To the west loomed the Kaikoura Range, on whose soaring slopes Hutton's shearwaters breed. More than thirty petrel species use these waters, in contrast to seven around Britain and a similar number on the rich fishing grounds of Newfoundland. Northern Hemisphere waters have auks and puffins (Alcids), but far too few to balance the numbers.[33]

New Zealand birds often feed in Australia's temperate waters, and sometimes choose to stay. George Bass and Matthew Flinders found slopes 'almost white with birds' when they landed on Albatross Island, off Tasmania's north-west coast, in 1798. From an undistinguished geological past as a crag on the edge of a frosty plain, this pinnacle became a crowded seabird 'city' as rising seas at the end of the last ice age flooded the plain running north from Tasmania. The shy albatrosses nesting here lack the genetic variation of those found in New Zealand, showing this to be a young population formed by spread from New Zealand: shy albatrosses don't breed outside Australia and New Zealand.[34]

The DNA of Australia's little penguins suggests that they too came from New Zealand recently. Seabirds must have spread west from New Zealand after each ice age ended, when Australia regained the islands it lost, then retreated east when the next glacial period turned Australian seabirds into 'climate refugees' by depriving them of Australia's 20000 islands.[35]

Ending Isolation

Seabirds all over the world face problems because mammals, including people, keep claiming the islands they breed on. Predators were helped by ice ages to reach islands, and boats are helping them today. The mammals that do best on ships, rats, are now a threat to seventy-five seabird species. When rats troubled Captain Cook he tied an overnight line to shore to invite them to leave. On one South Atlantic island house mice have helped pushed the Tristan albatross close to extinction by nibbling their chicks. That mice could endanger albatrosses shows how vulnerable seabirds can be. Mammals have been kept off many New Zealand islands, and removed from others, but the petrels on the main islands were decimated by the Maori and their rats long ago, and then by European mammals. Globally, thousands of seabird islands have become mammal islands in a major shift in vertebrate distribution.[36]

People were a large part of this. More than a thousand years ago Polynesians ate their way across the Pacific, seabird colonies fuelling much of their colonisation. Forest birds suffered as well. 'An island overflowing with birds must have been a blessing after days or weeks at sea,' mused David Steadman in a revealing book, *Extinction and Biogeography of Tropical Pacific Birds* (2006). Tame birds could be eaten while people grew their first crops. Pigs could be fattened on birds as well. Almost every island in Oceania was peopled at one time, and no major island kept all its birds. Rails suffered the most, being ground-feeders that readily lose their powers of flight. If Steadman's speculations are correct, 400 to 1500 rail species have been lost. That could mean that the world's perching birds, before this tragedy, did not number more than half of all bird species, as they do today. The DNA of these rails suggests they evolved less than 400000 years ago, which means their spread, like their demise, was very rapid, and their reign extremely brief.[37]

From the 200 vanished seabird colonies that we know about, Steadman has inferred that thousands more were destroyed, mainly

those of petrels. But Tahitians and other Polynesians have been represented by men who were idealists as people in harmony with nature.
About Tahiti's past nothing is known, but nearby Huahine lost at least
thirty-two birds, including eight petrels, all of which Steadman lists.
Of the twenty-four tubenoses that once bred in East Polynesia, only
seven remain.[38]

These losses leave us wondering about the tropics. Is the Pacific a
desert, or just deserted? Unproductive warm water can explain why
tropical seabirds are not prolific, and so can past hunting, leaving us
unsure about the role of each. Polynesian innovations – the outrig-
ger canoe and double-hulled platform sailing canoe – did for boats
what long wings do for birds, increased their reach. But seabirds occur
widely enough that very few have gone extinct. New Zealand lost a
penguin, Tonga a shearwater.[39]

The appropriation of seabird protein was something Europeans
participated in as well. After the First Fleet reached Sydney a second
British colony was formed on Norfolk Island, 1700 kilometres away.
The seas around it teemed with fish, but boats went unbuilt to prevent
convict escape, and rations were soon halved to forestall impending
famine. Aid eventually came from on high, when seabirds were seen
dropping at dusk to burrows on Mt Pitt. Convicts and their keepers
took to climbing the hill each day to procure enough to eat. From the
tally kept by the Keeper of Public Stores we have a meticulous record
of the demise of Norfolk Island's providence petrels.[40]

In three months, 171362 of the drab brown birds were eaten.
Captain Hunter enthused about this 'Bird of Providence'. The meat
tasted fishy, while the eggs 'were excellent', providing a reason to cut
open live birds, a practice banned under what was the first law enacted
in Australia to prevent cruelty. The evening 'shower' of birds suggested
to Lieutenant Ralph Clark gifts bestowed by God: 'This account will
make the story of Moses being in the wilderness to be a little more
believed, respecting the shower of quails. Everybody here owes their
existence to the Mount Pitt birds.' The harvest averaged 2000 birds a

day for the first three months, or four birds for each person each day. Another officer, Lieutenant Bradley, saw how unsustainable it was from all the eggs and young birds that were destroyed, and from the ground 'being torn up', partly by grubbing pigs.[41]

After ten years the resource was gone. In 2005 Norfolk Island's national park manager, Brooke Watson, parted a curtain of bracken and mistweed to show me the old convict road up Mt Pitt, made to reach the birds. Drink bottles discarded by American soldiers during World War II lay at our feet.

Providence petrels did survive on another island that was never pillaged – Lord Howe – and in 1985, about twenty birds were found on Phillip Island, a denuded lump in the sea near Norfolk. On a high cliff facing dark water ranger Robbie Ward showed me their burrows, which, because the ceilings had been rounded by generations of hungry chicks swaying their heads to beg, looked like toy train tunnels. These petrels owe their existence today to two nubs in the sea, but they display typical seabird mobility. I have seen them from a boat near the Gold Coast and they even turn up near California. One hope is that funds will be found to fence off Mt Pitt from cats and rats, and so entice the petrels back.[42]

Seabird destruction was repeated on many islands around Australia. A gruesome stench met George Robinson in 1832 when he stepped ashore on Albatross Island and beheld thousands of albatrosses slain for their feathers. Many more became bait in crayfish pots, a fate that was also visited on penguins, gannets, shags and muttonbirds. Cat Island in Bass Strait saw the numbers of its gannets reduced from 18000 to eighty by 1957. On Macquarie Island, hundreds of thousands of penguins were clubbed and boiled down for oil. Rat Island in the Houtman Abrolhos lost all its seabirds – including more than a million sooty terns and noddies – to guano farming, cats and rats. Lady Elliot Island on the Great Barrier Reef also lost a massive colony to guano harvesting, then goats. Seabirds are missing from all Torres Strait cays except remote Bramble Cay, presumably due to past harvesting. Mammals are currently a serious problem on four oceanic islands ruled by Australia: Christmas, Macquarie, Norfolk and Lord Howe, with cats and rabbits on Macquarie Island having proved especially damaging before they were removed. North of Australia, seabirds have been lost from most islands in Indonesia and Malaysia.[43]

Despite all this, most Australian seabird colonies did survive, on islands too remote for canoes or too small to attract settlers or sailors. Three-quarters of all seabird breeding in Australia occurs in Bass Strait, where short-tailed shearwaters (muttonbirds) have survived intense harvesting pressure. Operating over two centuries, muttonbirders have killed millions of these birds for meat and oil, but some islands seem to have as many as ever. I suspect that when people reached Australia shearwater numbers were rebounding from an ice age low, so that a rising population muted the impacts of mass harvesting. Shearwater DNA shows little variation, attributed to a minuscule population from a time when, because of falling seas, they had no islands to nest on. They probably bred in small numbers on Tasmania's west coast, which had no people until about 4000 years ago.[44]

I have emphasised the need of seabirds for small islands, but there are rare exceptions, including little penguins, which in spite of their name are big enough to see off most mammals. Michelle Foale showed me how they live surreptitiously in Burnie, an industrial town in Tasmania: one had entered a gap in a fence meant to keep them out, crossed a car park, then passed a bright street light and two corners of a building to reach its nest in a courtyard. A nearby service station on a busy road is sometimes visited late at night.

Adaptability and Vulnerability

Seabirds show so little aptitude on land that sailors, upon finding how easily caught they were, gave them names such as 'booby' and 'noddy'.

In the Straits of Magellan Sir Francis Drake saw fit to dispatch 3000 penguins in one day. On Albatross Island Matthew Flinders found the occupants, 'being unacquainted with the power or disposition of man, did not fear him; we taught them their first lesson of experience'. In *The Life of Birds* a whistling David Attenborough persuaded providence petrels on Lord Howe to fall to his feet. On the *Aurora* I kept a shearwater in an open cardboard box for days without it trying to escape. I thought it was weak from striking the ship one cold night, until I held it up in strong wind and it flapped vigorously, then flew off to become a dark shape in the distance; deprived of wind it had not moved. Biologists have noted Australian penguins showing 'surprisingly inappropriate and maladaptive' responses to fire by not moving away. Like all birds, these seabirds had ancestors in forests that would have been adept at eluding danger, but such skills were lost when they ceased to serve much purpose.[45]

Marine birds usually return to their birthplace to breed, a habit that should free them from the need to assess safety on land, since their own safety when they were young should translate into security for their own young. Philopatry, as this behaviour is called, is analogous to conservative values in humans, to trust in tradition. The problem today is that many seabirds are now becoming lambs to the slaughter, so to speak, by revisiting islands that now have mammals. Philopatry is the reason why petrel classification is controversial. Each colony behaves more or less like a separate species, creating questions about where to draw classificatory lines when colonies show slight differences.[46]

If seabirds learnt as fast as crows and cockatoos they might survive better on islands, and die less often from longline fishing. They are not as cautious as they should be near boats, or about what they swallow. Global plastic production expanded fivefold between 1970 and 2000, and the biggest repository for plastic waste is the ocean, where it decays into fragments that petrels eagerly eat. More than a hundred seabird species take plastic in pieces that may remain inside them for up to two years, although usually without doing obvious harm.[47]

The society of petrels is simple and ritualised. Their mastery over liquid habitats is not matched by aptitude elsewhere. It cannot help that most petrel young (except those of albatrosses and giant petrels) grow up alone in the dark. One theory about passerines is that cognition improved in place of olfaction, and if that is so, petrels went the other way, becoming good sniffers instead of good thinkers.[48]

Seabirds' Services

Islands that lose their seabirds change. Oceans are salty because water on its journey from land to sea bears away nutrients, some of which are returned to land by birds. Marine bird droppings have been important to humans as farm fertilisers, gathered as guano. They remain vital to some plants. The pisonia trees (*Pisonia grandis*) on coral cays are ornithocoprophiles, plants that do best in soil enriched by seabirds. Their flat limbs suit the nests of black noddies, which donate fertiliser in return. On a bare coast in South Australia, by swimming out to a limestone platform, I once found a lush herb, leafy peppercress (*Lepidium foliosum*), sprawling around napping penguins and brooding gulls. Only when I learned that it was an obligatory ornithocoprophile could I fathom its absence from mainland outcrops a short distance away.[49]

One puzzle about ornithocoprophiles is why Europe's seabird colonies supposedly lack them altogether. I suspect they are not absent at all but go unnoticed, having spread long ago to nutrient-enriched farms and become Europe's weeds, in which guise they are now known. Such plants thrive on Australian seabird islands, spread there by gulls. Botanist Mary Gillham told of seabirds 'tilling' and 'manuring' to produce an 'arable' habitat suitable for weeds of stockyards and waste places. Many peppercresses are global weeds.[50]

Captain Cook's achievements included success against that

scourge of the seas, scurvy. At each anchorage he plied his crew with messes of seashore plants. The most effective was probably a New Zealand peppercress, now called Cook's scurvy grass (*L. oleraceum*). Cook filled a boat with these greens, but this plant is now endangered. Four of New Zealand's six seabird peppercresses have become endangered, and a fifth is extinct. A loss of bird manure has been offered as a reason why.[51]

When rats replace seabirds on islands, the nitrogen content in the foliage goes down. There are fears on Norfolk Island of a complete change in vegetation if nutrients carried by seabirds are not restored. Each hectare of land on the island once received, by one estimate, 300 tonnes of rich guano a year. On subantarctic Macquarie Island nitrogen is most concentrated in the kelp near penguin roosts. My first impression of this island from the *Aurora's* deck, after two weeks of grey sky and water, was surprise that the colour green could be so vivid: seabird manure explains that. The sea near the island station is tested for human sewage, but the bacterial counts are highest around penguin colonies.[52]

Antarctica is a continent on which birds largely determine where plants grow. The dominant plants, lichens, are most obvious around bird roosts, where they plaster cliffs and boulderfields. Snow petrel colonies hundreds of kilometres inland are lichen 'oases' in sterile terrain. Mosses grow today on ground used by penguins thousands of years ago.[53]

The plants that are fertilised by seabirds are often moved around by them as well. Macquarie Island is extremely remote, but some of its plants enjoy vast distributions, thanks to prickly or sticky capsules. The burrs of buzz burr (*Acaena magellanica*), for example, have been seen on a skua's breast, and not by coincidence is this plant found as far away as Patagonia and islands south of Africa. I left Macquarie Island with hundreds of its burrs in my gloves.[54]

On islands in warmer seas some capsules are so sticky – those of *Pisonia* trees – that seabird wings are sometimes encumbered, dooming the bird to a slow death. On some Queensland islands, in some years, noddies and shearwaters die in the hundreds, eventually becoming fertiliser for the trees that kill them. The victims even include frigatebirds, which have 2.5-metre wingspans. On Cabbage Tree Island in New South Wales the sticky capsules of the bird-lime tree (*P. umbellifera*) were falling on endangered Gould's petrels, and the trees had to be killed. The glue kills many insects as well. On Christmas Island I have seen, stuck on one large seed as if to flypaper, dozens of ants and flies of four species. Hundreds of these seeds had fallen uselessly to earth, evidence of how wasteful of life nature can be. But bird-lime trees grow from Madagascar to Hawaii, showing that often enough the glue serves its purpose.[55]

Some animals, as well as plants, also need seabirds. The most southerly reptile in Australia, the Pedra Branca skink, lives partly on scraps of fish dropped by gannets and albatrosses. The *Aurora* passed by its only home, a forbidding rock stack in the sea south of Tasmania.

Seabirds in Trees

Very few seabirds depend on plants, but those that do so are instructive. One island on earth stands out by having seven species breeding in trees – Christmas Island, which also counts as one of very few tropical islands to have lacked people before the nineteenth century. An extinct volcano, it is a relic of a lost past.

From all the barking and barfing noises in the rainforest you could think there were seals inside, but the calls issue from an endangered bird high in the trees, Abbott's booby. Once found widely in the tropics, it now survives only here. Because its chicks, on their maiden flight, need a long drop to engage their wings, Abbott's booby is the only seabird anywhere to need tall rainforest. It may go back to a time when many tropical seabirds used trees. Not a petrel, it belongs in a small order with frigatebirds, cormorants, gannets, and other boobies.[56]

Abbott's booby nests win no prizes. Made from leafy stems, they rot before the chick can fly, and for some months it has to sit on branches. A breeding cycle lasting more than a year, and a hot wet summer of decay contribute to this odd situation. Many boobies died on the island in the past when trees were bulldozed for phosphate mines, and in 1990 an additional problem emerged: fledglings plummeting to the forest floor. The mine pits were exposing nearby nests to wind turbulence. This was unusual – a seabird threatened by too much wind – but real enough to motivate earnest revegetation. In 2012 the boobies faced yet another problem, breeding failures, which were blamed on a dry spring causing nests to dry out and collapse.[57]

The other tree-nesters on the island (red-footed boobies, white-tailed tropicbirds, noddies, three frigatebirds) use smaller trees. One tropicbird was breeding so low down I could have grabbed it from its hollow right by the jetty where refugees disembark. These birds, along with the frigatebirds and common noddies, nest on the ground on some islands, but Christmas Island teems with predatory land crabs, and once had native rats, now extinct. Completely missing today, thanks to the crabs and rats, are tropical petrels. The point to note here is that some seabirds use trees when necessary, but petrels never do.[58]

Why does no petrel ever roost in trees, despite the security they provide? The answer is probably to do with their feet. All the birds on Christmas Island today have an opposable toe, whereas petrels have more extreme paddle feet, placed further back on the body. Petrel and passerine feet went in different directions. Passerines are superior at perching because they have tendons that lock their hind toes in place. Petrels, instead of hind toes, have gliding wings that lock in place. In isolation on islands, where trees did not matter, evolution deleted the hind toes, which were not only useless, but impeded swimming. What was lost at the same time, it seems, was the capacity to respond to danger on land. I mentioned earlier that flightlessness usually ends in extinction. Seabirds show that flight alone does not buy survival if there are no strategies to counter predation.[59]

Seabirds in wind are hypnotic to watch. I don't believe I've spent

as much time absorbed by anything wild as by the wandering albatross. They ride fierce winds with an ease that brings to mind clichés
about poetry in motion. But their problems on land show that mastery in one setting can translate into extreme vulnerability in another.

Eleven

A Continent Compared

Rudyard Kipling appreciated the power of comparisons: 'And what should they know of England who only England know?' Australian biologists were long held back by a passion for their wildlife so loyal it discouraged the comparisons needed to achieve a meaningful understanding of Australia. Globalisation has changed that by turning ecology into a science that values international comparisons. Now is my last chance to consider the ecology of Australia's birds, drawing on observations as well as comparative science, elaborating on some of the points made before, and visiting other regions of the world.

Before going any further I should make the important point that many birds in Australia are far from unusual. Gould saw this, telling his countrymen that Australia's 'Eagles, Hawks, harriers, and Owls play their accustomed parts; while Swifts, Swallows, Martins, and Flycatchers perform the same offices with us'. Because of past travel on wings, Australia and Europe share twenty or so native birds, including the peregrine falcon, coot, turnstone and great crested grebe, along with a good number of genera of such birds as terns, gulls, ibises and spoonbills. Since they don't stand out in any way, I have nothing to say here about such birds. Australians think their bold crows are special, but most continents have noisy crows, Europe perhaps more so than Australia, a point masked by terminology (jackdaws and rooks are just small crows). One crow in Australia *is* special – the bare-eyed crow of New Guinea – but only because it lives on fruit, substantiating an earlier point about plentiful rainforest fruit. Australia did not produce the first crows. The black kites that float above bushfires and country dumps also seem fundamentally Australian, but I have seen their dark shapes above Voltaire's chateau in rural France.[1]

Australia is considered the land of marsupials, but there are ordinary mammals to match the ordinary birds. Rodents and bats make up half the fauna, including bush rats (*Rattus fuscipes*) that are little different from the rats (*R. rattus*) that gave Europe the plague. Because of ancient travel from Asia, Australia has native *Rattus* where Europe does not.[2]

When we carry out comparisons we must apply some care. Outback waterbirds travel so willingly they soon find new pools in remote places, perhaps by detecting vibrations emanating from thunderstorms hundreds of kilometres away. Reaching speeds of 100 kilometres an hour, grey teal will put 1000 kilometres behind them to trade a drying pool for one that is rising. After satellite-tracking their nomadic moves, biologist David Roshier drew a sharp contrast with the vast waterfowl flocks on the migration flyways of the Northern Hemisphere, 'all heading in the same direction with approximately the same plan for breeding success'.[3]

He's right of course, but as we've noticed all too often, Europe and North America can't on their own tell us if Australia is unusual, because these lands are themselves extreme, with savage winters driving birds south. Most birds reside in the tropics, and the contrasts in behaviour they offer are less stark. One ornithologist has remarked that birds in the southern temperate zone of the world – read Australia – behave more like tropical than north temperate birds. South Africa has nomadic ducks, and Roshier's Australian work has been quoted there to guide thinking about African ducks.[4]

The portrayal of Australia as a land of boom and bust, with

opportunistic nomadic birds that chase the rains after every drought, is one that comes up when the outback is the focus, but I'm not sure how important nomadism is. A recent book, *Boom and Bust: Bird Stories for a Dry Country*, repeats what others have said: the evidence for it is limited.[5]

Much of the nomadism there is involves nectar rather than water. Patchy flowering has been invoked as a 'principal driver' of nomadism in Australia. In one fifteen-year period, spotted gums on the south coast of New South Wales flowered well only once. Bad droughts drive swift parrots from Victoria to Queensland. Blossoming events resemble drenching rains by attracting great hosts of feasting and breeding birds. Australia's waterbirds are probably more nomadic than those anywhere else, but I suspect this only became true a couple of million years ago when the climate grew more erratic. Some of those waterbirds roaming the interior are shared with other continents. On the Thames near London Bridge, I have seen the same species of great cormorant that breeds in the outback after rain. Africa, Asia and North America have it as well.[6]

The place that has taught me the most about Australia is southern Africa, another dry, flat, southern land with, in the east, low ranges harbouring pockets of refugial rainforest, and, in the south-west, protea heathlands animated by large nectar birds. All the similarities narrow down the possible reasons for the differences. South Africa also has world leaders in fire and savanna research, and park managers who claim success with mosaic burning inspired by Kakadu and Uluru. South African expatriot Antoni Milewski has done the most to compare the two lands, and we've enjoyed email exchanges about topics as arcane as wood densities and termite numbers.[7]

New Zealand, as we've seen, is informative as well, as a part of Gondwana that detached simultaneously from Australia and Antarctica before drifting north more than 40 million years before Australia did. It must have shared with proto-Australia its earliest birds, at least along the long boundary they once shared. By staying further south New Zealand remained cool and wet, yet became another land of giant nectar birds and parrots, of honeydew and giant ground birds. It looked more like Australia in the Miocene, when it had eucalypts and banksias, on which New Zealand parrots, which enjoy nectar today, would have fed.[8]

Another special place is that long island east of Queensland, New Caledonia, a portion of Zealandia that, after sinking then re-emerging 37 million years ago, gained plants and birds from Australia. Its poor soils carry a profusion of bird-pollinated plants – paperbarks, grevilleas, lillypillies, mistletoes, and many more – and its shrublands are dominated by strident honeyeaters. Its capital Noumea, like several Australian towns, has screaming rainbow lorikeets roosting in crowds in its central park. Grey-eared honeyeaters hop along the edges of the paths below, pecking at who knows what. New Caledonia reinforces the point that plants on infertile soil commit strongly to pollination.[9]

Living with Aggression

The feature of Australia that is most revealing, I am convinced, is all the aggression between species. After spending six weeks in Kenya and southern Africa in 1994 I scrawled a line in my notebook that proved the germ for this book: 'Bird aggression very little observed in Africa', and the phenomenon has held my attention since. What impresses me is how often I see it in the vicinity of eucalypts, banksias, and their relatives.

During six weeks on a 2012 Churchill Fellowship in Europe and North America, the only sustained bird aggression I saw was around a eucalypt planted in parkland near Los Angeles. Hummingbirds were defending its flowers, and a hooded oriole its lerp, although they were only attacking their own kind. The insect secreting that lerp (*Glycaspis brimblecombei*), by spreading to eucalypt groves in Europe and North and South America – we don't know how – may be increasing conflict wherever it goes. Is the human demand for eucalyptus wood driving up global bird aggression?[10]

The fights in Australia shape how millions of birds live, and there is more to this than I have mentioned so far. Spotted pardalotes, for example, nest in soil; they use creek banks, road cuttings and, very unusually for such tiny birds, sometimes dig into flat ground. They have made burrows in pot plants, in sand dumped at building sites, on an eroded sheep track, and a dune near high tide. Underground nests are seen as a response by these lerp-feeding birds to honeyeater attacks, which sometimes end in death. Some of the most sustained chases I have witnessed were honeyeaters on the tails of spotted pardalotes, which will dive below ground to save themselves. The noisy wing beats of many honeyeaters are probably part of their warning systems.[11]

Grey-crowned babblers pay differently to live among honeyeaters – they lose nests to them. So often do blue-faced honeyeaters evict them that biologists postulate severe effects on babbler breeding, worsening a decline blamed mainly on habitat loss.[12]

These babblers are themselves aggressive. They will smash the eggs and ruin the nests of smaller birds, including finches that eat very different foods. I have seen young golden-shouldered parrots, while seeking small seeds on the ground, harassed first by these babblers, and then by friarbirds, just weeks before the wet season, a time of shortage that sees many youthful parrot deaths. Here were endangered birds harried for no purpose that I could discern, since their attackers eat different foods. Most parrots can defend themselves, but grass parrots have small beaks with limited nipping power.

Australian songbirds can of course be aggressive without being honeyeaters, if they are group breeders or substantial in size. Apostlebirds have killed domestic fowl while raiding their food. Does life with honeyeaters help explain this, or are large songbirds combative because intelligence and group-living improve defence? New Zealand biologists, surprised that introduced magpies should attack and even kill their native birds, have talked of evolution 'in an Australian environment that rewarded such behaviour'. I suspect that the large group-breeders found outside Australia, such as the brown jays that harassed me in Mexico, would show more aggression if they found themselves in Australia.[13]

The hyper-aggression may go some small way to explaining why few Asian birds have made homes in Australia. Exhausted birds reaching the northern Australian shore might find themselves mobbed right from the start, especially if they look unusual.

Nectar Niches

Especially because of all the easy nectar, Australia has ended up with dietary niches found nowhere elsewhere. Honeyeaters combine a liking for nectar with anything from lizards to the eggs of other birds. Lorikeets obtain protein by digesting pollen – rather than by chasing insects as honeyeaters do – yet despite this remarkable specialisation, rainbow lorikeets in aviaries can be weaned onto a diet of seeds and water. They love soft fruit, as I found when I picked pears in an orchard. They are successful all the way from Victoria through New Guinea to southern Indonesia and Vanuatu, and the breadth of their diet, which no hummingbird or sunbird can come close to, is surely a reason why. Woodswallows will flock with real swallows, depleting skies of insects, then drop down to take nectar in trees, shrubs, and desert-pea blooms near the ground. I have seen flowering desert bloodwoods explode with woodswallows as my car passed, and they like lerp as well. Treecreepers are committed bark-foragers that sometimes take nectar, dabbing fringed tongues into eucalypt and banksia sprays. They lick sap as well.[14]

Australian treecreepers stand out in another, very important way. DNA studies put them in an exceptional place on the songbird evolutionary tree – on the second-lowest branch, one above lyrebirds and scrub-birds (see DNA tree on page 70). They matter to science hardly less than lyrebirds, with such features as a foot design unique among perching birds for lacking hind-claw ligaments. To reach insects they can walk under limbs upside down. They hang from trunks to sleep. Their hind toe rotates more freely than on most birds, the better to grip trees whose bark, as defence from bushfires, is often thick and rough. As for the asymmetrical muscles and other features of their syrinx (vocal organ), American Peter Ames found 'a striking departure from the basic muscle pattern that prevails throughout the rest of the oscines'. It is not in fact a departure, but an early design, as old as or older than the design in the bird that went on to spawn the world's songbird fauna. Treecreepers sounded different to me after I learned that, although in truth their penetrating whistles aren't that unusual; like lyrebirds and scrub-birds, they vibrate their torso when they call. The world's songbirds have one form of syrinx for Australian treecreepers (seven species), another for lyrebirds and scrub-birds (four species), and a third for the remaining 4500 bird species – the only design to find its way outside Australia.[15]

Treecreepers look to me like survivors (hanging on by their superior feet) from some early outpouring of songbirds, one that may well have included fulltime nectar feeders, doomed when honeyeaters evolved. Birds can be sent extinct by new competitors, but a bark-living bird that acquires special feet early on has an advantage over newer birds that show interest in bark insects. An American biologist concluded that relict songbirds often feed in unusual ways, and foraging on bark was one of his four categories. A presence in Australasia was another category he mentioned, and Australia does have other bark relicts (sittellas and the enigmatic ifrita).[16]

Honeyeaters and treecreepers look so similar that until Sibley ruled it out they were thought to be close relatives. Beaks made for probing flowers also suit bark crevices, and long toes suit coarse bark as much as they do dangling flower stalks – which means that flower birds may have a better chance of adapting to bark than do birds that hunt in foliage.[17]

For the insights they provide into the earliest songbirds, few birds in the world can match Australian treecreepers. Their feet and vocal muscles have unique designs that may date back tens of millions of years. The white-throated treecreeper, shown here, bears some resemblance to a honeyeater and sometimes licks nectar from eucalypt and banksia blooms.

One reason for mentioning all this is the evidence that many plants go back in time much further than the birds they nourish today. In other words, some plants achieved a winning design – one that remains successful – long before birds (or mammals) did. An amazing 50 million years separates the oldest grevilleas and waratahs from the oldest known honeyeaters. There must have been earlier honeyeaters than those revealed by fossils but they could not, as the plants did, have overlapped in time with dinosaurs. Those museum bones Walter Boles showed me strongly imply this by having skeletal features that would have preceded those found in honeyeaters.[18]

Lorikeets, the other main nectar birds, seem from their DNA to be even younger. In Australia, and in Antarctica as well, since it had similar Proteaceae shrublands, there must have been earlier birds attending flowers, including parrots like none Australia has today and birds from long-extinct groups. Eucalypts and banksias have left their own fossils that predate honeyeaters, from the warm, wet Eocene. Walter's Eocene bones imply that treecreepers do not go back that far, but they could be survivors from the first wave of nectar-loving songbirds coming later. The simpler anatomy of plants could explain why they achieved durable designs well before birds and mammals did.[19]

DNA studies place bowerbirds on the same branch as treecreepers, as their nearest relatives. An American who examined a satin bowerbird found 'some of the most peculiar and unique cranial features in the entire order of perching birds'. The lachrymal bone has a design unique to bowerbirds and lyrebirds (whether he examined treecreepers is unclear). Like lyrebirds, bowerbirds have more secondary wing feathers than nearly all songbirds, and their lives proceed very slowly, satin bowerbirds taking up to seven years to achieve adult plumage. Given their stark anatomical differences, doubts have been raised about whether treecreepers and bowerbirds can belong on the same branch. The genetic evidence for this is probably misleading, an example of two ancient lineages from very early on grouping together during testing because they are so different from other birds – a phenomenon called long-branch attraction. Treecreepers almost certainly constitute the second songbird branch and bowerbirds, because they have a typical syrinx, the third.[20]

Bowerbirds specialise on fruit, but when silky oaks (*Grevillea robusta*) offer their treacly nectar, regent bowerbirds feed heartily on that, dabbing beaks whose long curve suggests a deep connection to flowers. Of the 350 grevilleas found in Australia, the silky oak is one of only four bound to rainforest, and its pollen comes closer than any to resembling 70-million-year-old pollen grains found in Victoria. Very unusually for a grevillea, it does not hybridise with other species, indicating the large genetic distance you would expect of a relict line. Its nectar may count as an ancient food. Satin bowerbirds like it as well, though their bills are stout.[21]

The swift parrot could be another survivor from a flourishing of nectar birds, pushed into a corner of Tasmania by lorikeet success. It stands out by having a tiny breeding range and no close relatives, the nearest being parrots in the Pacific, including the striking horned parakeets of the New Caledonian rainforest. New Guinea has four longbills (in the family Melanocharitidae), which before Sibley's work were taken to be honeyeaters but which evolved slender beaks independently. Dick Schodde sketched for me their odd tongues, with their idiosyncratic fringes. They could have been important in the past, but longbills and swift parrots sit too high on DNA trees to be extremely old.[22]

Success for nectar birds, as we've seen, has suited the plants that made it possible. What the latter want from birds is mobility. The two engage in trade, with trees and shrubs paying with nectar and fruit flesh for transport of their pollen and seeds. Plants outlay more to have their pollen moved in Australia than anywhere else, as the nectar birds show by their sizes and numbers.

Unusual nectar niches hold true for plants as well as birds. Outside Australia and the Pacific region, bird-pollinated forests are scarce to non-existent. There are paperbark stands in New Caledonia and Indonesia, and *Metrosideros* forests on many high Pacific islands, marking the spread of Myrtaceae out of Australia and New Zealand respectively. There are also Zambezi teak (*Baikiaea plurijuga*) and protea woodlands in Africa, rhododendron forests in the Asian mountains, mangrove stands, and coconuts behind beaches. None begins to compare with Australia for nectar volumes and numbers of flower-feeding birds and mammals.[23]

That Australia has the world's largest fruit-eating birds shows that
payments made for the transport of seeds are also exceptional. Birds
of paradise and prolific fruit pigeons show this as well. But we should
not forget that birds harm as well as help plants: bell miners and cockatoos contribute to forest death, parrots destroy seeds as efficiently as
squirrels do, cockatoos and choughs dig up orchid tubers. Whether
helpful or harmful, Australia has strong bird–plant relationships, and
there is more that can be said about this.

Plants tend to do better from birds than from mammals, for two
reasons: wings and teeth. Bird wings spread more pollen and seed than
do animal legs, and beaks are usually kinder to seeds than are jaws
with teeth. There are exceptions of course, including pollen-brushed
bats and seed-crushing birds. Birds also aid plants more than mammals do by consuming many insects that damage foliage.

Teeth guarantee that more leaves are lost to mammals than to
birds. Leaves are best digested in guts much larger than those found
in the average bird, although moa plants show that birds can become
major enemies of leaves when they have large bellies with grinding
stones in their crops. What's interesting is that most birds that graze
and browse – some ducks, geese, landfowl, ratites – occupy the deepest branches of the avian tree. Ducks and geese digest nutriments from
grass quickly but less thoroughly than do cows and sheep, removing
far more leaves than their size might suggest, especially the soft shoots
that farmers value most. Thousands of Cape Barren geese were shot
for the pasture they ate. On Tasmania's Maria Island, when I visited
long ago, the geese patrolling the campground had left it as smoothly
clipped as a bowling green.[24]

Leaf-eating in such lines could reach back to the Cretaceous, when
dinosaurs were the main herbivores and mammals were petite insectivores of the night. In Guatemala I watched crested guans of turkey
size (the two are related) lunching on lofty rainforest foliage, and
browsing birds like these may have been commonplace before mammals evolved rumens (stomach compartments for fermentation) and

ever-growing teeth. Mammals have come to dominate the leaf-eating niche by processing foliage more efficiently. The largest mihirungs were usurped as herbivores by giant marsupials, or so it is thought.[25]

Leaves are important as well to bowerbirds. In winter, when fruit is scarce, satin bowerbirds graze in lively flocks, pecking grass, clover, nettles, chickweed, pennyweed, dock. At least six bowerbird species eat leaves. The tooth-billed bowerbird shows itself to be the most committed browser by having serrated jaws. The deep position of bowerbirds on the songbird tree makes it possible that leaves served as a major food of songbirds very early on, but this is only speculation, because we can't be sure what early birds ate. Very few songbirds eat leaves today.[26]

Absence as Opportunity

Marsupials in Australia left many openings for birds by not doing as well as the mammals on other continents. A poverty of mammals helps explain why birds on islands often control a large share of the ecological space, and Australia is the island continent. It has hundreds of mammal species, but there are limits to what they do, mammalogists having noted unusually few predators or very large herbivores, even when the many extinct species are taken into account. Foxes and feral cats have been able to wreak havoc in Australia because their predation styles were something new. We have seen how marsupials benefited birds in the distant past by not producing a large carnivore in the rainforest, and by failing to match them as consumers of seed or fruit, and there's more to it than this.[27]

Much about Australia reflects 'missing' mammals: a bird as the largest and most dangerous rainforest tenant, parrots that do 'everything', and all the birds in New Guinea that use fruit. Two flightless rails – the Tasmanian native-hen and New Guinea flightless rail – stand out dramatically as the only exceptions to a global rule that flightless rails keep to islands without land mammals. 'Tasmanian' native-hens were doing well on the mainland before Aborigines and then dingoes arrived, judging by all the bones found in Victoria, South Australia and Queensland, some of which are only 12000 years old. Australia had flightless rails (*Australlus* species) more than 20 million years ago, showing that predation rates have lagged behind other continents for a very long time. The rails are exceptional not only for being the oldest flightless rails ever recorded, but for being the oldest rails ever recorded.[28]

The main bird groups on earth headquartered in the Southern Hemisphere – penguins, petrels, parrots, Palaeognaths – succeeded because they escaped the influence of some or all land mammals. After the demise of the dinosaurs the first serious carnivores on earth were mammals (creodonts and mesonychids) that spread widely but did not reach Australia, Antarctica or South America. The reason the south was able to produce the most extreme marine birds was that Antarctica (and other lands such as Zealandia) gave penguins freedom from land mammals. It might be said that the world has one hemisphere weighted towards mammals and one towards birds, since one is mostly land and the other partitioned by sea.[29]

Australia's remarkable reptiles also fit into the picture of mammal poverty. With 960 species and more named each year, reptile diversity is nowhere higher. As a teenager travelling around Queensland on a bicycle I discovered two new lizard species. Competition from lizards may be one reason why inland Australia has far fewer birds than Africa that hunt insects on the ground. Reptiles could explain why Australia has the world's largest kingfishers, the kookaburras, since these often take reptile prey. A reputation as snake-killers got them taken to Tasmania and south-western Australia, where they thrive today. Australia's pythons and goannas operate in place of furred predators. One who knows them well, American Eric Pianka, sees large goannas as 'clearly ecological equivalents' of the kit fox and coyote. I've watched a scrub python in north Queensland swallow a quoll within a day of entering its territory, by perching on a branch where, because of quoll traffic, quoll scent would have been thick. Reptiles stood out even more when Australia still had *Megalania* – a goanna large enough to eat people, and the largest lizard ever – plus gigantic land turtles, and *Wonambi*, a snake that grew 6 metres long.[30]

Early naturalists thought it only logical that an island continent in the remote south should have 'inferior' mammals. The devastation that foxes and cats visited on many marsupials was interpreted by Albert Sherbourne Le Souef as evidence that they were 'far ahead of them in the evolutionary scale'. Marsupials are not considered inferior today and that's not what I'm saying. Niches went unfilled because a small isolated continent cannot produce mammals (or birds) for every occasion. Twice Australia's size, South America had placental mammals as well as marsupials, but they still failed to fill all the available niches. Terror birds increased in bulk over time to become the top predators, sharing this space with sabre-toothed mammals, until big cats padding down the Panamanian land bridge ended their reigns. Africa sets the bar for mammal success today, but only because it served as a welcoming receptacle, for big cats and hyenas coming from Eurasia, giraffes from Europe, and cheetahs and zebras (horses) from North America. Australia, needless to say, remained beyond the reach of these groups. It was, in the most important sense, an island.[31]

But humans breached its isolation, spelling doom not just for the Tasmanian native-hen on the mainland, but for another flightless bird, *Genyornis*. And mammals keep expanding their ecological space, as deer of six species invade forests from failing farms, foxes infiltrate Tasmania, feral pigs and horses multiply, and monkeys promise disaster in New Guinea. Multiplying deer are a problem today even in England, where the understorey they consume deprives nightingales of habitat. This problem, a serious one, has a counterpart in Australia's Sherbrooke Forest, where sambar deer are trashing the lyrebird habitat once visited by Katharine Hepburn.[32]

The question of what is missing from Australia is important because there are so many crucial ramifications, including those arising from the lack – until recently – of large social bees. Australia has thousands of native species but they lack the industrial efficiency of the honeybees brought out from Europe 200 years ago. These soon spread from managed hives into the wild, to achieve extraordinary success as an invasive species. All over Australia they can be seen thriving on the bounty of nectar and nesting inside eucalypt hollows. When I worked briefly as an entomologist entomologist, honeybees from managed and wild hives were the main insects in my net whenever I deployed it around eucalypt blossoms. They compete so strongly with birds today that one bottlebrush stand lost 30–50 per cent of its honeyeaters when bee numbers rose. Pollen mobility in Australia has probably dropped since bees arrived, and may drop further. The Queensland government recently abandoned its program to eradicate newly arrived Asian honeybees, after funding agreements with the honey industry and other states collapsed, and there are European bumblebees in Tasmania that will probably reach the mainland one day. I have watched them in Hobart buzzing around perturbed swift parrots.[33]

The birds that are absent from Australia are too many to list, but one group has to be mentioned: the woodpeckers. Like parrots they are 'tool-heads', with insulated skulls that license them to operate beyond the bounds of other birds. Because flight demands that skulls be light, most birds have weaker jaws than mammals of equal size, the exceptions being parrots, woodpeckers and some finches, which succeed alongside mammals as consumers of hard seeds or insects hidden behind wood.[34]

Country folk in times past mistook Australia's treecreepers for woodpeckers, because they also drum on wood for insects, albeit with less noise and success. Grubs in trees appeal as well to black-cockatoos, which instead of precision tapping tear trunks open as if with angry pliers, harming eucalypts when they remove borers. On plantations near Coffs Harbour some 40 per cent of saplings have been lost to winds toppling cockatoo-gutted trees. My bushland property looks as if a blunt axe has been taken to many young trees, including one

I found that was lopped off half a metre above the ground. Aborigines used axes to obtain the same food – witchetty grubs.[35]

Rather than cockatoos or woodpeckers, the rainforests in northern Australia have striped possums noisily opening rotting trunks for grubs. Eucalypts and wattles have various gliders incising bark to encourage sticky flows, in place of sap-farming woodpeckers. Australian birds often replace mammals, but here are the opposite – marsupials replacing absent birds, showing how special woodpeckers are. Madagascar also shows this, by having lemurs instead of woodpeckers consuming wood grubs and sap.[36]

New Caledonia has tool-using crows in place of woodpeckers or anything similar. Northern biologists have wondered why, inconveniently for them, the world's best toolmakers after *Homo sapiens* hide away inside rainforest on an obscure Pacific island. One of my birding highlights was to watch from close by one of these odd-looking crows as it pecked to enlarge a tree hole, then slid in its tool – the central axis of a pinnate leaf – to tease out a handsome reward: a large cricket. After losing that tool it quickly made another, removing a pinnate leaf from a branch then neatly snipping off each leaflet in an operation so deft and methodical it left me breathless. So young was that bird that it made begging calls the whole time it worked, which persuaded a nearby parent to offer up a treat it had found. The 'Extraordinary large brains' of these birds (to quote experts) show how the pursuit of riches behind wood can promote intelligence, a point I raised about Queensland's palm cockatoos. The other bird made famous by adroit tool use, a tanager on the Galapagos Islands, is called the 'woodpecker finch'. As for true woodpeckers, they have large brains and very versatile diets.[37]

Creeping through Malaysian mangroves one sweaty morning, while woodpeckers of three species worked trunks and stems around me, I wondered why Australia, with the same or closely related mangrove trees, has no bird or mammal attacking mangrove wood. Here was more evidence that island Australia does not have birds for every occasion. The image of vacant niches could not be stronger.[38]

This absence is made stranger by the fact that Australia leads the world for the number of bird species bound to mangroves. Fourteen species have been listed, compared to just one in Africa, one in South America, and seven in Malaysia. But Australia is not the mangrove metropolis these numbers imply. Other continents actually have more bird species in their mangroves, but they're not specialists; they use other habitats as well. I have seen, for example, those same Malaysian woodpeckers in rainforest.[39]

Concentrated in the north-west of the continent, Australia's mangrove specialists are thought to be rainforest birds that took to mangroves after a drying climate removed their main habitat. The largest by far is a skulking bird that sounds like a pig half screaming, half grunting – the chestnut rail. Queensland's mangroves, which are taller and more expansive than Western Australia's, have more birds in them but most of these use rainforest as well. Mangroves are classed by some botanists as a kind of rainforest, and some birds obviously agree.[40]

If you proceed further down the West Australian coast to North-West Cape and Barrow Island, there are animals that went through far more to survive aridity. The only freshwater fish are three species of blind white cave fish, found up to 70 metres underground. On a barren plain I looked down an old stone well upon small white gudgeon suspended in the cool water far below, oblivious to the sun pouring down the shaft. They touched my imagination more than most birds do.[41]

Recent Arrivals

Besides what it lacks, Australia has groups that came in only recently. To the starlings and grass birds mentioned earlier can be added owls, thrushes, crows, sunbirds, Asian storks, and several more. We saw that grasses helped birds succeed in Australia and mistletoes also did this, in a modest way.[42]

The industry of mistletoebirds was something David Attenborough thrilled to in his charming series about plants. He showed one of these

For most of its past Australia lacked owls. They are an ancient group with an excellent fossil record because they often roost in caves in which their bones survive. The lack of any Australian owl fossils going back more than a few million years suggests a real absence deeper in time. Australia only has three owl genera compared to eight in Malaysia and 11 in North America (above Mexico). Instead of small owls, Australia has owlet-nightjars.

tiny, black and red birds wiping an excreted seed onto a stem – where it might sprout and grow – not once or twice, but ten times. The mutualism between bird and plant has been lauded as 'perhaps the most highly developed example of this phenomenon among Australian birds'. These minute birds move as many as 66000 seeds per hectare per season.[43]

But Australia had mistletoes tens of millions of years before it had this special species. The mistletoebird was long ago seen for what it is: a newcomer, a member of the Asian flowerpecker group, a songbird invader. Guesses about its entry date range from a couple of million to under 12000 years ago. It must have appropriated its niche, outdoing other birds in contests for the glutinous fruits, perhaps sending something extinct. The painted honeyeater – a rare bird today – has been called the 'original mistletoe bird' because it needs mistletoe (the nectar as well as the fruit) while belonging to an old Australian group. It keeps to inland woodlands, raising questions about what, near the coast, ate mistletoe in the past. Many birds take the fruits, but not in great numbers.[44]

By flitting about everywhere from mangroves to rainforest to desert, the mistletoebird shows itself to be superior at distributing as well as swallowing the seeds. The fruits it ingests bypass its gizzard, going straight to the intestine for such quick processing that seeds can pass out in less than four minutes. No old Australian bird can match this efficient digestion. Australia, after this bird arrived, could well have found itself with more mistletoe and more aggressive birds sustained by the nectar. All the mistletoes growing in Asia explain how a bird could arrive in Australia with a body ready-made for these plants.[45]

Another example of recent success in a tiny package is the well-known silvereye, and others like it called white-eyes (genus *Zosterops*). The first of these little green birds burst onto the scene less than 2 million years ago, evolving at speeds unknown in birds before it, to produce a parade of eighty-plus species that spread from South-East Asia west to Africa, north to Manchuria, east to Samoa and south to Tasmania, where they have streamed past me through falling snow. Biologists have compared them to Vikings and Polynesians, because epic journeys gave way to settled living. White-eye success turns on confidence over water, flock travel, rapid breeding and flexible diets.[46]

For recent success in a large package, the brolga comes to mind. Asia's sarus crane is so similar they must have separated recently from a common ancestor, one that probably lived in Asia.[47]

Australia will keep acquiring new birds. The cattle egrets that wandered in from Asia in the 1940s, and the pied chats spreading across New Guinea show that birds suited for farms have an advantage. A tiny woodpecker that likes tall paddock trees may join them one day. In the recent geological past the Sunda pygmy woodpecker spread as far east as Alor Island, near Timor, not far from Australia, where it has made a home for itself in eucalypt savannas.[48]

Birds that Remain Surprising

Ecosystems are too complicated to be understood at more than a modest level. An ecological framework is like a newspaper cartoon –

a simple picture emphasising prominent features, such as fights over sugar or the large noses of politicians. Given all the forces at play in the natural world, grand narratives cannot hope to explain most of what we see. I like the Orians and Milewski paper 'The Ecology of Australia' not only for talking up honeyeaters, nectar, exudates, serotiny and cooperative breeding (linking them all, for better or worse, to infertile soil), but for listing 'Puzzles and Anomalies' that stumped them. Why does Australia have more psyllids than aphids? they wondered. And why is mistletoe so prolific when southern Europe misses out altogether?

After seeing how easily biologists succumb to hubris, I am keen to admit to my own puzzles and anomalies. I can't explain why wattles lack floral nectar, why sap hardly matters to parrots outside of New Zealand, or why Australia and New Zealand have so much honeydew when the sources are so different: woodland eucalypts and rainforest beeches. I wish I knew why some eucalypts only have insects and never birds at their flowers. Pollination expert David Paton told me that every eucalypt species in his state of South Australia attracts birds at least occasionally, and this seems true in the Northern Territory as well, but it's not the case in the south-eastern states, where some flowers are consistently ignored. Pollen brought from far away by birds can sire seedlings unsuited for the site, a wastage that may matter more to some trees than others, justifying flowers that only target insects as pollinators.[49]

I wonder as well about all the ways there are to be a honeyeater. Success has come to them in many roles that don't obviously suit them. A DNA test found that Macgregor's bird of paradise, a velvety-black subalpine inhabitant with big orange wattles, is actually a giant honeyeater, the world's largest. Living on Gondwanan conifer fruit (*Dacrycarpus*), it exists because New Guinea's highest and coldest forests, by lacking birds of paradise, pigeons and bowerbirds, left an opening for a honeyeater to become the main fruit bird. Compare it to a gibberbird – the lark-like honeyeater that patrols desert gravel (see colour section), pecking at insects and fallen seeds – and to any hone-yeater that makes its living from bark, and then to a typical flower-fond honeyeater, and the variation you'll see is far greater than that in most of the world's bird families. In Australia only parrots (family Psittacidae) begin to compare. Sunbirds and hummingbirds are deeply conservative by comparison, with lifestyles that do not deviate much from one species to the next.[50]

Impressed by honeyeater variation, biologist Allan Keast in 1984 graphed several African bird families by plotting bill length against wing length. His graph for sunbirds is close to a straight line, because an African sunbird's wing length (which reflects its overall size) tells you its bill length, more or less, since this hardly varies between species of similar size. The graph for honeyeaters looks like a giant egg. In other words, honeyeaters combine any number of different body sizes and bill lengths, and some are very large. None of the African families chosen by Keast, for overlapping in size or diet with honeyeaters, took up much space on paper, since they vary so little from one species to the next. He needed eight to ten African families to equal the variation in honeyeaters. This finding was all the more amazing because Keast left out New Guinea's sensational honeyeaters and Australia's short-billed 'chats', which weren't known to be honeyeaters when he did this work.[51]

Honeyeaters occupy a place on the avian tree that adds a twist to all this. They are part of the third or fourth surviving songbird branch (suborder Meliphagoidea), the one that came after treecreepers and bowerbirds. This branch is the oldest one to qualify today as a serious success, with some 280 species alive, many of them newly evolved and thriving in Australia and nearby lands. This number compares to twenty bowerbirds and very few treecreepers, lyrebirds and scrub-birds. The Meliphagoidea vary enough to go into four or five families, but those that are not honeyeaters are tiny, or at least small: thorn-bills, scrub-wrens, fairy-wrens, grasswrens, gerygones, pardalotes. The largest among them are three thrush-sized bristlebirds, which are doing so poorly today that two qualify as endangered or vulnerable.[52]

Because the third songbird branch is overwhelmingly dominated by small consumers of insects, we can surmise that the common ancestor of the group was a tiny insectivore. The first songbirds that turned up in Europe 25 million years ago were tiny, with bones like those of fairy-wrens and pardalotes, suggesting they were in this group. We know they left no descendants, because no member of the assemblage gets closer to Europe today than South-East Asia, where gerygones warble from trees in Singapore's busy streets. Their DNA suggests they evolved from Australian gerygones only recently.[53]

Honeyeaters broke through the size barrier that keeps their relatives small, and significantly, they did so multiple times. The DNA of the blue-faced honeyeater puts it at a distance from the other large honeyeaters, and so does its stand-out colour scheme. The related black-chinned honeyeater, one-fifth its size, has a matching pattern of green, black and white, plus blue skin beside the eye, making one a goliath version of the other, with far more bombast to go with its stronger beak and harsher voice. Their DNA shows them to be close relatives.[54]

Something else that evolved again and again was a lifestyle tied to bark. The honeyeater tree is altogether confusing, showing any number of changes from small to big and from one food to another. The strangest relationship is the one between Australian chats and New Guinea's subalpine giant, Macgregor's honeyeater – a bird that shared an ancestor with the desert gibberbird far more recently than with wattlebirds and friarbirds, which come closer to it in size and diet. We can suppose that liberal nectar and exudates freed honeyeaters from the small insectivore niche, fuelling larger sizes and wider diets, but success has come in ways that don't include nectar. My guess is that sugar fuelled so much variation that honeyeaters could succeed in ways that no longer require it. On other continents nectar kept hummingbirds and sunbirds small. The quantity and accessibility made Australia different, allowing for much bigger birds with less specialised beaks.[55]

The Australian magpie is something else I wonder about. The two books written about it do not convince me anyone has its measure, written as they were before its DNA showed it to be a butcherbird that took recently to life on the ground. The black butcherbird, a discreet tenant of tropical rainforests and mangroves, turns out to be closer to the magpie genetically than to other butcherbirds. Their DNA suggests these birds shared an ancestor a mere 4 million years ago, give or take a million or two, an ancestor that presumably had its home in rainforest. The magpie has been moved into the butcherbird genus to reflect its status. We should call it the 'giant ground butcherbird', for that is what it is. The move from forest to open ground made it bigger, fiercer and probably smarter.[56]

Experts in times past thought the different schemes of black and white in different regions of Australia showed that magpies came in multiple species. Only one is accepted today – because of the broad overlap zones – but the different forms do act differently. In Queensland they behave like ordinary Northern Hemisphere songbirds, with one pair to a territory, while in Victoria they breed cooperatively in group territories. In the south-west of Australia up to five pairs share one territory, practising group defence but not group breeding. These magpies are the most promiscuous birds on earth, and the only magpies with different male and female patterns, as if gender recognition matters more in the west. Tasmanian magpies were esteemed by the nineteenth-century pet trade as the best singers. Why magpies vary in these ways is unknown.[57]

Birds Around the World

Australia does not stand alone in having unusual birds. Each continent has its specialties, except for Europe; this joined Asia so long ago it is best thought of as one end of a supercontinent, Eurasia, which, because of input from South-East Asia and the Himalayas, is richer in the east. I have seen the same tits and great spotted woodpecker in

Korea that animate parks in England; and in a Vietnamese pine forest, the same jays that naturalist Gilbert White, Bishop of Selbourne, beheld in the East Hampshire countryside.

The proper units for comparison across the globe are faunal regions rather than continents. The Palearctic region is made up of Europe, north Africa and northern Asia, and it has a sister region in North America – the Nearctic – that shares extreme winters which drive mass migrations, spring singing, rapid living, simple pair bonding (most of the time), and seed-caching by jays. The shared birds include immense numbers of ducks, geese, gulls and breeding waders.[58]

Biologists could think the Palearctic and Nearctic determined what was normal because these vast realms lose so much heat in winter that birds over immense areas respond in common by concentrating their breeding into the warmest months before migrating south. But frigid winters are important in only two of the world's eight faunal regions, and while these are vast, they lack the richness of the tropics, which support more than 80 per cent of bird species. English ornithologist Ian Newton was very blunt about a (Northern Hemisphere) temperate bias among his colleagues: 'Often these behavioral ecologists and ornithologists do not realize that the conventional wisdom applies only to a select group of birds from temperate regions, birds that do not represent general adaptations of birds.'[59]

Songbirds in the north often turn to fruit in autumn when insects become scarce, fuelling up on lipid-rich berries before migration. Evolution has favoured fruits that ripen in the cooler months to suit these birds. In Australia, mild temperatures and evergreen leaves ensure there are more winter insects and little winter fruit, except in all the gardens planted with northern privets, cotoneaster, camphor laurel, firethorn, hawthorn and holly. These fruits are enhancing the winter survival of pied currawongs, which have become the main predators of small birds in eastern cities and towns. Nest predation is increasing, and a connection has been made to proliferating currawongs and all

the fleshy fruits of exotic shrubs and trees. Satin bowerbirds are entering gardens in Canberra because these fruits suit them better than the
leaves they turn to in winter. The plants are responding by spreading
into paddocks and bushland, becoming weeds wherever birds drop
their seeds.[60]

Troubled by this, one biologist has suggested three possible
responses: remove the garden plants, develop seedless varieties, or
'Intensively cull pied currawongs'. Another biologist, uneasy about
predation at robin nests, also counselled plant removal and currawong
culls. But the fruits are often so colourful that many gardeners don't
want them removed.[61]

Of the world's other regions, the Neotropical – comprising Central
and South America – is the richest of all for birds, with a plethora of
parrots, hummingbirds, suboscines, tanagers, and all the followers of
army ants, including antbirds that hunt no other way. I accepted a
few stings for the chance to watch woodcreepers snatching the man-
tids and other insects that were fleeing up rainforest trunks in order
to escape an ant phalanx. Australia has nothing like this, although
it does overlap in a different way, by having shared a reputation for
poor songsters. Darwin found the inherited calls of suboscines less
than musical, and singled out the mockingbird as almost the only
South American bird to 'take its stand for the purpose of singing'. The
Neotropical region is the headquarters for a passerid group that most
reminds me of old Australian songbirds – the New World blackbirds
(icterids), some of which are big and noisy (oropendolas, grackles),
and others of which breed in unusual ways (oropendolas, cowbirds).
Cowbirds (in North and South America) resemble miners by replacing other birds in damaged forest, but they use stealth rather than
violence, laying their eggs, cuckoo-like, in other nests.[62]

The Oriental region, taking in South-East Asia and India, has very
different rainforest birds from those in New Guinea and Queensland.
Asia has provided Australia with fewer birds than might be expected,
given how close it is and how rich in species. To cross the seas, birds

and plants rely mainly on bird wings, but pigeons have ensured that more kinds of seeds than birds made the journey. Biologists have noted that in general, plants are better than animals at reaching new continents, and the durability of seeds helps explain that.[63]

The Ethiopian region (Africa south of the Sahara) matches Australia in having a super-abundant food, although it is fatty rather than sweet. When humid weather beckons, flying termites issue like wispy smoke from mounds or holes in the ground. Termites are too fleeting to warrant defence by birds, so they do not provoke aggression. Hundreds of bird species take them, including kestrels in vast migrating flocks, as well as many mammals. I have watched a Wahlberg's eagle and a small antelope (red duiker) side by side on the ground snatching queen termites from holes at their feet, while other birds were taking them in flight. Termites have much to feed on in Africa, including all the vegetation damaged by elephants, huge dung piles rich in fibre, and carpets of leaves shed each dry season. Australia has finches and magpies that take emerging termites, but not in great numbers.[64]

While less than half the size of Australia, southern Africa is far more diverse. An average African national park surpasses all of mainland Australia for species of bustards, larks, pipits, swallows, finch-like birds, starlings and birds of prey. Southern Africa boasts twenty-six larks and thirteen bustards to one of each in Australia (and Australia's lark-like birds – quail-thrushes, gibberbird – fail dismally to make up the numbers). Africa is so rich because birds have entered from Asia and Europe, and because it has two arid zones to serve as evolutionary cradles – the Kalahari and the Sahara – separated by moist forests that retreat west during ice ages, permitting enriching exchanges.[65]

Southern Africa has a bewildering series of birds bound to rock outcrops, including a handsome woodpecker. But it only has six parrots, none of them truly common, and, like eastern Australia and for the same reasons (ice-age losses and aridity), southern Africa is not rich in rainforest birds.[66]

For variety of land birds Australia is well ahead of glaciated lands, such as northern Europe, but far behind South America, South-East Asia and Africa, even with New Guinea included. What singles Australia out is the large proportion of birds of special evolutionary or ecological interest. They reveal far more about Australia's past than European birds disclose about Europe's, because most Australian songbird families evolved *in situ*. A 2012 paper invoked an 'early filling of ecological space by ancient radiations'. Australia had lyrebirds and logrunners before songbirds had even reached southern Africa, and before Europe had the families thriving there today. All the differences and similarities convince me that Dick Schodde was right to describe the bird fauna of Australia as 'the most distinct and "different" in the world'.[67]

Twelve

People and Birds

During the twentieth century, as Australians became more urban, the bronzed surfer came to replace the World War I digger and the pioneer bushman as a symbol of manly power. Equating national vigour with skin tone gave Australians new opportunities to feel superior to the pasty English. On the best beaches, oil squeezed from birds was conscripted into the patriotic endeavour of darkening skin.[1]

Muttonbird oil was nearly the only ingredient in a product famous in the 1950s and '60s: Vita Tan, the brainchild of canny industrial chemist John Paterson. To a 44-gallon drum of seabird oil he added perfume and other additives, then took his place on a Gold Coast beach where he sprayed and rubbed the young and beautiful. The hype he blasted from a megaphone as he marched along the streets induced tourists to buy his product. The pith helmet he wore had a stuffed muttonbird on top, and a second bird adorned his illegally parked Rolls-Royce, which remained beside the beach for years, covered in ads for Vita Tan, until it was towed away as a rust-eaten eyesore. On Bondi Beach his product was supplied by local identities Basil and Stan MacDonald.[2]

Hundreds of thousands of muttonbird necks were snapped each year on Bass Strait islands, before their lifeless bellies were squeezed to release the oil that went into Vita Tan and other products. Paterson's

Ever the showman, John Paterson liked to pose beside signs for Vita Tan, and with young women also at his side. A stuffed muttonbird can be seen here dangling below his sign. By one account, Paterson claimed that no birds were harmed to produce his product, which was patently untrue.

product contained no sunscreen, but it did embolden beachgoers to imbibe more sun. Vulnerable white skins were better served when lotions with sunscreens took its place.

The Vita Tan story shows that Australia has served as a stage, not just for unusual birds, but for unusual interactions between birds and people. These included the 'emu war' in Western Australia, launched in 1932 to combat the tens of thousands of emus that were trampling wheat. The federal minister for defence was humiliated when the birds eluded the machine-gun detachment he'd dispatched, in full media glare, to mow them down. Snide comments were made in parliament about a gun that jammed when emus approached an ambush.[3]

Bird–human relationships have changed markedly over time. A couple of generations ago Australians were eating the parrots and kookaburras they lovingly feed today. Native birds have gone from being exploitable resources to sentient beings that are treated with pro-tective respect. From the nineteenth century onwards values changed all over the world, but they did so in Australia in distinctive ways.

My focus in this chapter is on Australia as a nation, leaving the island of New Guinea to one side.

The pages of *Emu*, Australia's main bird journal since 1901, provide one way to chart the changes. *Emu* is a journal of science today, but less than a century ago it ran a report on a trip that mentioned, merely in passing, the daily killing of all the birds of prey around a waterhole. Changes to values did not come easily. The relationship of humans to birds is so complicated I can be little more than a swallow skimming over the surface, hovering in a few places worth highlighting. My particular interest is in exploitation, conservation and study. One is about taking, one about giving, and the other includes both.[4]

Muttonbirds

Muttonbirds show how extreme exploitation can become when birds are crowded together and easy to catch. They induced Aborigines to risk their lives negotiating treacherous waters to reach the Maatsuyker islands south of Tasmania – the southernmost land trod by Aboriginal feet. Islands off the north-west of Tasmania were visited as well. In 1828 at Cape Grim, thirty Pennemukeer men, women and children were feasting on muttonbirds obtained on these islands when they were massacred and hurled from a cliff by shepherds avenging some dead sheep and a spear in a shepherd's leg. But those killed were the wrong mob, having come for muttonbirds, not mutton.[5]

Off the north-east of Tasmania, unreachable islands were respected as the land of the dead, and they became Australia's largest seabird sanctuaries, supporting more than 15 million muttonbirds, which, then and today, make up some two-thirds (by mass) of all the seabirds nesting around Australia. Their idyll ended in 1797 when the *Sydney Cove*, the first free-trader in penal Australia, was thrown by a gale onto one of those islands, which became known as Preservation Island. The survivors became the second group of Englishmen in Australia to live away from Sydney, and like the first, on Norfolk Island, they ate birds. For five months, muttonbirds plus a cup of rice served as their daily fare.[6]

Tasmania was at that time thought to be part of the mainland, but Matthew Flinders, one of these men's rescuers, surmised that Preservation Island fronted a large strait. In Sydney he conferred with George Bass, back from a great journey south in a whaleboat powered by six strong convict oarsmen (who in turn were powered by mutton-bird meat). When Bass and Flinders set out the next year to explore the passage, they were tailed by the roving seaman-merchant Captain Bishop, and where the explorers found a strait to bring them acclaim, Bishop found seals to make him rich. The next three years saw 200 sealers invade Bass Strait.[7]

When, by raid or barter, some sealers settling on Bass Strait islands acquired Aboriginal women, they gained more than they'd bargained for: stealthy hunters able to swim through difficult waters, then, naked as their prey, slither up to basking seals and club them to death. Collapsing seal stocks persuaded the few sealers who remained behind with their partners to develop muttonbirding. This is one of the few forms of commercial wildfowling to survive to this day, and it is central to the identity of Tasmania's Aborigines, most of whom claim descent from those doughty women.[8]

The birds supplied feathers, eggs, body fat and, most importantly, meat and oil. Muttonbird fat oiled Tasmania's progress, servicing coalmines, smelters, foundries, tanneries, and the fledgling timber industry, by helping logs in sawmills slide along wooden skids. It was blended with lye to make soap, and also fed to calves. All this ended about a century ago when the rendering down of adult birds for their fat was outlawed.[9]

Larger and fattier than their parents, the full-sized chicks were, and still are, squeezed after death for their red stomach oil, which found service as a medicine, railway lubricant, leather softener, lamp oil, and Paterson's sun lotion. High in omega-3 fatty acids, its main value today is as a nutrient supplement for racehorses, greyhounds, poultry, pet snakes and people.[10]

Most of the chicks killed these days go to New Zealand, where the similar sooty shearwater is an esteemed Maori food. In decades past they were canned and sold as 'squab in aspic'. Great piles of their feathers and down were stuffed into upholstery, pillows, quilts, mattresses (1600 birds providing one featherbed) and, more recently, sleeping bags. From a million chicks in the heyday of the early 1900s, the yearly harvest has dropped below 100000, due to falling interest in the products, rising regulations and costs, and the difficulty of incorporating a seasonal harvest into modern lifestyles. Giant tiger snakes in the burrows on some islands add a measure of risk.[11]

The islands support far fewer birds today than when Matthew Flinders encountered one flock so large it took an hour and a half to pass him, containing, he calculated, 151 million birds. Muttonbirders have not killed anywhere near enough chicks to account for there being less than 20 million today, and that suggests serious problems out at sea, including climatic anomalies depleting food in the Northern Hemisphere.[12]

Muttonbirds at one time held special importance to Australia's main bird group, the Australasian Ornithologists' Union (AOU). In the nineteenth century, probably because of goldrush wealth, Melbourne became the centre for bird interest and nature conservation in Australia – a position it retains today. Melbourne being close to Phillip Island, which has muttonbirds, these were the first attraction at the union's second annual meeting, in 1902. At the opening dinner in Miss Kissock's tea-rooms, the birds were served up in three different ways. The meeting went on to discuss bird sanctuaries and education about protection, although not for muttonbirds.[13]

Muttonbird egging was popularised by one of the AOU's founders, Archibald Campbell, who wrote glowingly in the *Australasian* of taking 250 eggs on Phillip Island, teasing them from burrows with a crook on a stick. His other method was to flip a bird over, pin it with his knees, then press the abdomen to 'deliver' the egg. This operation left the owner 'to all appearances delighted', readers were told,

Muttonbird egging was socially acceptable in 1901 when Archibald Campbell,
a founder of the Australian Ornithologists Union, photographed these professional
eggers on Phillip Island.

although how Campbell assessed the mien of muttonbirds was not
made clear. He boasted of egging in three states, and bragged as well
about shooting lyrebirds.[14]

To acquisitive naturalists, Phillip Island was Eden, with eggs double the size of chicken eggs sometimes laid on open ground. Campbell
wrote fondly about a camping trip by the Field Naturalists' Club of
Victoria: 'The amusement of egging was carried on during the day
chiefly to supply the larder. Mutton-Birds' eggs fried are a great deli-
cacy, and were enjoyed by all in camp.' Women joined in. 'Occasionally
you come across a small party of ladies, gloved and veiled, deftly using
their egg crooks,' he noted. Shooting as well as egging went on. All
this is recounted in the main nature journal of the time, the *Victorian
Naturalist*. At a later camp the state governor dropped by. When his
Excellency Sir Thomas Gibson-Carmichael departed on the government steamer with his basketful of eggs, enthusiastic birders cheered
him off.[15]

Campbell's muttonbird articles and book about eggs must have talked hundreds Australians into joining the harvest. Some eggs even found their way to Melbourne biscuit factories.[16]

The muttonbirding on Tasmanian islands had one leading bird-lover wanting it emulated in Queensland:

> The appointment of the Great Barrier Reef Commission with powers to investigate and develop the resources of that remarkable and valuable natural feature was a great step forward. It is probable that a commercial use will be made of the large supply of eggs of sea-birds. For well over 100 years an industry in connection with the Mutton-bird or Short-tailed Petrel has been conducted in the Bass Strait area. Even more valuable industries may be developed on the Queensland Coast. If properly regulated and controlled, such industries would be permanent.[17]

John Albert Leach, who wrote this, also founded the Gould League of Bird Lovers, whose members signed a pledge: 'I hereby promise that I will protect native birds and will not collect their eggs.' Archibald Campbell and Sir Thomas Gibson-Carmichael were patrons of the league. There was hypocrisy in this, to which I will return later, but less than you might think. Early bird-lovers worried about extinction, but not about the exploitation of common birds if it was neither cruel nor excessive. Early *Emu* articles recounted the exploits of AOU members who, in the name of bird study, shot and caged birds and took their eggs – another topic I will return to.[18]

The treatment of muttonbirds was extreme because they suited industrial-scale exploitation. By furnishing eggs, poultry and feathers in commercial amounts, a muttonbird rookery resembled a modern chicken farm. Harvesting of other birds was seldom this intense but it was diverse, with birds valued at different times for providing food, feathers, oil, guano, sport and pets. What's interesting is that natural history was also a form of exploitation, with ornithology operating as another form of harvesting.

Early Exploitation

Legends tell of the Maori finding New Zealand by following godwits, and birds may well have led people to Australia. If the first boats came from Timor, there would have been birds as large as pelicans showing the way. Humans became a new mammal for the birds to contend with. Indigenous people ate a much wider range of species, obtained by a wider range of techniques, than people do today. Their hunting was so imposing that early European visitors, including Joseph Banks, found the birds shy. Complaints about the ducks suggest they were warier than English ducks, which were themselves hunted.[19]

Flightless birds faced a special risk from the new predator. The largest bird at the time, *Genyornis*, was probably one victim. Other extinct birds that may have fallen to net and spear, although this is only speculation, include a dwarf cassowary, a giant malleefowl and a giant coucal. Seabird colonies near the mainland were probably decimated when seagoing watercraft, which had fallen out of use some time after human arrival, were reinvented a few thousand years ago. The lack of large colonies in Torres Strait, except on remote Bramble Cay, suggests past losses.[20]

Most birds proved adept at surviving the new predator. Reports of Aboriginal hunting techniques in the nineteenth century show that great skill was required – ducks were snatched by an underwater hunter breathing through hollow reeds; other birds were tricked by nooses lowered over their necks from a place of concealment, or hit with clubs and boomerangs, or spooked into nets strung between trees by a slab of bark winging past like a hawk. Perches smeared with sticky fig sap ensnared small birds, as did tangles of sticky herbs draped around pools. Crows and hawks were grabbed by a hunter stretched upon a rock in the sun, counterfeiting sleep, with a piece of fish or meat in the hand as bait. Emus were drugged at pools laced with narcotics, or lured towards spears by a fairy-wren fluttering on a string. How was the wren caught? James Dawson offered one explanation in 1881: 'Small birds are killed with a long, sharp-pointed wand by boys, who lie in thickets and attract them by imitating their cries. When a bird alights on a bush above their heads, they gently push up the wand and suddenly transfix the animal.'[21]

No wonder noisy miners complained when Aborigines appeared, voicing warnings understood by other birds. Hunting pressure must have increased greatly a few thousand years ago, if it is true that human populations rose then.

What I've said about mosaic burning bears repeating here: it aided many birds, including birds of prey that snatched anything small startled by smoke and flames. By carrying smouldering sticks, black kites extended fires.[22]

Wells dug in parched riverbeds, like today's farm dams, would have extended lives during drought. The explorer Charles Sturt watched birds crowd into one such well, dug 7 metres into the bed. Another explorer, Ludwig Leichhardt, thought pigeons at a well suggested 'flies around a drop of syrup'. When pools were laced with toxic plants to stupefy fish, kites took some of the bounty. Great bowerbirds took 'odds and ends' from campsites to help lure females to their bowers. Here is anthropologist Donald Thomson in 1935: 'Although the bowers of these birds are often built close to native camps, the birds are not molested. Bower-birds are not eaten and the aboriginal does not destroy bird life wantonly.'[23]

Aboriginal people in some regions continue to use birds. In the Northern Territory as many as 60000 magpie geese are taken each year.[24]

Taking by Europeans

Early Europeans saw in Australia a virgin land with a natural bounty there for the harvesting, its birds included. The birds gained a powerful new predator, one with guns and ships.

As we've seen already, island birds proved especially vulnerable to exploitation. Soon to disappear were the Norfolk Island kaka, the white gallinule on Lord Howe, and dwarf emus on King and Kangaroo islands. Each of them limited to one or two islands, these birds were scarce to begin with. At the other extreme were the island birds so bountiful they invited industrial exploitation.[25]

On Macquarie Island's cobble and black-sand beaches, sealers flayed fur seals, boiled down the blubber of elephant seals, then turned on the birds, herding penguins into pens to be clubbed and tossed into waiting digesters, which held up to 2000 at a time. As many as 150000 royal and king penguins were rendered down a year, to yield about a pint of oil each, much of which, burning in lamps, lit up England at night. Douglas Mawson lobbied to stop the carnage, succeeding when the island lease was not renewed. Penguin numbers have since rebounded, but the old digesters on the beaches are arresting reminders of that time: rusting steel monuments past which oblivious royal penguins file on trips to the sea.[26]

Guano left by multitudes of roosting seabirds was mined as fertiliser in Queensland, Western Australia and Tasmania. Raine Island, one of nine islands exploited on the Great Barrier Reef, was depleted by a hundred Chinese and Malay labourers taking out tens of thousands of tons of guano, while Lady Elliot, further south, suffered the loss of 2.5 metres of island surface. Rat Island in Western Australia, which hosted a very large seabird colony, lost its surface as well, and its birds.[27]

Pied imperial-pigeons resemble seabirds by repairing to islands in great hosts to breed, and they offered the advantages of good eating and a semblance of sport. Pigeon shoots were organised on passenger boats plying the Queensland coast, and there were other visits by commercial fishermen and weekend shooting parties. Aborigines could reach just a few of the pigeons' islands, returning with canoes full of eggs and chicks, while Europeans shot adults in superfluous numbers, leaving chicks to starve. Green Island near Cairns lost all its pigeons. Shooting lasted into modern times because it was difficult to stop. 'More and more fast speedboats equipped with freezers and iceboxes visit the unpatrolled islands,' wrote Stan and Kay Breeden in 1970. 'Automatic shotguns are used to shoot the birds as they return to their nests, heavy and slow with food for their young.' More than a thousand were shot on one island near Cardwell as late as 1967.[28]

White Australians once ate a wide range of wild birds. As in Europe, poultry shops offered strings of quail, snipe and pigeons (top-knots, bronzewings, imperial-pigeons), but you could buy wattlebirds as well, and quail-thrushes. In Bendigo poulterers sold malleefowl, and in Melbourne, live Cape Barren geese. The latter were also caught in South Australia when young and fattened to sell as the Christmas goose. When British officials in 1864 inquired about birds useful to the mother country, Cape Barren geese, despite their aggression towards dogs and fowls, were recommended along with malleefowl, emus and magpie geese.[29]

Many other birds besides pigeons were caught and eaten rather than sold. Parrot pie, an early favourite, retained a place in *Mrs Beeton's Family Cookery* until the 1950s. Gould declared budgerigars 'an excellent article of food', while in New Guinea a bird of paradise soup made by missionary W.G. Lawes proved to be 'very unparadisaical in its flavor'.[30]

In the 1970s emus became the first Australian birds fully domesticated as a food, although the market for them remains small. Their meat is very low in saturated fats and the fine, non-greasy oil suits cosmetics and may help against arthritis. Emu farming has spread to Europe, North America and more recently India, where major investment scams have emerged.[31]

Often birds were killed just for sport. On the voyage to Australia bored migrants shot albatrosses following the ship, or caught them on lines baited with fatty meat. Gould noted that 'hundreds are killed annually'. Kangaroo Island promised good sport to an English surgeon, Dr Leigh, who migrated there in 1836: 'Accordingly I purchased a splendid rifle, double-barrelled gun, single gun, pistols, powder and shot.' But the gun-toting doctor could find nothing to kill. No kangaroos survived around the settlement, and as for emus, 'there has no emu been seen these ten years', he lamented. Leigh did not know it, but he was documenting the extinction of the Kangaroo Island dwarf emu.[32]

Shooting became so rampant that by the turn of the twentieth century Queensland had brought in closed seasons for such small birds as silvereyes, blue wrens, fantails and thornbills. In a 1901 article that was unusually passionate for its time, Frank Littler vented his dismay: 'An inborn, insatiable desire to kill something is one of the worst traits of Australian youths. They take their guns into the bush and are not content with legitimate game but must try their prowess on all and sundry that come across their path. They appear to feel that it is dependent on them to "slay, and slay, and slay".'[33]

Those too young for guns used shanghais (slingshots) to deplete birds around towns, including the sparrows and other foreign birds released by acclimatisers.

Except in Tasmania, exploitation of wild birds is a modest enterprise today. In the Northern Territory some 20000 – 40000 magpie geese are shot each year by non-Aboriginal hunters. On Norfolk Island, following an old tradition, sooty tern eggs are harvested. Many Tasmanians participate in muttonbirding because shearwaters, for no obvious reason, began nesting decades ago on mainland Tasmanian beaches, often near towns. Closed seasons, regulations and occasional arrests do not stop poaching and ruining of burrows. French scientists studying a colony in 1998 were compromised by poachers taking forty of the eighty-nine chicks they were studying.[34]

Ducks are the main birds shot in Australia today. Banned in some states in the 1990s, duck hunting in Victoria draws criticism from the government's animal-welfare advisory committee, the Royal Society for the Prevention of Cruelty to Animals, and *The Age*. A 2007 poll saw 70 per cent of Victorians supporting a ban.[35]

In New South Wales the Game Council sought hunter rights to shoot malleefowl, brush-turkeys, bustards, pigeons and other native 'gamebirds', under a new model of bird conservation – breeding on game farms and releases for hunters – but the council was disbanded in 2013 for unsound administration.[36]

Gondwanan Pets

Past Australians, as well as eating more bird species, put more behind wire. Australia has produced no new livestock of any account, but it did provide the world with its most popular pet bird. As the land of parrots and intelligent songbirds, it provided residents with interesting pet choices. Sydney dealers sometimes offered bowerbirds and apostlebirds, but magpies were more popular, for reasons outlined by Alfred North:

> Young birds reared from the nest are in great request as pets, for they soon learn to speak and in time acquire an extensive vocabulary, besides imitating any familiar sounds such as the barking of a dog or crowing of fowls. If allowed their freedom and the run of a garden an old male is often as good as a watch dog in the day time, for it will immediately give warning of one approaching the place by its loud notes of displeasure, if it does not also savagely attack the intruder.[37]

One magpie laid eggs on a washstand in the nest it built in its owner's bedroom. Lyrebirds taken from nests before fledging had a reputation, as free-ranging pets, for being playful. Butcherbirds had 'whistling propensities and grotesque manners in confinement'.[38]

Australian birds that reached England included a golden-shouldered parrot and star finch that won prizes in a bird show at the Crystal Palace in 1902; there were woodswallows in the same show that missed out. Arthur Butler's *Foreign Birds for Cage and Aviary* (1910) included many Australian birds, not because they were deemed special but because, in the spirit of empire, everything was wanted.

Butler mentioned anything from a consignment of noisy miners that reached Bavaria in 1893 to the malicious disposition of crimson finches. London Zoo by 1906 had sixty Antipodean species, including the straw-necked ibis, apostlebird and regent honeyeater.[39]

But the most popular bird pets of all were parrots. At Junee railway station, Alfred North saw hundreds of young galahs 'packed one on the other like fruit in open cases', ready for travel to Sydney. In Sydney he also saw budgerigars packed like fruit:

> It is impossible to form any approximate estimate of the large number of this species exported annually to Europe and America, but it must run into many thousands of dozens. The mortality amongst them is great, for they are placed in cages with just sufficient perching room for the occupants to be tightly squeezed in. In one bird-dealer's place I saw about fifty dead birds that had been taken out of the cages in one morning.[40]

Because they are so easy to breed and keep, budgerigars, along with zebra finches, have become two of the world's main research birds. Of the birds I kept as a teenager, 'zebbies' were my best breeders. They have become the lab rats of the bird world, conscripted into studies of almost anything – song tutoring, personality stability, memory, social isolation, gene expression, mating preferences, sperm morphology, Diazepam sedation, detection of the earth's magnetic field – in journals that span the spectrum from *Ethology* to the *International Journal of Bifurcation and Chaos*. Some of the research – that related to autism, for example – has medical applications. The zebra finch was the second bird (after the chicken) to have its genome sequenced.[41]

Not by coincidence are budgerigars and zebra finches the most common seed-eating birds in inland Australia. When Desmond Morris recommended the finch as a lab animal back in 1954, what he praised, without knowing it, was a commitment to breed whenever the desert provides a flush of seeds: 'New birds, transported to the laboratory in small boxes, will begin to nest-build and court within minutes of their release into an aviary. There are no seasonal difficulties, as it breeds all through the year. The species is exclusively a seed-eater and the nestlings require no special diet in captivity.'[42]

I find it hardly believable, but each of these species, if temperatures are benign, can live indefinitely without water, meeting their needs from the meagre moisture in dry bird seed.[43]

Why should Australia produce the two easiest of cage birds to breed, when Africa has an arid zone that is far older and far richer in birds? Competition, or its absence, and boom–bust weather, are likely reasons. Tropical storms, in those years when they stray south into the deserts to produce great flushes of seeds, cause budgie and zebbie numbers to multiply. The next drought, as the bust after the boom, drives extreme selection by restricting survival to the very fittest, including those with the least need for water. It was remarked of these birds in 1932: 'They increase enormously in a favourable season like the present and die in thousands in a time of drought.' The high bird diversity in Africa, and all the predators, may have reduced the tempo of evolution by dampening population swings. The red-billed quelea achieves immense numbers, but its chicks require insect food and when caged it seldom breeds.[44]

The Australian Companion Animal Council estimates that 7.8 million birds live behind wire in Australia today. But the solitary budgerigar or canary in its tiny cage has become less common. Lorikeets are no longer 'trapped in thousands and condemned to screech unhappy lives away in cages', as observed by Alec Chisholm in 1922.[45]

Protecting Assets

Although not exploitation, asset protection was something else that took many avian lives, and still does. On the land, it was the habit to reach for the gun whenever an eagle or hawk drifted over, until the message got through that rabbits and rats, not lambs, constitute their main prey. Government bounty schemes paid men to kill hundreds of thousands of birds, including emus, currawongs and bee-eaters (for eating bees). Bounties were paid out even on bee-eater eggs. Alfred North told of 10 000 emus killed in one year around Wilcannia, of 1500 of their big eggs broken on one station near Cobar in 1877. He was by no means disapproving, for emus ate grass, broke fences, and startled pregnant sheep.[46]

Birds also died when other pests were targeted for control. In 1910, because of rabbit poisons, 'rotting carcases of birds polluted every watercourse'. A naturalist who accompanied a Victorian trapper over three months in 1976 was appalled to see that thirty-four lyrebirds had been killed in steel-jawed traps, compared to just eight vermin (foxes and wild dogs). The birds had apparently been attracted by insects lured to the meat baits. A concern today is that birds may die when locusts are sprayed to prevent plagues.[47]

The strangest persecution of birds was the slaughter of peregrine falcons to protect racing pigeons. On a ridge near Hobart, biologist Nick Mooney led me to a rocky platform where wire from an old trap remained in place. From the 1940s into the 1970s this site, like many others, had a 'cocky cage', or wire milk crate equivalent, with a pigeon inside, surrounded by steel-jawed rabbit traps. Every day or so it was visited and everything caught inside it killed, the problem for the trapper being all the possums, devils and other birds of prey that had to be 'cleaned out' first, because 'falcon-hawks' (peregrines), as specialist aerial hunters, usually arrived last. Pigeon-racing clubs advertised bounties on peregrines until the 1970s, when birds of prey gained legal protection. Nick told me of one trap in north-west Tasmania that over two years killed 300 birds of prey, including only two peregrines, the vast majority undoubtedly migrating harriers, which pose no risk to pigeons. Peregrines do take pigeons (we found a numbered ring from one under a peregrine nest nearby), but the way pigeons are raced all but trains peregrines to catch them.[48]

Birds continue to die for the sake of crops and fish farms. All the emus culled each year demonstrate that the emu war never really ended. Some 317 000 were killed in Queensland under a bounty scheme when they attracted blame for spreading prickly pear, and culling in Western Australia continues to this day. Records show that in Queensland alone more than forty bird species are legally killed each year, including wedge-tailed eagles, bustards, spoonbills, honey-eaters, magpies and kookaburras, with cockatoos dying in the highest numbers, up to 7500 a year.[49]

Study as Taking

Bird study in the late nineteenth century entailed so much killing it can be construed as another form of exploitation. Natural history was blurred with sport, and study with acquisition, as men shot birds and took their eggs. Success was marked by a large collection. As good binoculars did not exist, many birds could only be identified by killing them, but even birds that could be identified were still shot. To quote historian Tom Griffiths: 'The gun had been, for many, the only intermediary with nature. It offered a way for early naturalists to get a long close look at the objects of study, so its use was often sympathetic and scientific.'[50]

Archibald Campbell, patron of the Gould League, whose members pledged to protect native birds and not collect their eggs, boasted of killing seven lyrebirds on one outing. His book about eggs and nests conveyed the thrill of the hunt: 'On you creep, every yard nearer, so that with the excitement your heart increases in palpitation til it throbs so loudly that you fancy the bird will hear it.' John Gould saw nothing amiss with telling readers about the crimson chat in these terms: 'As may be supposed, the sight of a bird of such beauty, which, moreover, was entirely new to me, excited so strong a desire to possess it that scarcely a moment elapsed before it was dead and in my hand.' A very good ornithologist, Gould was also a good shot. And it was a member of the Royal Australasian Ornithologists Union (RAOU), G.A. Heuman, who in 1926 casually admitted in *Emu* to the slaughter of birds of prey around a waterhole: 'I cannot remember the different species, but I generally shot a dozen before breakfast.' His interest was in finches so he probably saw virtue in culling their enemies.[51]

Egg and nest collectors had fancy terms to dignify their deeds: oology, caliology, nidiology. 'By their fruits ye shall know them', was their dictum. Cameras captured men perched beside nests high in trees. One article about a field trip conveys their mindset in the title: 'An Oologist's Paradise'. Egging on RAOU trips lasted into the 1930s: 'A participant recalls the row of tents from which in the evenings one heard continuous gurgling noises while eggs of waterfowl were being blown. Every now and again an oologist would stagger out of a tent to discard a bucketful of yolk and water.'[52]

Egg and nest collecting were once so respectable that the Australasian Ornithologists Union saw fit to publish this image in its journal in 1909. It shows the union's youngest member, Master Alfred White, with the nest and eggs of a raven.

A camp-out of such bird-lovers would have depleted breeding over
a wide radius. I once organised a small egg display at the Queensland
Museum, drawing on a private collection from this era, and soon
found myself in thrall to the elegant curves, lustrous surfaces, pastel
hues, delicate tracings and frantic freckles. Eggs invaded my dreams
at night. They are objects of precious beauty that took skill and often
courage to acquire. Here is Arthur Mattingley depicting the ascent to
a lofty egret nest:

> Puffing, panting, and perspiring at every pore, he now reaches
> the limb on which the nest is situated. Trembling in every mus-
> cle with such unwonted exertion, as well as with the enervating
> effects of excitement, he works his way laboriously along the hor-
> izontal bough until he reaches the smaller twigs in which the nest
> is placed. Ha! What a picture is before his eyes! 'Eureka!' Four
> delicately shaded blue-tinted eggs.[53]

The collections seldom served any purpose beyond gratifying their
owners and motivating them to know nature better. Occasional articles in *Emu* listing dimensions and colours fall short of showing that
oology was science. Campbell sounded like nothing more than a collector when he remarked that one lyrebird egg a year, over ten years,
was an 'ample and sufficient reward to satisfy any working oologist'.
Seeing through it all was Alfred North, with his jibe that oologists
'seek to conceal their own faults by loudly denouncing the plume-
gatherers and the subsequent innocent wearers of them'.[54]

Values changed with new generations. When a motion to end private collecting was put at an RAOU meeting in 1922, president Edwin
Ashby railed at the 'astounding absurdity' of 'revolutionary' resolu-
tions that would outlaw 'promiscuous collecting', and leave killing
to recognised ornithologists adhering to a clear course of research.
Seeing nothing less than an assault on manhood, Ashby used his pres-
idential address to extoll the 'formative effect on character' that came

from shooting birds and preparing their skins. The growing mind of a youth develops qualities, he said, that cannot otherwise be developed. Killing birds made the man, he was saying. This comment from Ashby, a man in his sixties, drew a savage rebuke from the talented Alec Chisholm, twenty-nine years his junior:

> Seeking to develop character in our boys, we will not present them with guns and skinning-knives for use on valuable birds; we will equip them with cricket bats and footballs by day and good books by night. And if a feeling for bird-study develops in a particular boy, we will indicate to him, firstly, that birds are the property and pride of the nation and that the person who kills birds or robs nests without governmental permission is corrupt to his citizenship, untrue to himself, and a source of irritation to better-balanced people.[55]

Chisholm was harsher than this in an earlier attack, depicting the average private collector – read Ashby – as a perversion of civilisation and 'relic of sin, masquerading under the honoured name of Science'.[56]

Chisholm prevailed. Photography had by his time become a gentler pursuit for those with the itch to collect. This conflict proved a turning point, as did an RAOU camp-out at Marlo, Gippsland in 1935, when, during breakfast and in front of the group, the Victorian Museum's George Mack aimed coolly at a scarlet robin attending its nest and pulled the trigger. It was a provocation, a snub to those who disapproved of killing for study. Mack had a permit to collect, but even so, all the New South Wales delegates left in disgust.[57]

Accounts of the incident appeared in the *Sydney Morning Herald, Sun, Argus* and *Melbourne Herald.* The fallout saw private collections in Victoria seized. Mack submitted specimens after this trip but none of a scarlet robin, a bird of uncommon beauty that was not, however, uncommon. So divisive was this event that Libby Robin's history of Australian ornithology has a chapter called 'Beyond Marlo'.[58]

Conservation and the Lyrebird

The interests of the nation, invoked by Ashby and Chisholm, were important to conservationists earlier on, in the years following federation. Birds were 'nature's police force, patrolling the face of the land, and keeping undesirables in check', wrote L. Harrison in 1908. 'We must try and win the farmer and his household, the squatter and his riders, to the side of their truest friends and best allies, the birds,' implored Linnean Society president A.H.S. Lucas at his annual address the same year, to which he added: 'For the sake of the land we love we need to train the children to love the tree and the bird.' Jessie McMichael, whose money kickstarted the Gould League, feared a 'disastrous toll on crops' if birds were killed; the League's first president was Prime Minister Alfred Deakin. 'Birds, in truth, are factors in keeping the earth habitable and the soil productive enough for man and beast,' declared Archibald Campbell in 1905, and, 'By an all-wise Creator birds have been fore-ordained to assist human intelligence towards its own highest ideals. Why spurn the gift?' There was a belief that countries had come undone by not protecting their birds.[59]

Books went by such titles as *The Useful Birds of Southern Australia* (1907) and *Some Useful Australian Birds* (1921). In the first of these Robert Hall contended that lyrebirds helped maize and potato crops by removing insects in their vicinity. The peak of exaggeration was a claim by Chisholm that: 'Ladies should not be allowed to wear beautiful headgear at the expense of Queensland's native birds, the loss of which would in time mean the loss of all vegetation.' This was conservation as self-interest, and the argument put today, that birds provide essential 'ecosystem services', is a refashioning of that idea.[60]

While a chorus of voices helped convince Australians to value and cherish their birds, so did the voices of the birds themselves, especially those of one species, the superb lyrebird. Lyrebirds, along with emus and birds of paradise, have served as national icons without necessarily benefiting. Emus found a place on the coat of arms for being 'notable and noble', the kings of the Australian fauna, but were not treated well, as we've seen. Papua New Guinea's house of assembly seriously considered the export of birds of paradise in 1965, but the vote was lost when protests came from as far away as David Attenborough in England.[61]

As for lyrebirds, they suffered early on since their plumes served well as parlour adornments. Hawkers prowled Sydney streets with fifty or more tails in a basket. The RAOU reported in 1912 on one dealer who illegally exported 800 in a year. Fears were held in England that lyrebirds would be driven extinct. John Leach saw the irony: 'On one occasion, lunching at a coffee palace in the far north-east of Victoria, I said to the lady of the house, "I suppose there are not many Lyrebirds in this district now." She replied, "They have disappeared." I said, "That is what I expected, there are nine Lyrebird tails in this room."'[62]

Oscar Wilde declared that each man kills the thing he loves, and this held true about Archibald Campbell, the man who fathered the Ornithologists' Union of Australia, which grew into Birdlife Australia, when he went after those lyrebirds. He wrote afterwards: 'out of ten birds I fired at, I bagged seven, two of which I dropped while they were scampering off through the ferns. I am glad to see that lately the Governor-in-Council has been pleased to place them under the game law.' He wanted to see less killing, but wasn't prepared to set an example. It did not help that lyrebirds breed more slowly than any other songbird. The one egg laid each year can take eight weeks to hatch.[63]

World War I disrupted the plume trade and steered tastes away from the extravagance implied by feathers. A few men who cared set about promoting the lyrebird's wonders. After long sessions in ferny grottoes studying their ways, sleeping each night in a giant hollow log, naturalist Tom Tregellas championed lyrebirds in articles and talks. Ray Littlejohns took immense pains to capture them on celluloid. Australians got to see their special songster at the cinema, and hear it on the radio – the first bird call to be broadcast. A gramophone record followed, and a booklet with photos, *Lyrebirds Calling from Australia*, was sent to cheer up World War II troops. Australians realised they had a bird to do them proud.[64]

By the late 1920s public opinion had swung. 'Cars are parked on the edge of the forest,' wrote Charles Barrett, 'while scores, even hundreds, of eager people go stepping slowly among the trees and ferns.' Tourists were said to be 'spellbound' by the rush of melody. As did the koala, the lyrebird won protection by public opinion. When Australian currency went decimal in 1966 it was the only bird to merit its own coin, the ten-cent piece.[65]

But the affection ultimately went too far. Michael Sharland mused about an 'upsurge of sentiment on its behalf and a feeling that it must be given the protection which so celebrated a bird deserved. "Save it from extinction" was the catch-cry, though with extinction it could not be said to be threatened.' He was writing in 1944, long after the bird trade had ended. In an overreaction to fears about foxes some lyrebirds were freed in Tasmania, which lacked foxes (and lyrebirds). They thrived, and the concern today is all the litter and soil they move when they feed, and the ferns they destroy. One lyrebird in a year can shift 200 tonnes of soil and litter per hectare, a feat no other songbird in the world can begin to match. The Tasmanian government ranks them a high risk to the Tasmanian Wilderness World Heritage Area, worse than starlings, cats and rabbits, and put a fence around the only known colony of an endangered orchid (*Thynninorchis nothofagicola*) to stop them causing its extinction.[66]

When John Leach celebrated 'Australia's Wonder-Bird' and the shift in public opinion, he praised the men (including Campbell) who'd championed its cause: 'Not only have Australian zoologists developed a public opinion which has secured the full protection of the Lyrebird, but they have also procured full protection for other members of the Australian fauna.' Leach omitted to acknowledge the role played by the bird itself. While emus proved too dim to win much affection, this unusual bird, by displaying talents unmatched on earth, showed that Australia's 'primitive' creatures could command respect for their intelligence and skill.[67]

But public affection is fickle, and while lyrebirds remain popular they are the nation's favourite no longer. The koala has taken their place. Koalas proved so easy to hunt that half a million pelts were exported to the US in 1927, the year President Hoover halted the trade. How did the koala steal a nation's affections? Not by displaying intelligence or skill. Offered fresh leaves on a dish, a koala cannot fathom how to eat them. On roads they keep dying because, unlike possums, they never learn to dodge traffic. The meagre nutrients in gum leaves keep their brains at 60 per cent of the size justified by their weight. Experts conjecture that koalas tug at our heartstrings more than most animals because they share the head-to-body ratio of an eighteen-month-old child. A koala on a tree trunk is a toddler on its mother's hip. Lyrebirds are more intelligent and skilful, but they can't match that.[68]

Modern Relationships

Birds are cared about today in ways unimaginable a generation ago. A camera trained on sea-eagles in Sydney Olympic Park streams their nest antics to the world, and when it showed chicks high in their tree trapped in fishing line, BirdLife Australia ran a sophisticated rescue. A cherry picker was brought in to help rush one chick to hospital for surgery – a parent had fed it a fish containing a hook. The group that the gun-happy Archibald Campbell helped found has grown into one capable of treating individual birds with great care: white-bellied sea-eagles are not rare.[69]

The gestures take many forms. The Mareeba Wetlands in north Queensland are the brainchild of local resident Tim Nevard, who saw that unused irrigation water could create a wetland for ducks, brolgas and storks. The café by the lake has the ambience created wherever staff have a passion for nature. The wetland managers are trying to return Gouldian finches to the region.

In Capertee Valley, north-west of Sydney, more than 100000 trees and shrubs have been planted by bird-lovers who came from near and far, to improve the regent honeyeater's future. Australia's Threatened Birds Network, which does the practical work of preventing extinction, could not operate without the thousands of volunteers who donate time and often travel and accommodation costs.

Where a century ago birds were portrayed by bird-lovers as agents of human progress, loyal protectors of crops, today they evoke questions about the future of progress by often becoming its victims. American Rachel Carson's book *Silent Spring* (1962), by highlighting bird deaths from pesticides, helped launch modern environmentalism, and bird deaths today focus thinking on anything from tree clearing to oil leaks. Early bird-lovers accepted and participated in the exploitation and killing of birds, but after federation a more protective sentiment took over, as Australia went from being a frontier in which nature needed taming to a land in which nature increasingly required protection. Today birds are protected for their own sake – for they are held to have intrinsic value – and not for their benefit to us. A Phillip Island resident proudly showed me the muttonbird burrows she watches over in her front yard: that anyone would fry the eggs for breakfast has become inconceivable.[70]

When social scientists describe how nature came to be accorded intrinsic value they usually mention the increasing urbanisation and affluence of the 1960s producing a post-material society ripe for biocentric thinking. But intrinsic rights for birds were sometimes invoked much earlier. Writing in *Emu* in 1901, Frank Littler offered a spirited defence of the birds that took garden fruit: 'But have they no right? Are the fruits of the earth to be man's wholly and solely? I think not, more especially as man, in many instances, takes away from birds the opportunities of sustaining themselves on their natural food as was given them from the first.'[71]

In his 1907 book, *The Useful Birds of Southern Australia*, Robert Hall went beyond the instrumentalism of his title to urge compassion: 'The feeling of man towards birds should, on the whole, be a kindly one.' Alec Chisholm wrote a book in 1922 called *Mateship with Birds*.

One suspects that all those who stressed the economic benefits of birds also believed in their right to live for their own sake. Look at what Alfred Russel Wallace wrote about birds of paradise in 1869: 'their happiness and enjoyment, their loves and hates, their struggles for existence, their vigorous life and early death, would seem to be immediately related to their own well-being and perpetuation alone . . .'[72]

The Australian ethos of a fair go, a product of convict culture, made it easy for naturalists to argue respect for birds, and cheaper cameras and better binoculars made it possible to 'hunt' without harming. In the bosoms of men who no longer killed, empathy was free to flourish.

Symbolic hunting remains important to birders today. It is manifest when they commit to ticking off (seeing) as many species as possible. Writing in the *Sociology of Sport Journal* about the historical overlap between ornithology and hunting, sociologist Keneath Sheard proposed that birdwatching reproduced in a more 'civilised' form the pleasures and skills of hunting. Describing the tick as 'the symbolic kill', he also offered this:

> Not only might quickly finding and holding a bird in steady focus with binoculars be regarded as the equivalent of the hunter's good aim, but the skill of 'shooting' birds with a camera, and the later capturing and fixing of a bird's image on photographic paper, might be regarded as a more civilized alternative to the killing that 'real' hunting involves . . . [73]

Those who think of birdwatching as meandering about the park cooing at cuckoo-shrikes fail to reckon on the skill and zeal that go into accumulating a life list of 600 or 700 Australian species. A ticker who becomes serious can become obsessive about the symbolic wealth embodied by a long list. In one frantic year of twitching one birder flew 65000 kilometres and drove 80000 kilometres, creating a large carbon footprint. Sean Dooley wrote with humour and self-deprecation about twitching as something that can give you ulcers. Englishman Mark Cocker, in his book about his fixation, *Birders: Tales of a Tribe*, explained the value of finding a rare vagrant. 'It's a rite of passage into true hunter status,' he said. 'To repeat it over and over again is to be elevated even further.'[74]

Many years ago, when I kept my own list, the federal shadow environment minister I compared numbers with seemed crestfallen to find he was a dozen birds behind, and vowed to catch up soon. Geographer John Connell's study of Australia's birding scene refers to the 'career path' in twitching and its 'hierarchical social structure'. To adopt a non-hunting analogy, each rarity is like someone famous you've met whose name you can later drop. Someone I defer to is Graeme Chapman, a former CSIRO photographer, who has not only seen most of the grasswrens – some of the shyest of birds and confined to remote places – but taken their defining photos.[75]

Hunting today takes many forms. On a barren plain of a vast grazing property in north-western Australia, a long drive down a sandy track took me to a camp on remote 80 Mile Beach where a large group of birders from around the world, most of them self-funded, were hunkered down under flapping tarps reviewing the day's banding. The next morning I was in the lead car when we 'twinkled' (gently herded) a large flock of waders towards the cannon nets hidden in the sand. From the radio came a firm voice: 'We are armed and remaining fully armed.' After the firing of the net, scores of stints, plovers and sandpipers – including some precious broad-billed sandpipers – soon had numbered aluminium bands on their legs. Many years before this, I went banding in India with the Bombay Natural History Society, and the villagers employed to catch the birds used the same nets they use to catch waders to eat.[76]

The hunting analogy ultimately undersells the birding scene by ignoring other motivations, such as the acquisition of knowledge and the protective sentiments I mentioned before. When Janos Hennicke climbs high up rainforest trees on Christmas Island he looks like a hunter, but his motive, in fitting radio-trackers and bands to endangered Abbott's boobies, is to help save them from extinction. Janos and I help run the Christmas Island Bird Week together.[77]

Interest in birds today is seldom separated from concerns about their future, something that holds true for even the most ardent twitchers. Mike Carter, who has ticked more birds in Australia than anyone, has won a Birds Australia prize for advancing knowledge, mentoring young birders, and campaigning to save wetlands. Another prominent twitcher, Tony Palliser, told me about the conservation project in Thailand he donated to after witnessing dire habitat loss while pursuing an endangered pitta.[78]

Birds motivate a large body of voluntary science, often with a conservation flavour. The world experts on bowerbirds, Cliff and Dawn Frith, work for no institution or organisation. I marvelled at their vast library when I visited their home deep inside Wet Tropics rainforest. For three decades they have funded their academic research by selling nature books and photos. They spend more time observing birds than any academic can hope to.

Kevin Wood, a retired electrical engineer who won his Birds Australia medal for amateur research, told me about the miner's shed he bought in western Queensland in order to study striated grass-wrens, one of the shyest of birds and tied to the harshest of plants: spinifex. Many more like him deserve a mention but I will stop at one, American economist John H. Boyd, whose amazing website about bird phylogeny alerts me to new research (he is often sent papers ahead of publication). With more than a thousand references and much discussion about the correct names for birds and the order they should go in, his site is a remarkable expression of one man's fascination with birds.[79]

Many conservation campaigns inspired by birds benefit whole ecosystems. That is my reply to those who decry the emphasis placed on birds (and mammals) when we know that invertebrates and plants play larger ecological roles. National parks in the woodlands of Victoria and New South Wales grew from concerns about declining birds. In one influential article in 1993 Doug Robinson showed how much had been lost from a Victorian site by citing bird lists from decades before.[80]

Birdwatching is almost the only activity bucking the trend of people losing contact with nature because of urban living and technology. Richard Louv made this point in *The Last Child in the Woods*, his book about the alienation of children from nature. A report into bird tourism in Australia presented birdwatching as 'one of the most rapidly growing pastimes in the Western World'. Britain's Royal Society for the Protection of Birds (RSPB) now has more than a million members, and the number of American birdwatchers has more than doubled in two decades.[81]

In Papua New Guinea, where other kinds of tourism have declined or stagnated, bird tourism is growing. In the Central Highlands, Paul Arut told me birds were far from his mind when he established Kumul Lodge. 'Hotel, club, dancing, prostitution – that's all the idea I had.' His workers were eating birds of paradise when they built the lodge. 'It's all meat,' he told me. A visiting ornithologist explained to him the existence of birdwatchers. 'That was eye-opening for me.' He never looked back. Australian birdwatchers appreciate New Guinea's birds but are playing almost no role in their conservation, a situation that needs to change.

The Ecology of Feeding

However big birdwatching becomes, it will never catch up to bird feeding in popularity. This has become urban Australia's main gesture towards wildlife, second only to pet owning as a connection to animals. Surveys suggest that as many as a third of all households feed birds. Enough feeding goes on that birds solicit in places where they are never fed: a butcherbird has flown through my living room to visit my kitchen, landing on a bench to investigate while I stood by. The brush-turkey that came inside was probably expecting food as well.[82]

The pages of *Emu* are largely silent about feeding in the distant past. An exception was Catherine Langloh Parker, writing in 1902 about the dramas incited by bread at her inland homestead, which included assaults on parrots by 'self-assertive' apostlebirds, and magpies diving down like hawks, 'establishing a scare'. In cities and towns, all the slingshots in boys' pockets probably made it unwise to attract birds.[83]

The feeding that goes on today separates Australia from other countries in two interesting ways: the foods are different, and the act is discouraged by governments and bird groups. An international review of bird conservation in home gardens singled Australia out as the land where feeding is deemed bad for birds.[84]

In Britain the RSPB encourages feeding as nothing less than the best way to help birds. The 233 'bird food & feeding' products it sells online include cakes, biscuits, suet balls, mealworms live and dried, coconut halves, and a range of tasteful tables and dispensers, as well as guards to stop squirrels getting a share. These sales provide the RSPB with a large revenue stream denied Australian groups. The shop at its Bedfordshire headquarters has a room crowded with 12.75-kilogram sacks of seed and imposing boxes of 'Buggy Nibbles'. In the US, where the Cornell Lab of Ornithology recommends putting out peanut butter and jelly (although not together), an estimated 80 million Americans spend $4.5 billion each year on bird food and paraphernalia, putting out 450 million kilograms of seed.[85]

The closest to any of this on the BirdLife Australia website is an information sheet, *Attracting Birds to Your Garden*, which recommends *not* feeding birds, advising that it's better just to grow native plants and furnish water. Reasons not to feed include a supposed risk of bird seeds sprouting to become weeds.[86]

The New South Wales government takes a line so extreme it would be read as comedy outside Australia: 'Think twice before you feed wild animals – a moment's pleasure for you may lead to the animal you feed becoming addicted to junk food.' This sounds like anti-drug rhetoric from the 1950s, but was on the Department of Environment and Heritage website when I looked in 2013. Animals do not become addicted to food.[87]

Magpies, butcherbirds, currawongs and kookaburras are bold enough to take treats from the hand, and Australians give them meat ahead of seeds or cakes. In Tim Winton's acclaimed novel *Cloudstreet*, Dolly feeds urban birds as only an Australian could – with blood and beef juice on her hands. A thrill runs through her when a magpie stabs some meat and pricks her hand. The feeding of flesh confirms Australia's status as a land of large songbirds and giant kingfishers, of birds replacing mammals in some predatory roles.[88]

All the meat raises concerns in many quarters about bacteria at feeders and high cholesterol levels in magpies, but a larger problem is that small birds become extra meals for those coming for beef. A study in Sydney found that meat feeding was correlated with more birds in gardens but fewer small birds. The main native birds around Sydney – magpies, currawongs and noisy miners – all come to food provided by humans (although miners don't fancy meat), and all of them sometimes kill small birds. In Brisbane magpies and butcherbirds are the main birds fed by a very large margin, occasioning concerns there. In the Northern Hemisphere most birds at feeders are so small these issues don't arise.[89]

But seed feeding can also be problematic in Australia, when it attracts sulphur-crested cockatoos that then sharpen their beaks on benches and beams. The Yarra Ranges Council in Victoria has banned the feeding of these birds after a flood of complaints. More than 150 people attended one council meeting, with claims of property damage caused by cockatoos in excess of $240 000. One Upwey resident covered their whole house with fishing net. Some insurance companies refuse to cover cockatoo chewing. The Victorian government issues permits to cull problem cockatoos, showing again that food can indirectly kill.[90]

In Europe and North America feeding became popular after birds were dying in city streets during cruel winters, and though it goes far beyond winter aid today the RSPB still tells members that birds come to gardens for food and shelter 'when conditions in the countryside are especially tough'. Droughts drive some birds into Australian cities and towns, but seldom into feeding trays in gardens. The birds helped in Europe are largely a subset of the small birds found in local woods – tits, chaffinches, and the like – but Australia has highly skewed urban bird faunas. Instead of thornbills, fantails and whistlers there are large, loud songbirds and parrots, creating adverse surrounds for drought refugees.[91]

When Darryl Jones surveyed people who feed birds, environmental atonement emerged as a major motivation:

'Humans have done so much damage to nature, I am trying to give something back.'

'I want to try and undo some of the destruction caused by humans.'

'We have destroyed their habitats, I want to give something back.'[92]

These comments suggest another reason why feeding is less than helpful – it diverts energy away from genuine assistance to birds.

Growing native plants for birds is less contentious but still attracts concerns. Bred for long flowering seasons, hybrid grevilleas and bottlebrushes often end up sustaining noisy miners and other feisty honeyeaters, exacerbating the poverty of urban birds. In a previous book, *The New Nature*, I noted the irony: 'by vying for more birds, you get fewer'. Some gardeners choose to take these plants out, although this doesn't bring small birds back.[93]

Despite everything we know about the topic, 'wildlife-friendly' garden books perpetuate the notion that native flowers lift bird diversity. A commendable exception is Graham Pizzey's 1988 book, *A Garden of Birds*, which is amazingly blunt about noisy miners: 'The moral, for anyone contemplating establishing a bird garden, is to carefully check the whole vicinity for signs of their presence. If they are anywhere near your intended home, and particularly if the habitat of large open trees extends towards your intended location, tread carefully, and even consider living somewhere else, if you haven't already bought.'[94]

When I last consulted the Birds in Backyards website run by BirdLife Australia, the advice about miners was incoherent. One page encouraged me to grow banksias, grevilleas and eucalypts to attract birds – including miners – while another told me they 'aggressively exclude most small birds . . . creating areas with a low diversity of small birds'.[95]

There are opportunities in some gardens to assist declining birds, although it's not as easy as people think, and not possible at all in many places. Spinebills, for example, come to native gardens in Melbourne, but, as one journal article notes, this entails them running a constant gauntlet between the attacks of larger honeyeaters.[96]

Birds into the Future

The surging affection for birds has largely bypassed one species. Until the 1950s, Australian consumption of beef, mutton and lamb was the world's highest. Chicken was a treat for special occasions, and most eggs and white meat came from backyard pens. Consumption rocketed when chickens became a factory food, and Australians now eat 560 million chickens a year, a hundred-fold increase. Chickens are the nation's most popular animal food, with consumption rates almost twice those in Europe, although well behind the US.[97]

Farmers face pressure to treat chickens poorly because a few processors have captured the market and driven down returns, allowing supermarkets to promote chickens as bargain items to lure customers. Ethical issues for the industry include extreme crowding (twenty birds per square metre), debeaking (to prevent stressed birds striking out at each other), infections induced by overcrowding, artificial light, and antibiotics in the food to speed growth. Turkeys, ducks and gamebirds are similarly mistreated.[98]

The conditions chickens live under would attract umbrage if they were applied to lyrebirds or malleefowl, but while hunter-gatherer societies often practised respectful exploitation, according honour to the animals they ate, white Australians reserve their respect for those birds they don't harm. They practise respectful non-exploitation and disrespectful exploitation. Researcher Jane Dixon found that while today's consumers approve of free-range chickens, their motives are mainly selfish. Said one interviewee: 'It's not so much a matter of a good life and death, 'cos if you're only grown to be eaten it wouldn't matter, but from a health point of view.' The European Union banned battery hens in 2009, but Australia's free-range farmers have been undermined by the Australian Egg Corporation arguing that 20 000 hens per hectare – more than two birds per square metre – should count as free-range.[99]

Chickens were domesticated from the red junglefowl, an Asian pheasant that barely survives today as a pure-bred bird due to growing contact with village chickens, which are free-ranging despite animal welfare not rating highly in tropical Asia.

As for the Australians who care most about birds, they tend to occupy two camps that do not, on key issues, agree. A shared dislike of duck hunting does not mean that conservationists shun supermarket chickens, or that animal activists condone culls of feral cats to save endangered birds. A debate I was part of at Melbourne University, opposite philosopher Peter Singer, confirmed a deep divide about feral animal control. Writing as someone biocentric and close to Indigenous culture, anthropologist Deborah Bird Rose has proposed a bridging position, whereby animals that are killed are afforded respect before and after death.[100]

While animal-rights campaigners might be optimistic that welfare in Australia will improve, at least to European standards, conservation is going backwards. Governments have found they can cut funding for threatened species and shed rhetoric about sustainability without losing elections. A belief in rights for nature is in many circles looking like a passing fashion, one that is losing sway to a 'wise use' doctrine which sees natural assets, including national parks, as tools for human advancement.[101]

In Britain, support for conservation investment remains strong across political divides and birds remain a key priority. I saw rare breeding cranes, bitterns, and a careening red kite at Lakenheath, a wetland constructed at great cost by flooding carrot fields and planting reeds. The Ouse Washes were nearly empty on my visit except for the congregation of waterbirds and waders drawn to one paddock by the pools and rises created by heavy equipment, by an electric fence to keep out predators, planted vegetation, and a water table kept high to trap worms near the surface. My ecologist guide that day, Malcolm Ausden, told me that future wetland design will accommodate birds expected to colonise Britain in the wake of climate change. It is worth mentioning here that the Mareeba Wetlands were conceived by an Englishman who moved to Queensland.

In Australia a recent senate inquiry heard from BirdLife Australia that although the critically endangered western ground parrot is down to 110 birds, there is no government funding for captive breeding or other emergency action. As for the Norfolk Island white-chested white-eye, which is now talked about as extinct, warnings of its likely disappearance 'evoked no action by responsible authorities'. The demise of this bird is not altogether certain, but except for occasional visiting birders, no one is looking for it. One islander I phoned recently, Honey McCoy, assured me they survive. 'At the moment they're hard to find. They're down to about two.' The meagre effort that went into saving this bird reflects poorly on Australia as a wealthy regional leader and role model. Cranes, bitterns and red kites have their largest populations outside Britain, whereas ground parrots and Norfolk white-eyes are Australian responsibilities alone.[102]

Another example is Sweden spending almost as much on the white-backed woodpecker, a bird found all the way from Europe to Japan, as Australia is spending on all its threatened birds put together. Some of Australia's rare birds are helped properly, but a couple get nothing, such as the buff-breasted button-quail I mentioned in chapter nine. BirdLife Australia noted that most states have recently cut their 'already lean environment departments'.[103]

Britons are doing more than Australians for their birds for two obvious reasons: massive habitat loss from farm intensification, and more people to fund conservation (a circumstance that does not apply to Sweden). Something else at play, I suspect, are the very different bird experiences in the two countries. England's garden birds are mostly small and rather ordinary. In Canberra, for example, residents receive visits from boisterous sulphur-crested cockatoos, gang-gang cockatoos, king parrots, crimson rosellas, eastern rosellas and, in some suburbs, satin bowerbirds dining on winter fruit. Reports about endangered parrots can seem a little suspect to public servants who have brazen parrots stealing not only their garden fruit, but their morning sleep. What are not obvious are all the birds in decline. Writing recently in *Wildlife Australia*, former diplomat Doug Laing summed up the situation:

> To an extent we have been mesmerised by the bold and the beautiful birds around our human spaces every day. What we have not seen until it is almost too late is the slow fade of the meek and the mild – those birds of our woodland remnants – which have not yet adapted, perhaps never will adapt to the world we are making for them in the 'bush capital'.[104]

Well to the north, pest controllers in Brisbane are called to homes to remove brush-turkeys that scratch up vegetable beds with their oversized feet and rake bark chips into big musty nest mounds. Says one website about turkey removal: 'save your garden, save your sanity! Call today for a free quote.' The wire netting I keep over part of my lawn has not stopped turkeys digging one hole that is deep enough for a foot to disappear inside. Rather than deterring these birds, dogs and cats are robbed of food by them. Yet to Noel Jack, who documented Brisbane's birds fifty years ago, brush-turkeys were little more than memories – shadowy inhabitants of the early rainforests lingering on as an occasional wayfarer on forest fringes. Aiding them today are leafier gardens with compost heaps and a community that long ago stopped killing birds for sport and food.[105]

Brush-turkeys were unknown around Sydney when convicts and early naturalists arrived. The Illawarra scrubs to the north had them 180 years ago, until cedar cutters shot them all out. One turkey seen near Sydney in 1980 was probably an escapee. But so dramatic is their advance today that the *Sydney Morning Herald* ran a headline claiming 'the brush-turkey is eating Sydney'. One pest control website mentioned 800 dollars' worth of garden mulch lost by a resident.[106]

A century ago small birds dominated around Sydney (fairy-wrens, small honeyeaters, yellow robins), not the large birds of today (magpies, currawongs, noisy miners). Biologists who deduced this by analysing museum specimens found that, as a group, parrots have done best from urbanisation. Sydney residents cannot be expected to know that much from the past is missing: regent honeyeaters, swift parrots, red goshawks and diamond firetails, to mention a few.[107]

Parrots have prospered in Melbourne as well, with 'dramatic increases' reported of cockatoos, corellas and lorikeets. A 2011 article by two Melbourne experts noted more 'large iconic species' and fewer small birds. In the Brisbane region as well, biologists have commented that large species such as magpies and cockatoos are thriving.[108]

The story since white settlement has thus been one of winners as well as losers, of smart, aggressive birds enjoying the changes people have made, often at the expense of other birds (including some smart and aggressive ones). The IUCN ranks Australia thirteenth out of more than 170 nations for the number of threatened bird species, but most Australians notice not the declines, but the large songbirds and parrots that are flourishing. The evidence declares itself everywhere. A farmer driving past his paddocks to the letterbox can stir up clouds of rosellas and corellas and set magpies and peewees complaining.[109]

Contempt for climate-change warnings shows that many people trust their gut feelings ahead of science, and just as frost on the grass invites doubts about global warming, so do galahs cavorting under sprinklers undermine claims about bird declines. To those who trust their eyes, birds in Australia are thriving. The book that helped launch modern environmentalism, *Silent Spring*, evoked a birdless world, not one in which raucous parrots appear to be partying at our expense. To be effective, any discussion of declining birds must include the winners people see or it will fall on ears turned elsewhere, because Australia is a land in which some of the world's loudest, smartest and most colourful birds are thriving.

Notes

Important sources, or those that are drawn upon more than once, appear in the bibliography, but many minor references are mentioned in these source notes in an abbreviated form (with article and co-author names not mentioned, and book subtitles often omitted).

As the first prescient observer of Australian birds, John Gould is cited many times in the book and because the reference is always Gould (1865) he is not mentioned each time in these notes. The seven-volume *Handbook of Australian, New Zealand and Antarctica Birds* is the major reference to Australian birds – as Marchant and Higgins (1990), Marchant and Higgins (1993), Higgins and Davies (1996), Higgins (1999), Higgins and Peter (2002), Higgins et al. (2006), and Higgins et al. (2001). Rather than cite specific volumes, I have shown 'HANZAB' when this was the source.

As the internet makes journal articles and reports so easy to locate, I have left out some citations that are marginal to the story where these can easily be found with key words. For legends about the Maori following godwits to New Zealand, for example, 'Maori' and 'godwit' are enough. The website jboyd.net provides a comprehensive list of genetic papers. URLs were checked between December 2013 and March 2014.

Introduction

1   Leach (1911). He said the division was based on mammals, but that Australia's birds were more interesting than these.
2   Sibley – Sibley & Ahlquist (1985), Sibley & Ahlquist (1990); Astounded –
    See Schodde (2000), Christidis & Boles (2008). Topic explored in chapter 3.
3   Orians & Milewski (2007).
4   Jones (2002); Newspapers – Quest Newspapers in South East Queensland.
5   Popular – Kaplan (2004), Dooley (2013) *Australian Birdlife* 2(4): 26–29; Stab eyes
    – Horsburgh et al. (1992) *Medical Journal of Australia* 157: 756–759; In Toowoomba –
    The Chronicle (Toowoomba) 22 & 24 September 2011; Bird, D. (2004) *The Bird Almanac*,
    Firefly Books, Richmond Hill, Ontario.
6   Interspecific Aggression high levels – Ford (1989), Dow (1977); MacNally et al. (2012).
7   Carbohydrate abundance – Orians & Milewski (2007) & chapter 1; Causal link –
    Infertile soil encourages pollen mobility ahead of seed mobility (Hopper 2009), and
    birds are best at providing that (Southerton et al. 2004), but often fight over nectar
    (Orians and Milewski 2007), uttering harsh calls – see chapter 1 for more references;
    Parrots prosper – see chapter 5.

8   Schodde (2006). See also Newton (2003).

9   See chapter 3.

10   Australia wetter – Martin (1998), Byrne et al. (2011); Platypus fossils – Pascual et al. (1992) *Nature* 356: 704–05; Eucalypt fossils – Gandolfo et al. (2011).

11   I have defined the Pleistocene in accordance with the Subcommission on Quaternary Stratigraphy.

Food Worth Defending

1   Rowan (1898) *A Flower-Hunter in Queensland & New Zealand*, J. Murray, London; Harvey (1988). *The Contented Botanist*, Melbourne University Press at the Miegunyah Press, Melbourne.

2   Gould (1865)

3   Balmford (1981) *Victorian Naturalist* 98: 96-105; Long (1981) *Introduced Birds of the World*, Reed, Sydney, Rolls, E.C. (1969) *They All Ran Wild*, Angus and Robertson, Sydney; McCoy – Tiffin (2007) quoted page 171.

4   Harsh (covering a wide frequency band) calls are recognized as hostile by myriad species of bird (and mammal) so it is not a subjective concept – see Morton (1977) *American Naturalist* 111(981): 855-69; Recent guide – Goodfellow and Stott (2001).

5   More nectar – Orians and Milewski (2007), Woinarski et al. (2000).

6   One book – Proctor et al. (1996); Ortega-Olivencia et al. (2005) *Oikos* 110:578-590.

7   Best pollinators – This is best established for banksias, eg. Krauss et al. (2009), Lorens et al. (2012) *Molecular Ecology* 21: 314–28, Coates et al. (1992) *Heredity* 69:11-20. Collins et al. (2008) *Australian Journal of Botany* 56: 119–30 mention other Proteaceae as well and Hopper & Moran (1981) *Australian Journal of Botany* 29: 625-38 mention eucalypt examples. Golden wattle – Vanstone & Paton (1988). The state floral emblems have clear attributes showing that natural selection has favoured birds as visitors, including the red of waratahs and desert peas, and the shape of pink heath and kangaroo paw flowers; Brown honeyeater – HANZAB.

8   Hummingbirds seldom visit trees – Stiles (1978) *American Zoologist* 18: 715-27. (They do readily visit eucalypts, reinforcing the point that Australia is different); Nectar bird weights – HANZAB, del Hoyo et al. (2008). My size assessments only compare specialised nectar birds. Larger birds may feed periodically on nectar, including cockatoos, crows and Neotropical icterids, but they are not nectar specialists and like other biologists I do not include them. Wattlebird hunting – Hindwood (1944) *Australian Zoologist* 10: 231-51, Longmore (1991), Leach (1928) *Emu* 28: 83-99.

9   80 per cent of birds – Chris Tzaros pers. comm.; 24 birds – Tzaros (2005) *Wildlife of the Box-ironbark Country*, CSIRO, Melbourne. See also McGoldrick and MacNally (1998), and, for a different example of domination, Tullis et al. (1982) *Australian Wildlife Research* 9: 303-09; 24 honeyeaters – Tzaros et al. (2005); Chisholm (1922); Red wattlebird flocks – HANZAB, Longmore (1991).

10   Fights over flowers – Orians and Milewski (2007), Olsen and Joseph (2011), Paton (1986), Beehler (1994) *Biotropica* 26(4): 459-61 & references therein.

11   Male & Roberts (2002) *Biotropica* 34(1):172–76 record aggression over fruit while noting it is usually nectar that motivates this.

12 A simple qualitative comparison of canopy densities can be conducted by use of Google Images. It shows that oaks, beeches, elms and spruce have thicker crowns than eucalypts. Birches and some conifers are comparable.

13 Many visits – Olsen and Joseph (2011); New Holland territories – HANZAB.

14 Sugar feeds mammals – Woinarski et al. (1997, Van Dyck and Strahan (2008), Orians and Milewski (2007).

15 Drummond – Meagher (1974) *Rec WA Museum* 3: 14-65; Dawson (1881). The 'parakeets' he refers to would have been lorikeets. South Africa came closest to Western Australia because protea nectar was harvested as syrup (bossiestroop).

16 Manna, etc. – Paton (1980); Thrive on lerp – Beveridge (1884) *Journal and Proceedings of the Royal Society of NSW* 17:19–74.

17 Orians and Milewski (2007). See also Hopper (2009).

18 Sugar funds pursuit – Tullis et al. (1982) *Australian Wildlife Research* 9: 303–09; Recher et al. (1970) *Emu* 70: 90.

19 The Hobart painting was by F. Maurice.

20 Swift parrot numbers – Garnett et al. (2011).

21 Biography – Tree (2003) *The Bird Man*, Ebury Press, London.

22 Malaria – Thompson (1970) *Annals of the Association of American Geographers* 60(2): 230-44.

23 Blue gums for paper – Foelkel (2009) *Papermaking Properties of Eucalyptus Trees, Woods, and Pulp Fibers*, Associacao Brasileira Tecnica de Celulose e Papel (online book).

24 International plantings – git-forestry.com/download_git_eucalyptus_map. htm; Africa's tallest tree – http://git-forestry-blog.blogspot.com/2008/07/tallest-tree-in-africa-is-you-guessed.html; Europe's tallest tree – git-forestry.com/KarriKnight-GiantEucalyptusdiversicolor.htm.

25 Zacharin (1978).

26 International consumers – Smith (1974) *Ibis* 116: 155–64; Firecrowns – Roy et al. (1999) *Oryx* 3: 223-32, Colwell (1989) *Ibis* 131: 548–66.

27 McKinnon et al. (2001); McKinnon et al. (2004).

28 Blue gum visitors – Hingston et al. (2004) *Australian Journal of Botany*, 52: 353–69; Many birds take nectar – HANZAB, Franklin & Noske (1999) *Emu* 99: 15–28.

29 Climate workshop – Climate change, species and ecosystems – identifying key science questions for Australia, 18–19 October 2007; Byrne – in Steffen (2009); fossil pollen – Dunlop & Brown (2008), Byrne (2008) *Quaternary Science Reviews* 27 (27–28): 276–85, Dunlop and Brown (2008); Low (2011). See also Nevill et al. (2014) *Annals of Botany* 113: 55–67.

30 Southern Hemisphere different – Markgraf, et al. (1995) *Trends in Ecology and Evolution* 10(4): 143–47, Hopper (2009); Adapted to soil – Hopper (2009), Butt et al. (2013) *Ecology and Evolution* 3(15): 5011–22; Soils vary etc. – Morton et al. (2011); Low (2011).

31 Four eucalypts occur in southern Indonesia and one reaches Mindanao. Mobile eucalypt pollen – Potts and Wiltshire (1997; Southerton et al. (2004; Barbour et al. (2008); Viable –Barbour et al. (2008); Heathland pollen – Hopper (2009); Flowering erratic – McGoldrick and Mac Nally (1998), Paton (1986).

32 Swift parrots invoked – Wallis et al. (2011).

33 Sydney eucalypts – Beadle et al. (1982). Mingling of 6–8 species does not occur widely
    but is commonplace in South East Queensland, for example. Not all pollinated by birds
    – Andrew Hingston pers. comm.

34 Hybridise – Potts and Wiltshire (1997), Barbour et al. (2008), Steane et al. (1998)
    Australian Systematic Botany 11: 25-40; Potts – McKinnon et al. (2004).

35 Flowering continuity – Woinarski et al. (1997), Paton (1986); Thinking Honeyeater &
    woollybutt – Woinarski et al. (2000).

36 Near Perth – Whelan & Burbidge (1980) *Australian Journal of Ecology.* 5: 1–7.

37 Thousand flowers – Ford et al. (1979). Highest flower density, old landscape –
    Hopper (2009).

38 Proteaceae in SW – Burbidge (1960) p 109; Macadamia pollination – pers. obs &
    Heard & Exley (1994) *Environmental Entomology* 23(1): 91–100.

39 Carpenter and Jordan (1997).

40 Dettmann and Jarzen (1998).

41 Rift Valley Proteaceae – Dettman & Jarzen (1990) *Review of Palaeobotany and
    Palynology* 65: 131–144, Dettman & Jarzen (1991); Botanists have described – Hill et al.
    (1999). *Adenanthos* is today confined to heathlands.

42 Birds as better pollinators – Southerton et al. (2004), Cruden (1972) *Science* 176:
    1439-40; Krauss et al. (2009). Eucalypts are also noted for high outcrossing rates
    (Southerton et al. 2004).

43 Fossil feathers – Talent et al. (1966); Flying foxes recent – Hall and Richards (2000);
    No large social bees – Orians and Milewski (2007).

44 Hummingbird flowers 'young' – Proctor et al. (1999); Fossil – Mayr (2004) *Science* 304
    (5672): 861–64; Banksia – McNamara & Scott (1983) *Alcheringa* 7: 185–93.

45 Bird weights – HANZAB and del Hoyo et al. (2008). Wherever there are bird flowers
    there are larger birds than these that sometimes take nectar, including crows and
    cockatoos, but they do not qualify as nectar specialists.

46 Tasmania is not fertile, but it comes ahead of Western Australian kwongan, some of
    which is on white sand.

47 Sugarbird weight, diet – del Hoyo et al. (2008). When I say 'protea specialist' I include
    Leucospermum, also in Proteaceae; Cape & south-west – Hopper (2009); 300 fynbos
    flowers – Orians and Milewski (2007).

48 Hawaii's giants & their DNA – Fleischer et al. (2008); 'Ohi'a fed birds (including kioea)
    – Perkins (1903); Perkins – cited by Scott et al. (2001) *Evolution, Ecology, Conservation,
    and Management of Hawaiian Birds*, Cooper Ornithological Society, Camarillo.

49 Percy et al. (2008) *Proceedings of the Royal Society of London* B 275: 1479–90.
    Lobelias with tubular flowers were significant nectar sources in Hawaii but the first
    honeycreepers had short beaks and would have required an accessible nectar source,
    namely 'ohi'a.

50 Two main nectar sources – Higgins et al. (2008), Longmore (1991), Woinarski et
    al. (2000), Franklin and Noske (2000), Saunders et al. (2003) *Corella* 27(1): 1–12,
    McGoldrick and MacNally (1998), del Hoyo et al. (2008). Some lorikeets, including
    purple-crowned & little, are eucalypt specialists.

51 Leach (1928) *Emu* 28: 20–42 for lorikeet dominance. Visiting in 2013 it was very obvious

that grey-eared honeyeaters dominate maquis. The website endemia.nc has pictures showing many Myrtaceae, Proteaceae and Cunoniaceae, among others, with obvious bird flowers.

52　Proctor et al. (1999)

53　Brush-like – Proctor et al. (1996); Flower colours – Franklin and Noske (2000); Accessible –Franklin and Noske (1999).

54　Hidden nectar, white-eyes avoid proteas – Anton Pauw, University of Stellenbosch, pers. comm. Nectar can be hidden on small-flowered grevilleas.

55　It could be argued that pollinating marsupials in Australia are another reason for accessible nectar, but it is not plausible that non-flying mammals provide good enough cross pollination to influence floral evolution at this scale. *Metrosideros* in Hawaii has open nectar in the absence of parrots, but the genus evolved its floral form in the presence of parrots in New Zealand. But that said, parrots are more likely to have influenced floral structure in Proteaceae than Myrtaceae. Africans says 'fynbos' rather than 'heathlands'.

56　Protea farms – Bomford & Sinclair (2002); Cockatoo help – Lamont & Van Leeuwen (1988) *Journal of Applied Ecology* 25(2): 551–59. Lamont et al. (2007) mention high damage to banksia inflorescences by parrots without explaining that grubs are the target, although their sources show this.

57　Diets – HANZAB, Higgins et al. (2008), Longmore (1991). Spiny-cheeked honeyeaters may eat leaves to obtain water rather than food.

58　Longmore (1991)

59　Tongues – Olsen & Joseph (2011); Fleischer et al. (2008); Paton & Collins (1989) *Australian Journal of Ecology* 14: 473–506; Wattlebird 120 bristles – Higgins et al. (2008)

60　Del Hoy et al. (2008), Christidis and Boles (2008), Joseph et al. (2014).

61　Proteaceae adapted to infertility from the start – Hill (1998). See also Hill et al. (1999) page 279; Wind-pollinated Proteaceae – Hill et al. (1999).

62　Tough leaves – Hill (1998); Dense wood – Kraft et al. (2010) *New Phytologist* 188: 1124–36.

63　Mistletoes long flowering – Napier (2014) *Emu* 114(1): 13–22; Grey mistletoe – Reid (1990) *Australian Journal of Ecology* 15: 175–90; Nests in mistletoe – Cooney et al. (2006) *Emu* 106(1): 1–12.

64　Christmas tree DNA – Wilson & Calvin (2006) *American Journal of Botany* 93(5): 787–96; Most Australian heaths were previously included in family Ericaceae. *Brachychiton* has an old fossil record and is bird-pollinated today.

65　Vanstone and Paton (1988).

66　Honeyeaters on mountain ash – Ashton (1975) *Australian Journal of Botany* 23: 399–411, Loyn (1985) *Emu* 85(4): 213–30; California's giant sequoias are larger and also benefit from fire assisting germination, but they sometimes sprout without it, as I have seen along roadsides, and they are not flowering plants (angiosperms).

Forests that Exudes Energy

1   Maiden (1889; Trease and Evans (1978).

2   Explosives – Cribb & Cribb (1981) *Useful Wild Plants in Australia*, Collins, Sydney.

3   Mundy (1852).

4   Sweetened flour, sugar-plum, and icing on a wedding cake – Henderson (1851).

5   Queen's Domain manna mentioned – Tasmania (1855). Manna sugar – Basden (1966).

6   Donkin (1980), Angas (1847).

7   Manna in the past – Donkin (1980).

8   Donkin (1980), Maiden (1889), Morris (1898); question of origin – Anderson (1849); Basden (1966).

9   Anderson (1849).

10 Presented in 1880, Wooster's article appeared in 1882 in 1 (4): 91–93; See also Paton (1980).

11 Exhibition – Tasmania (1855), Lerp industries – West, T. (1858) *The Sydney Magazine of Science and Art* p. 75, Manna plantations – Simmonds, P. L. (1861) *The Technologist* 1: 225–29.

12 (1980).

13 Ibid.

14 Moore – Reprinted in Moore (1884), *Almanack* – Morris (1898) p 283, Ryan (1951) *Emu* 51: 175–76;

15 Woinarski (1985b), Woinarski (1984), see also Longmore (1991); Roaming flocks – HANZAB, Longmore (1991).

16 Weebill – HANZAB, Braced jaws – Bock & Morioka (1971) *Journal of Morphology 135:* 13–50, White-naped – 2008 PhD thesis by Megan Harper, Flinders University.

17 Woinarski et al. (1989); Psyllids common – Woinarski & Cullen (1984) *Australian Journal of Ecology* 9(3): 207–32.

18 Livingstone (1857)

19 Botswana – Herremans-Tonnoeyr, D. et al. (1995) *Babbler* 29–30: 34–35.

20 Californian researchers – Lockwood and Gilroy (2004); Sacramento birdwatcher – Tim Manolis, pers. comm.

21 California's lerp epidemic – an online search on 'lerp' and 'California' produces many papers, Cottony cushion scale – Low (1999).

22 Dahlston – *Laredo Morning Times* 3 Sept 2000 p 16A.

23 – HANZAB.

24 Large clusters – Brookfield Conservation Park 20 April 2011.

25 River red gum – Collett (2001) *Australian Forestry* 64(2): 88–95; year-round supply – Paton (1980).

26 – Latin America – Latta et al. (2001) *Oikos* 93: 250–59; New Zealand – Gaze and Clout (1983). I did the Honeydew Walk in Nelson Lakes National Park; Greenberg et al. (1993).

27 Epicormic strands, thin bark – Burrows, G.E. (2013) *Australian Journal of Botany*, 61: 331–49, bark photosynthesis – Cernusak & Hutley (2011) *Plant Physiology* (155): 515–23, Shed bark – Woinarski et al. (1997), Holmes and Recher (1986).

28 Australian mammals – Orians and Milewski (2007), Woinarski et al. (1997), Van Dyck & Strahan (2008). Honeyeaters do occasionally take wattle gum. The importance of exudates to lemurs probably says more about absent woodpeckers than infertile soils (see chapter 11).

29 Psyllids replace aphids – Orians and Milewski (2007).

30 Long-lived leaves – Orians and Milewski (2007); Smooth barked eucalypts stay cooler – Gill & Ashton (1968) *Australian Journal of Botany* 16(3) 491–98; In cold regions winter cold is enough to kill most adult insects, but deciduous trees are also common in the tropics and subtropics. Northern Australia has some deciduous eucalypts.

31 Manna, honeydew scarce in northern Australia – Franklin (1997) *Corella* 21: 48–54; Mountain ash exudates – Loyn R. H. (1985) *Emu* 85(4): 213–30, Experts have remarked – Higgins et al. (2008).

32 Pardalotes killed – Dow (1977); Bark birds driven away – Paton (1980).

33 Sap – HANZAB, Higgins et al. (2008), Longmore (1991), Loyn (2002); In central Queensland – Gurulmundi State Forest; gliders – Chapman et al. (1999) *Emu* 99: 69–72; *Lophostemon* – Ravenswood region, October 2011.

34 Roth (1899), Cribb & Cribb (1974) *Wild Food in Australia*, William Collins, Sydney.

35 agric.wa.gov.au/PC_92527.html.

36 Wattle nectar – Ford & Ford (1976) *Australian Journal of Botany* 24(6) 793–95, *Acacia mangium* – At Mission Beach in 2006.

37 Traill thesis for Monash University (1995). See also Driscoll & Lindenmeyer (2010) *Ecography* 33: 854–865 (for white-plumed), Tran (1990) *Australian Journal of Ecology* 15: 207–17; Tischler (2013) *Austral Ecology* (2013) 38: 809–19.

38 Territoriality usually species-specific – Dow (1977), Greenberg et al. (1994) *Bird Conservation International* 4: 115–27, Ford (1989) notes that interspecific competition is important among honeyeaters. Magpie – HANZAB.

39 Bell miners – HANZAB, Loyn et al. (1983), Multiple papers in *Australian Forestry* 8(4) in 2005; Leave insects – Haythorpe and McDonald (2010) *Austral Ecology* 35: 444–50.

40 Loyn et al. (1983) and (1987) *Natural History* 96(6): 54–60.

41 Ibid.

42 Clarke and Schedvin (1999).

43 BMAD – Bell Miner Associated Dieback Working Group (2004).

44 Multiple papers in *Australian Forestry* 8(4) in 2005 & BMAD Working Group (2004).

45 Ibid.

46 Helmeted honeyeater – Pearce et al. (1995) *Wildlife Research* 22:633–46, Garnett et al. (2011).

47 *Ecology* paper – Mac Nally et al. (2012), Clarke et al. (2007) *Victorian Naturalist* 124(2): 102–05.

48 Dow (1977), Piper and Catterall (2003), Mac Nally et al. (2012), HANZAB.

49 Barakula – Maron (2008) *Emu* 109: 75–81.

50 One expert – Chapman (1969).

51 Spread into the sandstone country – Low (2002).

52 Dow (1977).

53 Recognition of a positive butcherbird–noisy miner relationship has been slow in

coming, but see Roberston et al. (2014) *Austral Ecology* 39(3): 255–66; Atherton Tableland – Nielsen (1996) and pers. comm, Victoria – Loyn (2002). [An ecological paradox: More woodland predators and less artificial nest predation in landscapes colonized by noisy miners.]

54 Noisy miner society – HANZAB.

55 Ibid.

56 Trees suffer – Grey (2008) PhD thesis, La Trobe University, Melbourne. The other endangered birds are the forty-spotted pardalote and swift parrot.

57 Cricket Ground – Leach (1929); Regent pursuits – Ford et al. (1993) *Emu* 93:277–81.

58 'Rich patch nomads' – Ford (2011) *Pacific Conservation Biology* 17: 303–09; Miner activity spaces – Dow (1979) *Ibis* 121(4):423–36, HANZAB.

59 Climate reports – mainly Low (2011).

60 Buloke – Maron (2007) *Biological Conservation* 136: 100–07.

61 600 metres – Clarke & Oldland (2007) *Wildlife Research* 34: 253–61; Leafy wattles – Hastings & Beattie (2005) *Ecological Management & Restoration* 7(2): 105–12; Brochure – The Noisy Miner: Challenges in managing an overabundant species, at http://www.latrobe.edu.au/zoology/dept-of-zoology-documents/Final-low-res-LR-LatrobeUni-Birds-Broch.pdf; Canberra – Doug Laing pers. comm. and pers. observ.; Adelaide – David Paton pers. comm. and pers. observ.

62 Michael Clarke, Deakin University, pers. comm, Maron *Emu* (2011) **111:** 40–51.

63 DNA trees – Driskell and Christidis (2004), Nyari & Joseph (2011) *Emu* III: 202–11; Changed less – See Hawkins et al. (2005) *Journal of Biogeography* 32: 1035–42.

64 Nature resort – Kumul Lodge, which is where my I made my observations, and where the bird guide, Max Mal, worked.

65 Miner DNA – Christidis & Holderness (1998) *Nature Australia.* 25(12) 32–39; intense predation from grey goshawks could have driven a shift from grey to green to disguise bell miners against leaves in wet eucalypt forest. Examples are known of intermediate species diverging the most in appearance – see Norman et al. (2002) *Proceedings of the Royal Society of London* B 269: 2127–33; Schodde (2006).

66 Fed small items – Piper and Catterall (2003).

67 Miner foods – HANZAB, Habitats I have seen them without eucalypts included Hyde Park and fragmented rainforest.

68 Dow (1977) struggled & failed to find analogues for noisy miners, and no other biologists have reported any since.

69 Division of labour – Arnold et al. (2005) *Behaviour* 142: 1577–90; Mynas – Haythorpe et al. (2012) *Emu* 112: 129–36; Lowe et al. (2011) *Journal of Ornithology* (2011) 152:909–21; Parsons et al. (2006) *Austral Ecology* 31: 217–27.

70 Ford et al. (2001).

The First Song

1 Hepburn biography – Mann (2006) *Kate,* Holt & Company, New York.

2 Hartshorne (1973), Messiaean's piece – L'Oiseau-Lyre et la Ville-Fiancee (*The Lyrebird and the Bride-City*).

3   Haffer (2004) *Journal of Ornithology* 145: 163–76.

4   Mayr (1944) *Bulletin of the American Museum of Natural History* 83(2): 123–94.
    See also Mayr (1944).

5   Mayr (1963) *Emu* 63(1): 1–7.

6   bid. & Robin (2001) p 193.

7   Mayr (1972) *Emu* 72: 26–28, Cracraft (1972) *Emu* 72: 171–74.

8   Fossil feathers – Talent et al. (1966).

9   Mayr dismissed the feathers – Frith and Calaby (1976) page 98; Robin (2001)
    pages 235–37.

10 Vickers-Rich (1975) *Emu* 75(3): 97–112; Vickers-Rich & van Tets – In Rich and
    Thompson (1982) page 340.

11 Sibley and Ahlquist (1986), Schodde (2000), Corben & Brush (1999)
    Auk 116(3):806–14.

12 Sibley and Ahlquist (1985) Sibley et al. (1988) Sibley and Ahlquist (1990).
    Often wrong – Ericson (2011).

13 Pleasure about Australia – Schodde (2000); Australian radiation – Sibley and
    Ahlquist (1985).

14 ibley et al. (1988), Sibley & Ahlquist (1986).

15 Ibid. & Ericson et al. (2002).

16 Some European biologists call all members of order Passeriformes 'songbirds',
    a usage avoided here. Tendon – Raikow (1982) *Auk* 99(3): 431–45; Dippers are the
    only passerines that feed underwater, although there also songbirds such as water
    red-starts that will pluck insects from the water's edge.

17 Suboscines singers – Saranathan et al. (2007) Two unrelated species are mentioned.

18 Wallace (1856) *Annals and Magazine of Natural History* 18(2): 193–216 at
    biodiversitylibrary.org/itemdetails/19544.

18 Mayr (1989) *Auk* 106: 508–12.

19 Les Christidis pers. comm.

20 Ibid.

21 Ibid.

22 Christidis and Schodde (1991). With the Passerida clustered within the Corvida,
    the latter can no longer be said to exist.

23 'Anomalous', Fürbinger, Sharpe – Sibley (1974); Bird guide – Leach (1911).

24 Newton (1893); Sibley (1974).

25 'Primitive' invalid – Omland et al. (2008) *BioEssays* 30: 854–867; 'Sister' is used in many
    of the DNA papers I mention, for example Barker et al. (2002), Ericson et al. (2002).

26 Christidis & Schodde (1991).

27 Feduccia & Olson (1982) *Smithsonian Contributions in Zoology* 366: 1–22.

28 Newton (1893). Stephens Island wren – Donald et al. (2013) *Facing Extinction*,
    Christopher Helm, London. New Zealand had other flightless wrens that vanished
    after rats arrived but before European arrival, and Teneriffe had a flightless bunting
    which also did not last into historical times, so this wren became the world's only
    flightless perching bird.

29 Barker et al. (2002), Ericson et al. (2002).

30   Barker et al. (2004), Agence France Presse story 21 July 2004.

31   Agence France Presse story 21 July 2004. The 45 million year date may be too high, going on fossil evidence.

32   I have not come across any biologist questioning this since the papers that appeared in 2002 and 2004. Unlike some genetic findings it has been embraced by palaeontologists, eg. Mayr (2013).

33   47 per cent – Jønsson & Fjeldså (2006). Surveys in Britain Yalden & Albarella (2009) Table 8.1. Three estimates for the number of seabirds are 700 million, 900 million and 1189 million (Riddick et al. 2012); 1500 million Quelea – Cheke et al. (2007) *Journal of Applied Ecology* 44: 523–533; Other statistics – Kemp et al. (2001) state that 22 of the 26 most common birds in Kruger National Park, which is half the size of Switzerland, are songbirds, with queleas make up 80 per cent of all birds in the park. In field surveys in Australia and on most continents songbirds outnumber other birds, although not in South America or Antarctica (which lacks many birds). Australia's importance for bird ancestry increases when parrots and pigeons are included. Songbirds do not dominate by biomass because most are so small.

34   Ericson et al. (2002), Barker et al. (2004), Ericson et al. (2003) *Journal of Avian Biology* 34: 3–15. The scenario can also include one early colonisation of the Northern Hemisphere by suboscines, to account for broadbills, pittas and asities.

35   Songbirds missing – Mayr (2005) *Biological Reviews* 80: 515–42; Mayr (2009).

36   Boles (1995) *Nature* 374: 21–2, Boles (1997) *Emu* 97: 43–50, Mayr (2013).

37   Doubts Mayr (2013). The bones could belong to zygodactylids; The site – Boles et al. (1994) *Alcheringa* 18: 70; Bat, salamanders – australianmuseum.net.au/Murgon.

38   Lyrebird & logrunner remains – Boles (1995), Boles (1993) *Emu* 93: 44–9; Perching birds in Europe & Africa – Mayr (2013) *Systematics and Biodiversity* 11(1): 7–13; Manegold (2008) *Ibis* 150: 377–87, Mayr (2009); Marsupials – Luo et al. (2003).

39   Newton (2003).

40   3 km – HANZAB (Albert's lyrebird); startled chick – Reilly and Chambers (1988).

41   HANZAB.

42   HANZAB.

44   'Astonishingly sharp' – Olsen & Joseph (2011); Prince Philip – Robin (2001).

45   Hartshorne (1952) *Victorian Naturalist* 69(5) 73–74.

46   Ibid. & Hartshorne (1973) *Born to Sing*, Indiana University Press, Bloomington.

47   Syringeally primitive – Ames (1971) *Peabody Museum of Natural History Bulletin* 37: 1–194.

48   Hartshorne (1952) *Victorian Naturalist* 69(5) 73–74.

49   Lyrebirds are probably globally unique in having very sophisticated songs as well as highly ornate plumage. Nightingales, by way of contrast, are very drab.

50   Song & speciation – Brambilla et al. (2008) *Journal of Evolutionary Biology* 21: 651–57.

51   From *Descent of Man*.

52   Gray et al. (2001) *Science* 291 (5501): 52–54.

53   Hartshorne (1968) *Journal of Aesthetics and Art Criticism* 26(3): 311–15.

54   Learning calls – Gahr (2000) *Journal of Comparative Neurology* 426:182–96.

55   Pacific bazas, bronze-cuckoos and malkohas are examples of non-songbirds that

search foliage for insects. Birds vanished – Mayr (2006) *Ibis* 148: 824–27.

56   Good nests – Collias (1997). Link to intelligence – Hall et al. (2013) *Biology Letters* 9: 20130687. Mousebirds build reasonably sophisticated nests. A good number of songbirds do nest in cavities, including pardalotes and (most) starlings. Uniquely among parrots, South American monk parakeets build rough roofed nests of sticks.

57   Lyrebirds and scrub-birds diverged first, which means the evolutionary distances in songbirds are largest between them and anything else. Bowerbirds and treecreepers diverged next, which means they are also far removed from other songbirds, including lyrebirds and scrub-birds.

58   Choughs – Heinsohn (1991) *Animal Behaviour* 41: 1097–100, Boland et al. (1997) *Behavioral Ecology and Sociobiology* 41: 251–56, HANZAB.

59   Fairy wrens – Cockburn et al. (2013) *Emu* 113: 208–17; Magpies – Hughes et al. (2003).

60   Cockburn (1998), Heinsohn & Double (2004) *Trends in Ecology and Evolution* 19(2): 55–57, Boland & Cockburn (2002) *Emu* 102: 9–17; Skutch (1935) *Auk* 52: 257–73, HANZAB. Theories about its function are more complex than I have outlined.

61   Varian-Ramos & Webster (2012) *Animal Behaviour* 83: 857–64. Rowe & Pruett-Jones (2013) *Emu* 113: 218–31.

62   Weak relationship – Jetz,& Rubenstein (2011) *Current Biology* 21: 72–78.

63   Russell (1989) *Emu* 89: 61–62, Cockburn (2003).

64   Cockburn (2003*)*. Satin bowerbirds and some birds of paradise, although promiscuous, sometimes roam in flocks.

65   Slater and Mann (2004).

66   Martin (1996) *Journal of Avian Biology* 27(4): 263–72.

67   Stutchbury and Morton (2001).

68   Scrub-wren study – Magrath et al. (2000) *Auk* 117(2): 479–89; Woinarski (1985a); see also Ford (1989).

69   HANZAB.

70   Iwaniuk & Nelson (2003) *Canadian Journal of Zoology* 81: 1913–28.

71   HANZAB.

72   Diamond and Bond (2003); Kaplan (2004); Chough tools – Hobbs (1971) *Emu* 71: 84–85; Babblers – Boehm (1955) *Emu* 55: 159–60.

73   Sandpaper – Pizzey et al. (2007).

74   Poem 'Apostle-birds' in Wright (1994) *Collected Poems*, Angus & Robertson, Sydney.

75   Babbler names – HANZAB. African babblers are cooperative breeders with some strikingly similar calls, although they are not related and look different.

76   'Explosive, obnoxious' – Howell & Webb (1995) *A Guide to the Birds of Mexico and Northern Central America*, Oxford University Press, Oxford; Skutch (1935).

77   Jurisevic & Sanderson (1994) *Emu* 94: 69–74, Rooke & Knight (1977) *Emu* 77: 193–98.

78   Small birds sparse in suburbia – Ford (1989), Parsons et al. (2006), Catterall et al. (2010); Large songbirds were tallied by noting weights in HANZAB & the *Handbook of the Birds of the World* series, eg. del Hoyo et al. (2008). I chose 120 grams as the cut-off point. Australia's large songbirds are 2 lyrebirds, 8 bowerbirds, chowchilla, 2 wattlebirds, 2 friarbirds, 2 butcherbirds, magpie, 3 currawongs, chough, apostlebird, 3 birds of paradise, figbird, 5 crows & ravens. The number increases greatly when

New Guinea is included, with many more bowerbirds and birds of paradise, also honeyeaters and butcherbirds.

79 *Corvitalusoides* – Boles (2006) *Alcheringa Special Issue 1*: 31–37; Lyrebird – Boles (1995); Big songbird – Nguyen et al. (2013). *Kurrartapu johnnguyeni* was the size of black butcherbirds, which weigh 180 grams.

80 85 per cent – Jones (2002); Amputated – Janelle Miles story in *The Australian* 3 November 2008; Horses, etc. – Jones (2002).

81 Distinguish adults – Jones (2002); Cholesterol – Ishigame et al. (2006); Bennett (1860).

82 Morgan et al. (2005) & Morgan et al. (2006) *New Zealand Journal of Zoology* 33: 17–29. Attempts to eradicate – Jones (2002).

83 'Piebald mass' – Rowley (1975); 'Rampage – Heinsohn et al. (2000) *Proceedings of the Royal Society of London B* 267: 243–49.

84 The comment about 50 million years was made before doubts grew about the use of Zealandia to calibrate molecular clocks (see chapter 7). These songbird groups are unlikely to go back that far. Some bowerbirds are not large; the golden bowerbird is only 75 grams.

85 Ford (1989). Butcherbirds are closely related large songsters. Pets – North (1901–04).

New Guinea: Australia's Northern Province

1 A core part – Beehler et al. (1986) is an additional reference to those cited later that specify that New Guinea is part of Australia; Torres Strait opened 8000 years ago – Reeves et al. (2008) *Quaternary International* 183: 3–22. Bass Strait opened 14,000 years ago – Lambeck & Chappell (2001) *Science* 292 (5517): 679–86. Savanna rainforest line – Heinsohn & Hope (2006) page 85.

2 Bird of paradise history – Frith et al. (1998); Gilliard (1969); Frith and Frith (2010).

3 Best-selling book – Oliver Goldsmith's *A History of the Earth and Animated Nature*; Prayer book – Muzinic et al. (2009) *Journal of Ornithology* 150: 645–9.

4 Trade – Swadling et al. (1996; Frith and Frith (2010); Court dress – Marshall (1966).

5 Role models – Frith & Beehler (1998); Attenborough (1966) *Wildlife Australia* 3(2): 47.

6 *Wallace (1862).*

7 Dinsmore (1970) *Auk* 87(2): 305–21.

8 from *The Descent of Man.*

9 Diamond (1986) *Annual Review of Ecology and Systematics* 17: 17–37.

10 Burger et al. (1993) *Environmental Toxicology and Chemistry* 12: 1291–6.

11 *Jakarta Post* – Article by Simon Sinaga, 12 June 2001.

12 Abundant fruit & display – Primack and Corlett (2005); Pearson (1977) *Condor* 79: 232–44, and see Bell (1982); Beehler & Dumbacher (1996) *Emu* 96: 81–88.

13 Ibid. & Primack & Corlett (2005); New Guinea marsupials – See dietary accounts in Flannery (1995).

14 Kemp and Burnett (2003) available at indopacific.org/macaca.asp.

15 DNA – *Irestedt* et al. (2009) *BMC Evolutionary Biology* 9:235; Cordillera – Hall (2002) *Journal of Asian Earth Sciences* 20: 353–431, Hill & Raza (1999) *Tectonics* 18(6): 950–66; Wuster et al. (2005) *Molecular Phylogenetics and Evolution* 34: 1–14.

16 Clock date wrong – See chapter 7; Nearest relatives – Aggerbeck et al. (2014), Jønsson et al. (2011).

17　Australia dried – Byrne et al. (2011).

18　Schodde (2006), also Heinsohn & Hope (2006).

19　Schodde (1989).

20　Hocknull et al. (2007); Flannery (1995).

21　Read et al. (2001) *Australian Journal of Botany* 49: 321–32; 'ancient and distant' – Hill (2001) *Australian Journal of Botany* 49: 321–32.

22　Conifers in New Guinea – *Flora Malesiana*, pngplants.org; *Dacrydium* lasted – Moss & Kershaw (2007) *Palaeogeography, Palaeoclimatology, Palaeoecology* 251: 4–22. It lasted until the LGM on the Atherton Tableland and on Stradbroke Island near Brisbane; Other New Guinea relicts include frogs (*Mixophyes, Lechriodus*) and the pygmy possum (*Cercartetus*).

23　112 birds – Williams et al. (1996) *Pacific Conservation Biology* 2: 327–62; 165 species – Bell (1982); More than 700 – Beehler et al. (1986); Schodde (2006).

24　Barker et al. (2004), Aggerbeck et al. (2014). There is a second 'berrypecker' family, that of the painted berrypeckers (Paramythiidae).

25　Clarifying paper – Aggerbeck et al. (2014); Basal Passerida – Aggerbeck et al. (2014) call them 'transitional oscines', meaning transitional to Passerida; Walk back – Schodde (2006).

26　Jønsson et al. (2011).

27　Schodde and Christidis (2014).

28　DNA study – Jønsson et al. (2010) *Ecography* 33: 232–41; Fossil – Boles (1999) *Alcheringa* 23 (1–2): 51–56; New Zealand's extinct piopio was yet another oriole. The Americas lack true orioles, using the name for some of their New World blackbirds.

29　*Toon et al.* (2012) *Molecular Phylogenetics and Evolution* 62: 286–95.

30　Mayr (1953) cited by Heads et al. (2001) *Journal of Biogeography* 28: 893–925; Driskell et al. (2011) *Molecular Phylogenetics and Evolution* 60(3):480–85.

31　Warblers – Gardner et al. (2010); Bellbird & whistler – Norman et al. (2009). Laughing kookaburras do occasionally enter Queensland rainforest.

32　Spencer – Eback (2012) *Zootaxa* 3392: 1–34.

33　Recent DNA studies – See Malekian et al. (2010) and papers cited therein; Honeyeaters & allies – Gardner et al. (2010); Take home message – Norman et al. (2009).

34　Fish – Humphries & Walker (2013) *Ecology of Australian Freshwater Fishes*, CSIRO, Melbourne.

35　Dawkins (2004).

36　Heinsohn and Hope (2006).

37　New Guinea's woodlands – Bell (1982) *Ibis* 124: 252–73, Beehler et al. (1986).

38　Cape York's New Guinea birds – Beehler et al. (1986), Heinsohn & Hope (2006), Schodde (2006); Recent spread possible – Legge et al. (2004) *Wildlife Research* 31: 149–61; Rowe 'Torres Strait Holocene' Rowe (2007) *Palaeogeography, Palaeoclimatology, Palaeoecology* 251: 83–103.

39　Heinsohn (2008); North Queensland was wetter 5000 years ago – Rowe (2007) *Palaeogeography, Palaeoclimatology, Palaeoecology* 251: 83–103.

40　Dumbacher et al. (2008) *Molecular Phylogenetics and Evolution* 49: 774–781, Dumbacher et al. (1992) *Science* 258: 799–801.

41   Ibid.

42   The drongo cuckoo in South East Asia is the best mimic I know about. Diamond (1982) *Auk* 99(2): 187–196.

43   Diamond (1986) *Emu* 87: 201–211, Bell (1983) *Emu* 82: 256–75.

44   MacGregor's honeyeater – Cracraft and Feinstein (2000); Panama – Beehler (1980) *Wilson Bulletin* 92: 513–9. See also Beehler (1994) *Biotropica* 26(4): 459–461.

45   Malaysian location was Gunung Brinchang in the Cameron Highlands, contrasted with Kumul Lodge in the Central Highlands. Forest structure was very similar; Honeyeaters blamed –Woinarski (2000) *Northern Territory Naturalist* 16: 28–31.

46   Sahul – Davidson (2013) *Quaternary International* 285: 1–29.

Land of Parrots

1   Mayhew (1864) pages 69–72.

2   NSW parrot numbers – HANZAB, del Hoyo et al. (1997), Hindwood and McGill (1958). I excluded feral populations. Australia's common parrot species seem to reach much higher densities than the most common species in Africa and Asia (pers. obs.), so that the differences in parrot densities are even greater than species numbers indicate.

3   White (2004) *Living Bird* 24(3): 8.

4   Fletcher – Crome and Shields (1992).

5   Shields – Crome & Shields (1992).

6   Muscles – Tokita (2004) *Journal of Morphology* 259(1):69–81.

7   Olives, pandanus – HANZAB; Peregrines – Nick Mooney pers. comm, Penny Olsen pers. comm.

8   'Pugnacious' – Crome & Shields (1992).

9   Wallace (1865) & see also Primack & Corlett (2005); 50 species – Beehler et al. (1986), 29 parrots – de Hoyo et al. (1997) page 292. It can't be claimed that all parrots are analogues for rodents and monkeys. Lorikeets are not.

10   Majority takes nectar – HANZAB.

11   Termites – Perry et al. (1985) *Australian Forestry* 48(1): 46–53.

12   Pepperberg (2006a), Pepperberg (2006b) *Journal of Comparative Psychology* 120(1):1–11, Emery (2006). Many other papers exist, and the two books about Alex by Pepperberg are presumably unique as books written about one individual bird. Marler & Slabbekoorn (2004) *Nature's Music*, Elsevier Academic, Amsterdam.

13   Patel et al. (2009) *Current Biology,* 19: 827–30, nature.com/news/2008/080625/full/news.2008.914.html.

14   Bird play – Diamond & Bond (2003); Rowley (1990); Lorenz (1952) *King Solomon's Ring*. Cromwell, New York.

15   Butler (2008) *Brain Research Bulletin* 75(2–4): 442–49.

16   8 times – Fitch et al. (2010) *Neuron* 65: 795–814. The evidence that a couple of suboscines learn their calls suggests the real number may be higher, unless the skill was lost by all but a few suboscines lost.

17   Amniote – Chiari et al. (2012).

18   Reiner (2004) *Journal of Comparative Neurology* 473:377–414; Emery & Clayton *Current Biology* 15 (23): R946–50.

19 Cockatoo drumsticks – HANZAB, Crome & Shields (1992).

20 Contact calls – Hile et al. (2000) *Animal Behaviour* 59: 1209–18; Branches collapse – Crome & Shields (1992).

21 Budgerigar UV – Pearn et al. (2003) *Ethology* 109: 961–70.

22 Kare & Mason (1986) in Sturkie (Ed.), *Avian Physiology*, Springer-Verlag, New York.

23 Crops attacked – Caley (1966).

24 Jones (1853); Sentry duty – Crome & Shields (1992); Mundy (1852) *Our Antipodes*, Richard Bentley, London.

25 Parliamentary enquiry – Environment and Natural Resources Committee (1995).

26 Ibid.; Ruin Canberra sleep – Doug Laing pers. comm.; Melbourne Arts Centre – abc.net.au/news/stories/2008/02/15/2163908.htm?site=melbourne.

27 Caley (1966).

28 Environment and Natural Resources Committee (1995).

29 Condamine – Bomford and Sinclair (2002). Little corellas were mainly responsible; Galahs – Saunders & Ingram (1995) *Birds of Southwestern Australia*, Surrey Beatty & Sons, Sydney.

30 Kea damage. theguardian.com/world/2009/may/29/parrot-steals-passport; stuff.co.nz/national/8255617/Kea-steals-tourists-wallet.

31 Marriner (1906).

32 Keas Temple, P. (1996) *The Book of the Kea,* Auckland: Hodder; Giant eagles – Lee et al. (2010).

33 Anderson (2003) *Society & Animals* 11(4): 393–418; Bennett – in Crome & Shields (1992).

34 Early history – del Hoyo et al. (1997),Chansigaud (2010); Wallace (1862) *Journal of the Royal Geographical Society of London* 32: 127–137.

35 world-budgerigar.org/

36 Baretta – Bryant (1985); Background Briefing episode 'The Parrot Smugglers' 18 November 2012.

37 Vet - Wilson (1998) *Journal of the American Veterinary Medical Association.* 212(8): 1222–23.

38 Broinowski (1890) *The Birds of Australia,* Stuart, Melbourne.

39 Mattingley (1903) *Emu* 3: 122–23.

40 Karak – Richard Hill pers. comm, Garnett et al. (2011), Maron & Hill (2004) *Wildlife Research* 31: 311–17, Marginson & Ladiges (1988) *Australian Journal of Botany* 1: 151–70.

41 Flammable, cheap defences – Orians & Milewski (2007), Peeters (2002) *Austral Ecology* 27: 658–71 (cheap defences). Carbon-based defences are not usually as potent as alkaloids and other nitrogen-based defences, but they are more cheaply replaced.; Long-lasting – Orians & Milewski (2007), Morton et al. (2011); Legumes can fix nitrogen rather than extracting it from soil and they are more likely to have nitrogen-based defences.

42 Lamont et al. (1991), Lamont & Enright (2000) *Plant Species Biology* 15: 157–166, Witkowski et al. *(1991) Australian Journal of Botany,* 1991 (39): 385–97, Lamont et al. (2007). Paperbarks and bottlebrushes practise serotiny but parrots seem not to eat their seeds, perhaps because these are so small.

43   Marri – HANZAB.

44   Corellas (little and long-billed) – Burbidge (2008) *Australian Field Ornithology* 25:
      136–39; Grow in shade – Wardell-Johnson (2000) *Austral Ecology* 25: 409–21;
      Poisoning – ccwa.org.au/content/government-logging-agency-poisoning-thousands-
      cockatoo-habitat-trees-near-bunbury.

45   Protein & phosphorus – Lamont & Groom (1998) *Australian Journal of Botany* 46:
      387–406, Groom & Lamont (2010) *Plant Soil* 334:61–72.

46   Glossy feeding – Crowley & Garnett (2001) *Austral Ecology* 26, 116–26; HANZAB;
      More She-oaks – Martin & McMinn (1994) *Australian Journal of Botany* 42: 95–102.

47   Fire verses birds – Groom & Lamont (1997) *Australian Journal of Ecology* 22, 352–55.
      This is obvious because the seeds targeted by parrots are always protected by a thick
      layer of wood while other serotinous plants that grow beside them, while facing the
      same fire risk, have far less wood around their seeds.

48   Prehistoric parrot – McNamara & Scott (1983).

49   No serotiny in north – Williams et al. (1999) *Biotropica* 31(4): 626–36. Limited serotiny
      can explain why there are fewer cockatoo species in the north and competition from
      finches can explain fewer small parrots (see chapter 9).

50   Pine seeds – Schwilk & Ackerly (2001) *Oikos* 94: 326–336; Cockatoo use of –
      HANZAB.

51   Carnaby's – Garnett et al. (2011). They use *Pinus radiata*.

52   Eyre Peninsula – Department for Environment and Heritage (South Australia) (2008)
      Eyre Peninsula Yellow-tailed Black-Cockatoo *Calyptorhynchus funereus whitei* Regional
      Recovery Plan.

53   Corellas & murnong – HANZAB; Missionary – Francis Tuckfield cited by Gott (1983)
      Archaeology in Oceania. 17: 59–67; Orians & Milewski (2007).

54   Kakapo – HANZAB.

55   Rodent arrival – Rowe et al. (2008) *Molecular Phylogenetics and Evolution* 47: 84–101.

56   DNA studies – Schweizer et al. (2011; Wright et al. (2008) and references therein.
      A recent paper advocates dividing parrots into six families, by splitting the main
      parrot family into three.

57   Tiger parrots – Joseph et al. (2010) *Molecular Phylogenetics and Evolution* 59(3): 675–84.

58   DNA studies – Ericson et al. (2006; Hackett et al. (2008). The palm cockatoo stands
      out in Australia for having territories.

59   Often don't work well – Philippe et al. (2011) *PLoS Biol* 9(3): e1000602; Tests more
      recently –.McCormack et al. (2013), Kimball et al. (2013) *Molecular Phylogenetics and
      Evolution* 69(3): 1021–32, Suh et al. (2011) *Nature Communications.* 2(443): 1–7, but
      see Pacheco et al. (2011) *Molecular Biology and Evolution* 28(6):1927–1942 & Pratt et
      al. (2009) *Molecular Biology and Evolution* 26(2): 313–26. The Pratt study produced
      idiosyncratic results, including placing the parrots as sisters to all other Neoaves,
      and is unlikely to be correct.

60   Moult pattern – Pyle, P. (2013) *Condor* 115(3): 593–602; Messelasturids – Ksepka &
      Clarke (2012) *Journal of Vertebrate Paleontology* 32(2): 395–406. The Phorusrhacids,
      which are considered stem seriemas, also have raptor-like beaks (Mayr 2009, page 153).

61   Zygodactylids – Mayr (2009); Falcon tree – Griffiths et al. (2004) *Molecular*

Phylogenetics and Evolution 32: 101–09. Falcons are missing from Europe's early fossil record.

62   Brains different – Jarvis (2007) *Journal of Ornithology* 148 (Supplement 1):S35–S44, Ball (1994) *Brain, Behavior and Evolution* 44: 234–46. A couple of suboscines appear to learn their calls (Saranathan et al. 2007) and it is perhaps possible that this a retained facility among suboscines rather than newly evolved.

63   Capuchins – Alfaro et al. (2010) *Journal of Biogeography* 39(2): 272–88; Alfaro (2012) *American Journal of Primatology* 74:273–286; Primate studies – Finstermeier et al. (2013) *PLOS One* 8(7): e69504, Chatterjee et al. (2009) *BMC Evolutionary Biology* 9:259; Perelman (2007) *PLOS Genetics* 7 (3): e1001342 The 20 m date refers to crown rather than stem primates.

64   Department in Vienna – cogbio.univie.ac.at/labs/kea-lab/; Keas trump crows – Auersperg et al. al (2011) *PLoS One* 6(6): e20231. The crows were better at some tasks; 4 types of play – Diamond & Bond (2003).

65   Kaka play – Diamond et al. (2006) *Behaviour* 143: 1397–1423, Diamond & Bond (2004) *Behaviour* 141: 777–98, Kakako play – Hutton and Drummond (1909), Diamond & Bond (2004) *Behaviour* 143: 1397–1423.

66   2011 study – Schweizer et al. (2011). Their clock dating is plausible; Apes split from monkeys – Chaterjee et al. (2009) *BMC Evolutionary Biology* 9:259, Finstermeier et al.(2013) *PLoS ONE* 8(7): e69504. Gibbon aptitude – Fedor et al. (2008) *Journal of Comparative Psychology* 122(4): 403–17. Gibbons are surprisingly poorly studied, limiting the potential for comparison. New Zealand had a parrot of kea size in the early Miocene, that could have been as smart as one.

67   Miocene perching bird – Boles (1999); Bartlett (1867) *Proceedings of the Zoological Society of London* 1867: 688–89; Lyrebird brains – Iwaniuk et al. (2006) *Brain Behavior and Evolution* 68:45–62; Bowerbird branch – Barker et al. (2004), mimicry, paint – Keagy et al. (2011) *Animal Behaviour* 81: 1063–70, HANZAB; Bowerbird brains – Franklin et al. (2014). Some bowerbirds are modest in size, notably the golden.

68   Emery (2006); Capuchin diet – Emidio & Ferreira (2012) *American Journal of Primatology* 74:332–43.

The Last of the Forest Giants

1   Kofron (1999).
2   Tuite *Courier Mail* 3 October 1990.
3   Vietmeyer, Noel (19??) The Great Cassowary Games *Geo* undated detached article.
4   Kofron (2003) *Memoirs of the Queensland Museum* 49(1): 339–42.
5   *The Australian* – 29 May 2006.
6   Low (2007) *Australian Geographic* 85: 104–18.
7   AAP – 25 May 2006, *SMH* – 29 May 2006.
8   *Innisfail Advocate* 24 March 2005.
9   Kofron (1999).
10   Ibid.
11   Dawson (1881), Thomson (1935), Olsen, P. (2005) *Wedge-tailed eagle*, CSIRO, Melbourne; New Guinea eagle – Corlett & Primack (2005).

12   In Europe – Chansigaud (2010), Frith & Frith (2010) *Birds of Paradise*, Frith & Frith,
      Malanda; Locke – Walmsley (1992) *Studies in Eighteenth-Century Culture* 22: 253–67.

13   Cassowary meat – Chansigaud (2010).

14   Fossil – Miller (1962).

15   Toddler trampled – Pressey, A. (ed.) (1990) *Australia's Dangerous Creatures*,
      Reader's Digest, Sydney.

16   Murray and Rich (2004), Wroe (1999–2000) *Nature Australia* 26(7): 56–63;
      Riversleigh riddled – Archer et al. (1996).

17   Mihirungs are ratites – Elzanowski and Boles (2012); Murray and Rich (2004).

18   Painting – Gunn et al. (2011) *Australian Archaeology* 73: 1–12; Dawson (1881),
      Smith (2009).

19   Eocene extinctions – Mayr (2009); Terror birds – MacFadden et al. (2007)
      Geology 35:123–26; Human extinctions – Flannery (1994).

20   Eggshells – Smith (2009), Miller et al. (1999) *Science* 283: 205–08, Webb (2013)
      Corridors to Extinction and the Australian Megafauna, Elsevior Science, Burlington;
      Fiercely contested – Wroe et al. (2013) *Proceedings of the National Academy of Sciences
      of the United States of America* 110 (22): 8777–81. Late survival of some megafauna
      until the Last Glacial Maximum can't be ruled out, and many extinctions occurred
      prior to human arrival, for example at Mt Etna. Claims that this glacial was no worse
      than previous maxima jar with some evidence, for example the extinction then of
      Dacrycarpus.

21   St Bathans – Worthy et al. (2006) *Proceedings of the National Academy of Sciences of
      the United States of America* 106 (51): 9419–19423; Crocodiles & Turtles – Chiari et al.
      (2012), Iwabe et al. (2005) *Molecular Biology and Evolution* 22:810–13.

22   Arldt, Gentilli – Gentilli (1949).

23   Molecular studies – Harshmann et al. (2008) *Proceedings of the National Academy of
      Sciences of the United States of America* 105(36):13462–67, Phillips et al. (2010) *Systematic
      Biology* 59(1):90–107.

24   4 orders – worldbirdnames.org/ioc-lists/family-index/.

25   Southern origin questioned – Mayr (2009).

26   Westcott et al. (2005) *Oecologia* 146: 57–67.

27   Gold Coast – Low (1996) *Nature Australia* 25(4): 16–17. The rainforest formations with
      these plants extend a short way into Queensland, including Tallebudgera Valley, which
      is where I first saw them; Weber – Honours thesis, University of Queensland, October
      2011, & Weber et al. (2014) *Journal of Biogeography* 41: 222–38; Dwarf cassowary –
      Miller (1962), Worthy et al. (2014) *Integrative Zoology* 9: 148–66.

28   Wire plants – Dawson & Lucas (2000), Lee et al. (2010). My tugging experience was
      with a *Coprosma*. Some wire plants may not be as tough; Emus & goats – Bond et al.
      (2004) *Oikos* 104: 500–08.

29   Outback saplings – Johnson (2006); Madagascar – Bond & Silander (2007) *Proceedings
      of the Royal Society of London B* 274:1985–92.

30   Dawson and Lucas (2000); Lord of the Rings – for example when hiding from Nazgûl
      and meeting Ents.

31   Westcott et al. (2008) *Diversity and Distributions* 14: 432–39.

32 Quinine – Clifford & *Monteith* (1989) *Biotropica* 21: 284–6; Nitre – Noble (1975)
    Journal of Ecology 63 (3): 979–984; Orchards – Hynes & Chase (1982) *Archaeology in
    Oceania* 17(1): 38–50. Lady apple looks like a *Genyornis* fruit because it grows mainly
    in woodland yet has a fruit far too large for emus to swallow. It is common in Top End
    woodlands, thousands of kilometres from the nearest cassowary population. It is an
    important Indigenous fruit.

Australia as a Centre of Origin

1   Cracraft (2001; Edwards and Boles (2002). By saying Chicxulub took out the dinosaurs
    I'm obviously not including birds, which are dinosaurs that survived. Chicxulub also
    gets called an asteroid and bolide.

2   West Antarctica – Cracraft (2001), Johnson (1993); Tasmania – McLoughlin et al.
    (2008) *American Journal of Botany* 95(4): 465–71; New Zealand – Johnson (1993), Vajda
    & McLoughlin (2004) *Science* 303: 1489, Vajda & Raine (2003) *New Zealand Journal
    of Geology & Geophysics* 46: 255–73. Asian volcanoes that preceded KPB are believed to
    have contributed to global extinctions.

3   Ericson (2012). The Australavis part of the theory suggested itself to readers of Hackett
    et al. 2008) when this study appeared.

4   Gentilli (1949).

5   Simpson (1940) *Proceedings of the Sixth Pacific Science Congress of the Pacific Science
    Association* University of California Press, Berkeley & Los Angeles 2: 755–68, footnote
    4; Snide language – Nelson & Ladiges (2001) is a good example; A Mess – Nelson &
    Ladiges (2001).

6   Dawkins (2004); Parrot beak – Dyke & Mayr (1999) *Nature* 399: 317–18; 'Parrot' an
    ibis – Mayr & Bertelli (2011) *Palaeobiodiversity and Palaeoenvironments* 91:229–36.

7   Most fossil hunting – Barker et al. (2004), Cracraft (2001).

8   Kioea – Fleischer et al. (2008).

9   [Magpie goose] made differently – Rowley (1975); 'Old' belongs in inverted commas
    because we shouldn't suppose that the magpie-goose, as a species, is older than any
    other species, although it may well be older than most.

10 Galloanserae – Hackett et al. (2008), Mayr (2009); Screamer – Elzanowski and Boles
    (2012).

11 Screamers earliest – Hackett et al. (2008), Mayr (2009).

12 Gonzalez et al. (2009).

13 Waterfowl numbers – Australia's total comes from Christidis & Boles (2008) with the
    New Guinea Salvadori's teal added. The British list is taken from Svensson et al. (2009)
    and the Alaskan from the 2011 Checklist of Alaskan Birds produced by the Alaska
    Checklist Committee and published by the University of Alaska (Fairbanks). I have
    included non-breeding seasonal visitors but excluded vagrants.

14 Bag like a lizard – Vancouver (1984) *A voyage of discovery to the North Pacific Ocean
    and round the world, 1791–1795*, Hakluyt Society, London; Courting males –
    HANZAB.

15 Gonzalez et al. (2009). The freckled duck is another oddity that appears to be
    a significant relict (see jboyd.net) but its DNA was not included in this study.

16   Worthy (2009) *Zoological Journal of the Linnean Society* 156: 411–54, Gonzalez et al.
     (2009), and Worthy pers. comm.

17   Cenozoic birds (*Anatalavis, Anserpica*) – Worthy & Scanlon (2009) *Journal of Vertebrate
     Paleontology*, 29(1): 205–11, Mayr (2009); Wyoming Neoave – Elzanowski and Stidham
     (2011); The pink-eared duck's diet of microscopic items may be globally unusual.

18   – Fowls outweigh – Yalden and Albarella (2009); Interbreeding – Peterson & Brisbin
     (1998) *Bird Conservation International* 8:387–394.

19   I say 'peacock' rather than 'peafowl', the true name of the species, because the species
     is better known from this name; Lower branches southern – Cracraft (2001).

20   Palaeognaths branch off between reptiles and landfowl, and that makes it impossible
     that megapode reproduction was retained from dinosaur ancestors; Chicks can fly –
     HANZAB; Outback mounds – Noble (1999) *Australian Zoologist* 31(2): 396–402 ;
     Small brains – Corfield (2008) *Brain, Behavior and Evolution* 71(2): 87–99.

21   Southern African landfowl – Numbers taken from Sinclair & Ryan (2003) *Bird of
     Africa, South of the Sahara* (2003), Princeton University Press, New Jersey; Northern
     stem landfowl, northern origin – Mayr (2009). Megapodes have colonised many
     remote islands, showing good overwater travel. Their absence from mainland Asia
     has been attributed to competition from pheasants or predation by carnivores,
     see Olsen (1980) *Emu* 80: 21–24.

22   Briskie & Montgomerie (1997) *Journal of Avian Biology* 28(1): 73–86. A few birds have
     independently evolved penis-like structures. Malleefowl have lost theirs, see Brennan
     (2008) *Journal of Avian Biology* 39: 487–92.

23   Ibid.

24   McCracken et al. (2001) *Nature* 413: 128–29, McCracken (2000) *The Auk* 117(3): 820–25

25   Metaves & Coronaves – Fain & Houde (2004) *Evolution* 58 (11): 2558–73, Ericson
     (2008); Southern past proposed – Kaiser (2007) *The Inner Bird: Anatomy and evolution*,
     UBC Press, Vancouver; Metaves not confirmed – Ericson (2008), Morgan-Richards et
     al. (2008) *BMC Evolutionary Biology*. 8: 20.

26   Pigeon remains – Worthy (2012); In Europe – Worthy et al. (2009) *Auk* 126: 649–656.
     This paper records an early Miocene fruit pigeon; Genetic studies: Johnson et al. (2010)
     Molecular Phylogenetics and Evolution 57: 455–58, Periera et al. (2007) *Systematic Biology*
     56(4): 656–72.

27   Pigeon genera from Svensson et al. (2009) & del Hoyo et al. (1997).

28   The partridge-like pigeons are the partridge pigeon, squatter pigeon and spinifex
     pigeon (HANZAB, Crome & Shields (1992). Like partridges, they fly from danger
     if running is an insufficient response. The flock pigeon also nests on the ground;
     Lyrebirds – Bennett (1860). New Guinea's rainforest rails are further evidence of
     gamebirds doing poorly.

29   Unclear – The DNA tree in Pereira et al. (2007) implies an Australian origin for more
     pigeons than does Lapiedra et al. (2013) *Proceedings of the Royal Society of London B*
     280: 20122893. All analyses imply an Australian origin for fruit pigeons and a clade that
     includes *Geopelia* and *Gallicolumba*; Wallace (1865).

30   *Shapiro et al.* (2002) *Science* 295: 1683.

31   Genetic evidence – Hackett et al. (2008); anatomical Mayr (2007) *Journal of*

Ornithology 148 (Supplement 2):S455–S458; *Parargornis* – Mayr (2005) *Biologist* 52:12–16 & Mayr (2009); Wallace (1863) *Zoologist* see *www.wku.edu/~smithch/wallace/S077.htm*. New Guinea has the moustached treeswift. Treeswifts, which are confined to South East Asia and New Guinea, belong in a separate family from regular swifts and one that has deviated less from the ancestral form (they are less aerial).

32 Mayr (2009) page 213; Early hummingbirds ate insects – Mayr (2009); Australian owlet-nightjar eats ants – Doucette (2011) *Bird Observer* 868: 6–8; Because fossil swifts and hummingbirds go back to the Eocene, the owlet-nightjar lineage must be extremely old, and have survived in Australia only because it was not colonised by small owls.

33 1990s version – Feduccia (1995) *Science* 267, 637–38.

34 Dyke and Van Tuinen (2004).

35 Waders – Baker et al. (2007) *Biology Letters* 3: 205–09; Many fossils – Fountaine et al. (2005) *Proceedings of the Royal Society of London B* 272: 289–94; Robertson et al. (2004) *Geological Society of America Bulletin* 116 (5–6):760–68.

36 Cracraft (2001); Barker et al. (2004); Vegavis – Clarke et al. (2005) *Nature* 433: 305–08; but see Mayr (2013); Waimanu – Slack et al. (2006).

37 Major paper – Jetz et al. (2012) *Nature* 491: 444–48; Mayr (2013).

38 European fossils – Mayr (2009), Mayr (2005) *Biological Reviews* 80: 515–42. The fossils are mainly of stem rather than crown members of these groups.

39 Marsupials – Luo et al. (2003); Megapodes – Mayr (2009) page 36; Frogmouths – Nesbitt et al. (2011) *PLoS ONE* 6(11): e26350.

40 Flamingoes – Boles (2006), *Miller (1963) Condor* 65 (4): 289–299 & australianmuseum. net.au/Phoeniconotius-eyrensis/; Extinct parrots – Boles (1998) *Emu* 98: 32–35. I say 'reported' on the assumption that additional remains have been found, but information about them has yet to be published.

41 Storks & Conservatism – Boles (2006). 17 families – From Boles (2006) plus Nguyen et al. (2013): Casuariidae, Megapodiidae, Dromornithidae, Accipitridae, Rallidae, Turnicidae, Columbidae, Cacatuidae, Psittacidae, Aegothelidae, Apodidae, Halcyonidae, Menuridae, Meliphagidae, Ornithonychidae, Oriolidae.

42 Europe's extinctions – Mayr (2006) *Ibis* 148: 824–827; Could not compete – Mayr (2005) *Biological Reviews.* 80: 515-42, Mayr (2006) *Ibis* 148: 824–27.

43 Rail – Worthy and Boles (2011); Darter – Worthy (2012) *Auk* 129(1):96–104; Swift – Boles (2006); Eagle – Early grebes, stone curlews – Boles (2006); Cormorants – Worthy (2011) *Zoological Journal of the Linnean Society* 163: 277–314; Kingfisher – Boles (1997) *Memoirs of the Queensland Museum* 41(2): 229–34.

44 The giant kingfishers are four kookaburras & the shovel-billed kingfisher; Ploughs soil – Coates (1985).

45 Kingfisher success – Newton (2003) page 137; Kingfisher tree – Moyle (2006) *Auk* 123(2): 487–99; Ericson (2012) proposed that kingfishers arose in Africa, but I find the Afroaves theory unconvincing in light of the fossil record. An early proliferation in Asia would help explain a strong presence in Australia.

46 Missing European birds – Mayr (2009). Grebes and bustards were also missing.

47 Mayr (2009), Ksepka et al. (2011), Worthy et al. (2011), Waterhouse (2006) *Historical Biology* 18(2): 227–38.

48   Whales – en.wikipedia.org/wiki/Evolution_of_cetaceans. (The site says 'The pakicetids
      were digitigrade hoofed mammals that were the earliest whales.')

49   Unambiguous parrots – Mayr (2009); Mayr (2014) *Palaeontology* 57(2): 231–42;
      German parrots – Mayr & Göhlich (2004) *Belgian Journal of Zoology*, 134 (1): 47–54.

50   Gondwana during crucial time – see Lawver & Gahagan (2003) *Palaeogeography,
      Palaeoclimatology, Palaeoecology* 198: 11–37.

51   Mawson's pollen – Truswell (1983) *Palaeontographica Abt. B* 186: 121–74; Prydz Bay –
      Macphail & Truswell (2004) *In* Cooper et al. (Eds) Proceedings of the Ocean Drilling
      Program, Scientific Results 188: 1–43.

52   Hooker (1859). Darwin had speculated about southern plants floating on icebergs but
      Hooker provoked interest by providing a detailed discussion; Hutton Hutton (1896);
      Paris Academy – Hutton (1896).

53   Buttercups & beeches – Ashworth & Cantrill (2004) *Palaeogeography,
      Palaeoclimatology, Palaeoecology* 213: 65– 82.

54   Endured ten weeks – Read & Francis (1992) *Palaeogeography, Palaeoclimatology,
      Palaeoecology* 99: 271–90; Davis darkness: antarctica.gov.au/about-antarctica/fact-files/
      weather/sunlight-hours.

55   Shared plants – For conifers see Wang & Ran (2014) *Molecular Phylogenetics and
      Evolution* 75:24–40; Beetles – Moore & Monteith (2004) *Memoirs of the Queensland
      Museum* 49(2): 693–99.

56   Chulpasia – Sige et al. (2009) *Geobios* 42 (2009) 813–23; Snakes – Scanlon (2005)
      Memoirs of the Queensland Museum 51(1): 215–35, australianmuseum.net.au/
      Alamitophis-tingamarra; Eucalyptus – Gandolfo et al. (2011); Directions of travel
      can be inferred from concentrations of relatives (eucalypts have all their close relatives
      in Australia) and sometimes from the first ages of fossil appearances. Some plants went
      the other way, including probably *Papuacedrus*.

57   Reached Seymour Island – Reguero et al. (2002) *Palaeogeography, Palaeoclimatology,
      Palaeoecology* 17:189–210; Mayr (2009).

58   Closest relatives – Gonzalez et al. (2009) & Fain & Houde (2007) *BMC Evolutionary
      Biology* 7:35; There could have been more journeys than those mentioned here, for
      which the evidence no longer survives because of subsequent spread to additional
      continents; The Cape Barren goose is mostly confined to islands in southern Australia
      that lack kangaroos and wombats, although they readily coexist with introduced
      livestock.

59   'Explain NZ' – Trewick & Gibb (2010).

60   McGlone (2005).

61   An online search on 'Moa's ark' shows how popular the term is; Zealandia – Wallis &
      Trewick (2009), Trewick & Gibb (2010).

62   Too Close – Wallis and Trewick (2009), Trewick & Gibb (2010), McGlone (2006)
      Ecological Studies 186: 17–32; 52 million – eg. Tennyson (2010), citing Gaina et al.
      (1998) Earth Interactions 2: 1–23. The Eocene perching bird in Queensland is 54 million
      years old, with skeletal structures suggesting it predates the evolution of songbirds,
      making it very unlikely than any songbirds reached New Zealand on a landbridge.

63   Lyrebirds – Barker et al. (2004); Cockatoos – Wright et al. (2008); Ksepka et al. (2011).

64 Parakeets – Garnett et al. (2011); Shirikai (2002) *The Complete Guide to Antarctic Wildlife*. Princeton University Press, Princeton.

65 The other Norfolk parrot is the endangered Tasman parakeet (Garnett et al. 2011).

66 Trewick and Gibb (2010); Mayr Mayr (2013); One 2011 study – Schweizer et al. (2011); Spiders (*Nephila*) that blew from Australia are (or were) displayed at the Auckland Museum. Winters are too cold for them to breed there; Egrets, plovers – HANZAB; Birds arriving recently – Trewick & Gibb (2010), Low (2002); Parrots crossed sea – Schweizer et al. (2011).

67 Bats – Hand et al. (2013) *Journal of Vertebrate Paleontology* 33(6): 1442–48; Wright et al. (2011) suggest the New Zealand parrots evolved outside Zealandia. Birds on mammal-free islands often have weak flight so it shouldn't be supposed that the New Zealand wren ancestors had weak wings; Sanatorium – Apart from Trewick & Gibb's (2010) verdict about birds, McGlone et al. (2001) *Journal of Biogeography* 28: 199–216 say the flora is best characterised as a subset of the Australian flora. See also Tennyson (2010).

68 Jønsson and Fjeldså (2006); Africa a recipient – Cuckoo-shrikes, parrots – Jønsson et al. (2010), bush-shrikes – Kearns et al. (2012). Some of the African birds have their nearest relatives in Australia or the Sunda Islands, which is taken as evidence of direct travel to Africa, perhaps via Indian Ocean islands that no longer exist.

69 Wallace (1865); Matthew (1915).

70 Reprinted until 1974, modern critics – Nelson and Ladiges (2001).

71 Matthew (1915).

72 Matthew (1915); Gentillli (1949).

73 Agassiz (1850) *Christian Examiner and Religious Miscellany* 48: 181–204, accessed at people.wku.edu/Charles.smith/biogeog.

74 Passerida, Crows, vireos, bush-shrikes – Jønsson et al. (2011; Aggerbeck et al. (2014), minivets – Jonsson et al. (2010). There are many groups for which we cannot identify a continent of origin, for example kingfishers and quail, that could alter the balance towards Asia.

75 The ancestors of parrots, if it lacked a crushing beak, might not have benefited from the absence of rodents, but it could have benefited from an absence of placental mammals.

76 Ericson (2012); Northern fossils – Mayr (2009); Mayr (2009) classifies zygodactylids as stem-perching birds but Ksepka et al. (2011) question this; Fossil information missing – see Boles (2006).

77 Fantails – Nyári et al. (2009) *Zoologica Scripta* 38: 553–61; Monarch – Filardi & Moyle (2005) *Nature* 438: 216–19; Articles – Bellemain & Ricklefs (2008) *Trends in Ecology and Evolution* 23(8): 461–9; Heaney (2007) *Journal of Biogeography* 34: 753–57.

The Forest Makers

1 Claims about ancient rainforest can be found on any number of Daintree tourist websites. Details about the documentary will be made available to researchers who contact the author.

2 Explorers – Mason & Mason (1993) *Kurangee: Cape Tribulation Pioneers*, Pineapple Publications, Mareeba; Charcoal – Hopkins et al. (1996) *Journal of Biogeography* 23: 737–45.

3   In mountains & plateaus – See Rainforest Conservation Society of Queensland. (1986)
    Map 5 on page 147, showing primitive angiosperms concentrated in inland elevated
    locations. See also map 4; Three biologists – VanDerWal et al. (2009).

4   Nielsen (1996).

5   Nutmeg family – Jessup (2007) Myristicaceae, *Flora of Australia*, 2: 18–57; Fossil record
    – Sniderman & Jordan (2011) appendix 2.

6   All have far high species diversity in Asia, for example *Licuala* has something like 28
    species in New Guinea & scores of species in Asia.

7   Hooker (1859).

8   Burbidge (1960).

9   Emigration – Sniderman and Jordan (2011).

10 Burbidge (1960).

11 Relict genera in south-eastern Australia include *Telopea* and other Proteaceae and
    Tasmanian conifers. New Guinea relicts were mentioned in chapter 4. In drier habitats
    grevilleas and eucalypts can be considered Gondwanan relicts.

12 Specht (1988) *Proceedings of the Ecological Society of Australia* 15: 19–30. Kangaroo grass
    and black spear grass also occur in Africa, and black spear grass has reached Hawaii,
    proving it does not need land connections to travel immense distances. I have found
    the seeds of each to sink within a few days when soaked in water. *Themeda* is far more
    diverse in Asia than Australia.

13 I have tested scores of seed species by placing them in water and only seashore plants
    floated for more than a week.

14 3 million – Minton & Rogers (2004) *Stilt.* 46:35–35.

15 Flinders (1814).

16 del Hoyo et al. (1997).

17 Corlett (1998). Some pigeons swallow fruit but digest the seeds. The fruit pigeons
    of interest here are clade Ptilinopinae, composed of genera *Ducula, Gymnophaps,
    Hemiphaga, Lopholaimus, Ptilinopus.* Worthy et al. (2009) reports an early Miocene
    fruit pigeon from New Zealand, close to *Hemiphaga,* strengthening the evidence for an
    Australia origin.

18 Minmi – Tiffney (2004); When brushtail possums steal fruit from kitchens they behave
    like major fruit consumers, but dietary studies indicate otherwise; Stomachs made for
    leaves – Corlett (1998) notes that leaf monkeys avoid fruits because simple sugars upset
    their digestion.

19 Flocks of 3000 Frith (1982), also Campbell (1901); Frith (1977) The Destruction of the
    Big Scrub. *Parks and Wildlife* 2: 7–12.

20 Frith (1982).

21 Ridley (1930); Flight muscles – del Hoyo et al. (1997).

22 Pigeons in history – del Hoyo et al. (1997); To Brazil – del Hoyo et al. (1997);
    Macquarie I – parks.tas.gov.au/fahan_mi_shipwrecks/journals/Scientists/soabeline13.
    pdf; Domestic pigeons digest most of the seeds they swallow but a small number
    probably pass out intact.

23 Torres – Frith (1982).

24 Jardine (1904) *Emu* 3: 181–85; Banfield (1908); Johnstone-Need (1984) *Spinifex and*

Wattle, The Author, Cairns; Thomson (1935).

25 HANZAB; Mace rich – Crome (1975) *Emu* 75: 189–98.

26 HANZAB, Frith (1982). Many early *Emu* articles voiced concerns about the slaughter; Green Island – Atherton & Greeves (1984) *Emu* 85: 261-63 (quoted) & Frith (1982).

27 New Guinea paradox – Primack and Corlett (2005), Gressitt (1982) *Biogeography and Ecology of New Guinea*, W. Junk, The Hague; One theory – Sniderman & Jordon (2011).

28 *Pittosporum* – Chandler et al. (2007) *Australian Systematic Botany* 20: 390–401; *Melicope* – Harbough et al. (2009) *Journal of Biogeography* 36: 230–41, Hartley (1986) *Telopea* 2(6): 619–30. Australia probably exported many tree genera but it is difficult to be certain. There is clear evidence for *Ilex, Dacrycarpus, Casuarina, Melaleuca*.

29 Dipterocarps were blamed for the poverty of frugivorous pigeons and parrots in Malaysia by Janzen (1974) *Biotropica* 6: 69–103; Pearson (1977). Bell (1982) *Emu* 82: 24–41 also emphasised a poverty of mammals as the reason for exceptional numbers of frugivorous birds.

30 Wallace (1865).

31 Wallace line & pigeons – Wallace (1865) uses pigeons to demarcate the Indian region from the Australian (though he did not coin the term 'Wallace's Line'); Borneo fieldwork at Danum Valley, Sulawesi in Wartabone National Park & Tangkoko Nature Reserve; pigeon numbers from del Hoyo et al. (1997).

32 Switched diet – Primack & Corlett (2005), Corlett (1998). Pigeon phylogenies, eg. Pereira et al. (2007), Gibbs & Penny (2010) *Molecular Phylogenetics and Evolution* 56: 698–706 show that the seed-dispersing pigeon genera group together. Seed-grinding pigeons occasionally disperse live seeds (Corlett 1998). Largest genus – del Hoyo et al. (1997); Pigeon milk – del Hoyo et al. (1997).

33 in Maluku – del Hoyo et al. (1997); The other birds that can spread seeds between islands, *Aplonis* starlings and white-eyes, are far smaller.

34 The pied imperial-pigeon as defined here, following Christidis & Boles (2008), is often split into two species, by defining the pied imperial-pigeon as those populations extending from Asia to western New Guinea, and the Torresian Imperial-pigeon occurring further east. There is very little difference between these putative species and no certainty about reproductive isolation. Both 'species' behave as supertramps over part of their range, and presumably had a supertramp ancestor.

35 Mayr and Diamond (2001); Drinks seawater – Frith (1982), Campbell (1901). It nests in very small numbers on the Australian mainland, mainly in mangroves.

36 New Guinea habitats – Coates (1985).

37 More rainforest – Banfai et al (2007) *Diversity and Distributions* 13: 680–691; Biologists have noted – McKenzie et al. (1991) *Kimberley Rainforests of Australia*, Surrey Beatty and Sons, Sydney, page 384.

38 40 per cent shared – Russell-Smith & Lee (1992) *Biotropica* 24 (4): 471–87; Shapcott (2000) *Australian Journal of Botany* 48: 397–407, Shapcott (1998) *Journal of Tropical Ecology* 14:595–614, Shapcott (199) *Biotropica* 34(4): 579–90.

39 Calendar tree – Banfield (1908). All groups would not have had the same calender tree so I am guilty of simplification.

40 The nests of various starling species are regularly described in books as 'rough' or 'crude' and never as 'neat'. Often they are not bowls so much as dense piles of straw with a hollow in the middle for the eggs.

41 The yellow-eyed starling of New Guinea is the only other starling to build pendulous nests in outer limbs (Feare & Craig 1999). It is a rarely seen species, very similar to the metallic starling, that the latter, which may have evolved more recently, appears to be replacing.; One naturalist – Chisholm (1929) *Birds & Green Places*, J. M. Dent & Sons, London. Heavily massed weaver nests sometimes break limbs.

42 'Swarm' – *Lovette & Rubenstein* (2007) *Molecular Phylogenetics and Evolution* 44: 1031–56.

43 400 – Coates (1985); Three broods, large flocks – HANZAB. Pairs or small groups – Feare and Craig (1999). The Asian glossy starling can form very large flocks but only in rural landscapes in which some of them nest under roofs.

44 There may have been several penetrations of starlings into Queensland during wet interglacials.

45 A very small number of starlings in Cairns remain for winter. Africa lacks fruit migrants – Hockey (2000) *Emu* 100: 401–07.

46 Rich in cuckoos – Herberstein et al. (2014) *Behavioral Ecology* 25(1), 12–16. The number of genera as well as species is unusually high, and includes deep sisters. Channel-bill eggs – HANZAB. The channel-bill lineage could have been parasitising currawong relatives (cractids) long before currawongs evolved. Crows are new to Australia.

47 Costin (1979) lists global distributions.

48 Ibid. This is an old book and its plant list would no longer be up to date but there is no reason to suppose that a modern list would alter the conclusions reached.

49 Hooker (1859).

50 Small plants with northern relatives include mints (*Mentha*), carrots (*Daucus glochidiatus*), raspberries (*Rubus* species).

51 75% – Carlquist (1974) *Island Biology* Columbia University Press, New York. Violet – Ballard & Sytsma (2000) *Evolution* 54(5): 1521–32; waders – HANZAB.

52 Never landing on the sea – Phalaropes are the sole exception, but do nothing to negate the importance of other waders as seed vectors; Godwit – Gill et al. (2009) *Proceedings of the Royal Society of London B* 276(1656): 447–45; 340 hours Proctor (1968) *Science* 160(3825): 321–22.; Missing toes – Green & Figuerola (2005) *Diversity and Distributions* 11: 149–56.

53 Frith et al. (1977). The label to his picture names the grass.

54 Non-stop flight can be deduced from the lack of reliable sightings from South East Asia (see HANZAB); Snow Mountains – Coates (1985).

55 Europe and eastern Asia share species of duck, bittern, spoonbill, ibis, grebe, crake and many songbirds that would sometimes move seeds short distances in an east-west direction.

56 Other waders that could bring seeds include little curlew, common sandpiper, curlew sandpiper, red-necked stint, pectoral sandpiper.

57 Plants in Asia that can be attributed to bird spread north, given the far greater diversity of the genus in Australia and lack of potential for wind dispersal, include heaths

(*Acrothamnus suaveolens, Leucopogon malayanus*), a baeckea (*Baeckea frutescens*), a dodder laurel (*Cassytha pergracilis*), flax lilies (*Dianella* species), sawsedges (*Gahnia aspera, G. javanica*), *Gonocarpus chinensis,* goodenias (*Goodenia koningsbergeri, G. pilosa*), a sword-sedge (*Lepidosperma chinense*), lobelia (*Lobelia loochooensis*), triggerplants (*Stylidium tenellum, S. uliginosum*) and various grasses. See van Steenis (1985) *Brunonia* 8: 349–72; *Drosera* – Rivadavia et al. (2003) *American Journal of Botany* 90(1): 123–30; Tasmania & NZ share – Jordan (2001). Some of the shared plants have seeds that would have floated across.

58 Freshwater plants – Les et al. (2003) *International Journal of Plant Sciences* 164(6): 917–32, Green et a. (2008) *Freshwater Biology* 53: 380–392: 917–32; Figuerola & Green (2002) *Freshwater Biology* 47: 483–84, Kristiansen (1996); Duckweed (*Lemna triscula*), loosestrife (*Lythrum salicaria*) and hornwort (*Ceratophyllum demersum*) are plants that otters and platypuses would both encounter. Snail – Ponder (1981) *Proceedings of The Linnean Society of New South Wales* 105:17–21; *Aglaoreidia* – Sanchez Botero et al. (2013) *Alcheringa* 37: 415–419.

59 Oaks moved north – Clare et al. *Bioscience* 48(1): 13–24; Blue jays – *Journal of Biogeography* 16: 561–71.

60 Medicines – Low (1990) *Bush Medicine* Collins/Angus and Robertson, Sydney. All of these plants are in the same genera as their northern namesakes.

61 Two biologists – Givnish & Renner (2004) *International Journal of Plant Sciences* 165(4 Suppl.):S1–S6.

62 Hill (2001) *Australian Journal of Botany* 49: i–ii.

63 *Caustis* paper – Vollan et al. (2006) *Australian Journal of Botany* 54: 305–13. Crossing the equator is very difficult for aerial seeds because winds do not cross from one hemisphere ot the other.

64 Floating rafts – Metcalfe et al. (2001) *Faunal and Floral Migrations and Evolution in SE Asia-Australasia*, A.A. Balkema Publishers, Lisse, The Netherlands; Ridley (1930).

65 Frith (1952) *Emu* 52: 89–99 & Frith (1982).

66 Frith (1982), see also Frith (1957) *Emu* 57: 341–45.

67 Judy Davies pers. comm.

68 Fenton (1891) *Bush life in Tasmania fifty years ago,* Hazell, Watson & Viney, London.

69 Weeds of National Significance – weeds.org.au/WoNS/; Meeting – I was on the Hymenachne National Management Group; Magpie Goose – Delaney et al. (2009).

Of Grass and Fire

1 Grass statistics – Strömberg (2011). The major grass crops include rice, wheat, maize and sugar-cane.

2 Thrive on destruction – Bond and Midgley (1995); Quick replacement - Chapman (1990), *Reproductive Versatility in the Grasses*, Cambridge University Press, Cambridge; Growing points – Harvey (2002) *The Forgiveness of Nature: The Story of Grass*, Vintage, London.

3 Mutch (1970), Bond & Midgley (1995), Zedler (1995) *Trends in Ecology and Evolution* 10(10): 393-95. Critical evidence for evolved flammability was provided recently by McGlone et al. (2014) *New Zealand Journal of Ecology* 38(1): 1–11.

4   Grasses the fuel – Murphy et al. (2013); Trees need – Bond et al. (2012) *Austral Ecology*
     37: 678–85.
5   Modelling – Bond et al. (2005).
6   Dinosaurs – Prasad, *et al.* (2005) *Science* 310: 1177–79; Gondwanan fossils – Strömberg
     (2011); Ecdeiocoleaceae – Bremer (2002) *Evolution* 56(7): 1374–87, Michelangeli et al.
     (2003) *American Journal of Botany* 90(1): 93–106.
7   Stages – Jacobs et al. (1999); Electrical storms (=lightning fires) – Russell-Smith et al.
     (2003), Woinarski & Legge (2013).
8   Australia lagged – Strömberg (2011), Jacobs et al. (1999), Martin (1990) *Alcheringa*
     14 (3): 247-55; Browsers – Johnson & Prideaux (2004) *Austral Ecology* 29: 553–57;
     Procoptodon – Prideaux (2009) *Proceedings of the National Academy of Sciences of the
     United States of America* 106 (28):11646–50.
9   Mayr (1944).
10   Drilling – Martin & McMinn (1994) *Australian Journal of Botany* 42:95–102.
11   DNA work - Alström et al. (2013) *Molecular Phylogenetics and Evolution* 69(3): 1043–56;
     Other birds include the grass owl, black-winged kite and all Australia's quail, which
     represent the only landfowl other than megapodes.
12   The phylogenetic position of cockatoo grass (*Alloteropsis semialata*) suggests spread
     from Asia to Australia rather than the reverse – see Ibrahim (2009) *Annals of Botany*
     103: 127–36. Africa is clearly the land where cisticolas evolved, with almost 50 species
     compared to only a couple on other continents.
13   The manikin genus in Australia (*Lonchura*) is nested within a clade of African manikin
     genera, which is nested within a larger clade dominated by Australian and a few Asian
     finches, which are nested within a larger clade dominated by non-Australian finches,
     which had a Passerid ancestor that left Australia - see Sorenson et al. (2004) *Systematic
     Biology* 53(1):140–53.
14   Swedish biologists - Ericson et al. (2003). Passenger pigeons were the dominant seed-
     eating birds in North America before their extinction, but 'finches' are far more diverse
     than pigeons in North America and otherwise more successful. Quelea - Cheke et al.
     (2013) *Pest Management Science* 69(3): SI 386-396; The thick-billed grasswren has a
     thick bill. There are three white-faces, and the two with the most grain in their diets
     have unusually small idiosyncratic desert distributions, suggesting a decline over time.
     White-faces may have done better before finches colonised Australia.
15   Hawfinches crack cherry seeds and crossbills tackle pine cones; Multiple invasions
     by grassfinches can be deduced from Estrildid phylogeny, including by mannikins,
     firetails, and by the Gouldian finch. There is no reason to suppose that strong-jawed
     passerids were kept out of Australia by competition from parrots because they are
     missing from South East Asia as well.
16   Changes in 22 years – Grant & Grant (2006) *Science* 313(5784):224–26. A new finch
     turned up and imposed competition during the drought; House finches – Badyaev et
     al. (2008) *Evolution* 62(8):1951–64; Bmp4 – Parsons & Albertson (2009) *Annual Review
     of Genetics* 43:369–88; Seed-eating evolved multiple times – DNA trees show this, eg.
     Jonsson & Fjeldsa (2006) *Zoologica Scripta* 35(2): 149–86.
17   Crown parrots predated Australia's separation from Antarctica according to all

molecular clocks. If we assume that crown parrots evolved in New Zealand rather than Australia it is still possible that banksia cones and gum nuts contributed to their evolution since New Zealand once had these.

18 Finches smaller for what they eat – Dietary lists in HANZAB show strong overlap between the hooded and golden-shouldered parrot and co-occurring finches; In captivity zebra finches and budgerigars can survive without water, but it is unclear if they can do so for long under desert climates.

19 I have ignored the potential of rodents and ants to compete with birds when they consume seeds, having not found much research to draw upon, and having noted that competition between birds alone provides a powerful explanation for observed patterns. Ants & birds eat more seeds in Australia than rodents, at least in the arid zone (Morton et al. 2011). Experts – Franklin et al. (2000); Hooded parrots – Garnett & Crowley (1994) *Australian Bird Watcher* 15(7): 306–09; Finches posing threat – Garnett et al. (2010); Biologist – Mark Holdsworth; New Zealand parakeets – Kearvell et al. (2002) *New Zealand Journal of Ecology 26(2):* 139–48.

20 Olsen (2007).

21 Elton (1930) *Animal Ecology and Evolution*, Clarendon press, Oxford; Savannas volatile – van Langevelde et al. (2003), Sharp & Bowman (2004) *Journal of Tropical Ecology* 20: 259–70; Fires remove – Bradstock et al. (2012) page 197.

22 Responsive, low cost – Bond (2008) *Annual Review of Ecology, Evolution and Systematics* 39: 641–59.

23 Its survival will depend on sensitive management of the two surviving populations.

24 McLennan in Olsen (2007).

25 5 per cent, Lakefield – Crowley & Garnett (1998) *Pacific Conservation Biology* 4: 132–48.

26 Crowley & Garnett (2000) *Australian Geographical Studies* 38(1): 10–26.

27 Ibid.

28 Unique discourse – Fire as a serious threatening process to birds does not come up in reviews outside Australia, for example Parr & Chown (2003) *Austral Ecology* 28: 384–95, Askins et al. (2007) *Auk* 124(3): 1–34. Changed fire regimes are considered very harmful for many birds outside Australia, but nearly all the afflicted birds remain too common to qualify for listing as vulnerable or endangered.; Partridge pigeon – From the Australian government's Species Profile and Threats Database at environment.gov. au/cgi-bin/sprat/public/publicspecies.pl?taxon_id=64441, 384–95.

29 Recovery plan – O'Malley, C. (2006) *National Recovery Plan for the Gouldian Finch (Erythrura gouldiae)*, WWF-Australia, Sydney and Parks and Wildlife NT, Department of Natural Resources, Environment and the Arts, NT Government, Palmerston; Grasswren – Perry et al. (2011) *Emu* 111: 155–61; 14 birds – environment.gov.au/cgi-bin/ sprat/public/publicthreatenedlist.pl?wanted=fauna; 47 birds – Woinarski & Legge (2013); Franklin (1999).

30 Some of the birds threatened by fire in non-grassy habitats include noisy scrub-bird, eastern bristlebird & black-eared miner Garnett et al. (2011); Some birds benefit – Woinarski & Legge (2013); Garnett et al. (2000) *Sunbird* 30(1): 18–22.

31 More rain in the dry season is probably one reason why Queensland has less burning.; Northern burning summarised – Russell-Smith et al. (2008) *Journal of Arid*

Environments 72: 34–47, Bowman et al. (2010), Yibarbuk et al. (2001), Three-quarters – Woinarski & Legge (2013); Arnhem Land – Yibarbuk et al. (2001); Leading ecologists – Fitzsimons et al. (2010).

32 Habitats avoided – For example Fensham (1997) *Journal of Biogeography* 24: 11–22 noted that no fires were lit in gidgee and very few in other inland habitats; Gammage (2011) asserted that fire management occurred everywhere without justifying this statement.

33 Reasons for fire – see for example Gould (1971) *Mankind* 8: 14–24.

34 Theories lack firm support – See for example Bowman (2000), Bird et al. (2013); Mt Etna – Hocknull (2007); The conifers have been recorded from central and south-western Australia, regions from which they vanished well before 50 000 years ago.

35 Lightning, 400 million – Scott (2000) *Palaeogeography, Palaeoclimatology, Palaeoecology* 164: 281–329; In Queensland – 2013; Banksias – (McNamara & Scott 1983); South-western Australia – Dodson et al. (2005) *Austral Ecology* 30: 592–99.

36 Flannery (1994), Gammage (2011). For a gentle critique of Gammage see Fensham (2012) *Australian Geographer* 43(3): 325–27.

37 Plumwood (2006); Low (2011) 'When is Nature Not?' in Freeman et al. (eds) *Considering Animals*, Ashgate, Farnham; Flammability preceded – He at al. (2011) *New Phytologist* 191(1): 184–96 & Crisp et al. (2011) *Nature Communications* 2: 193 trace serotiny back more than 60 million years in banksias & Myrtaceae respectively. Grass flammability could be as old. Lightning – Woinarski & Legge (2013).

38 Ecosystem engineer – Wright & Jones (2006) *BioScience* 56(3): 203–09; Low (2002).

39 Fraser et al. (2003) *Ecological Management & Restoration* 4(2): 94–102 & HANZAB.

40 History of burning unknown – Bird et al. (2013); Woinarski and Legge (2013).

41 Flannery (1994); Johnson (2006); Nineteen experts Mooney et al. (2011). See also Sakaguchi et al. (2013); There could have been some changes – Bird et al. (2013); *Flammable Australia* – Bradstock et al. (2012).

42 Extrapolated unwisely – Bird et al. (2013) note that Lynch's Crater is in too wet a region to be indicative of Australia; Lynch's Crater – Rule et al. (2012) *Science* 335(1483): 1483–86.

43 McGowan et al. (2012); Atchison (2009) *Vegetation History and Archaeobotany* 18:147–157 provides another example of late fire management, going back about 3500 years.

44 Intensification – Habberle & David (2004) *Quaternary International* 118–19: 165–79, Johnson & Brook (2011) *Proceedings of the Royal Society of London B* 278: 3748–54, Williams et al. (2010) *Journal of Quaternary Science* 25(6): 831–38, Ulm (2013) *Quaternary International* 285: 182–92; Indian immigrants – Pugach et al. (2013) *Proceedings of the National Academy of Sciences of the United States of America* 110(5):1803–08. Fire regimes could be young – Russell-Smith et al. (2003), Bird et al. (2013). Mooney et al. (2011) say there was no increase in burning a few thousand years, which is one reason why I do not embrace the opposite conclusion.

45 Esparza-Salas, R. (2008) Molecular Ecology of the endangered Gouldian Finch *Erythrura gouldiae*. PhD thesis, James Cook University.

46 Heuman (1926) *Emu* 25: 134–6; 2011 estimate – Garnett et al. (2011); In captivity – Forshaw & Shepherd (2012) *Grassfinches in Australia*, CSIRO, Melbourne; freefall – Franklin et al. (1999) *Australian Zoologist* 31(1): 92–109.

47 Do not turn to insects – Dostine & Franklin (2002) *Emu* 102: 159–64; Gouldian plight – Garnett et al. (2011), Woinarski et al. (2005), Dostine et al. (2001).

48 Woinarski & Legge (2013); Retired ranger – Greg Miles, interviewed on ABC radio Darwin 12 July 2012; Academic – Corey Bradshaw at conservationbytes.com/page/52/.

49 Declining expertise – Russell-Smith et al. (2003).

50 Woinarski & Legge (2013); Petty et al. (2007) *Kakadu National Park Arnhemland Plateau Draft Fire Management Plan*, Kakadu National Park & Tropical Savannas CRC.

51 Ecofire – Legge et al *Ecological Management & Restoration* 12: 84–92, Legge et al. (2012) *EcoFire 2004-2011 fire pattern analysis, central and north Kimberley*, Australian Wildlife Conservancy, Perth, WA; No one advocates wilderness – See for example Fitzsimons et al. (2010).

52 Garnett et al. (2011).

53 Barnard (1934) *Queensland Field Naturalist* 9(1): 3–7.

54 Brahmins etc. – Sharp & Whittaker (2003) *Journal of Biogeography* 30: 783–802; Feral cattle – Van Dyck & Strahan (2008).

55 Five best properties – Damon Oliver pers. comm.

56 Guide – NPWS (2002) *Plains Wanderer Habitat Management Guide*, NPWS Western Directorate Threatened Species Unit. Near Sydney, Melbourne, Adelaide – HANZAB.

57 Not natural – Benson et al. (1997) *Cunninghamia* 5(1): 1–37. Plains-wanderers avoid a scattering of trees because they fear raptors perching in them; Kangaroos do not graze low enough – Damon Oliver pers. comm.

58 Sclerophyll grass – Morton and James (1988), Bradstock et al. (2012) page 197. Spinifex is usually called xerophytic rather than sclerophyll but their meanings overlap Bowman (2000).

59 Figs, bamboo, palms and she-oaks are other plants that have specialised birds, but they do not dominate so much of the continents they occupy. Lizards – Pianka (1989) *American Naturalist* 134(3): 344–64, Morton et al. (2011). The theory that termites explain exceptional lizard diversity in Australia (Morton & James 1988) falters on the evidence that Africa has many more termites (see chapter 11), something Mileswki as well as Pianka has written about. Morton & James focus on numbers of termite species rather than individuals, which are what matter to consumers.

60 Spinifex birds – HANZAB. The spinifex pigeon is not as restricted to spinifex as its name suggests. Spinifexbird ancestors presumably went from soft Asian grasses to soft Australian grasses before spinifex was colonised.

61 Quarter of Australia – Groves (1994) *Australian Vegetation*, Cambridge University Press, Cambridge. There are spinifex grasslands plus woodlands with a spinifex understory, which together cover close to a third of Australia by some estimates; Black grasswren – HANZAB; Sniffer dogs – birding-aus.org/using-dogs-to-find-night-parrots/.

62 Fire a problem – Garnett et al. (2011); Perfectly designed – Pianka (1989) *American Naturalist* 134: (3) 344–64; Large town – Mt Isa.

63 Killing mulga – Bowman et al. (2008) *Journal of Arid Environments* 72: 34–47; Fire the enemy – Wright & Clarke (2007) *Australian Journal of Botany* 55: 709–24; 1986 fire – Bowman & Murphy in Sodhi & Ehrlich (2010) *Conservation Biology for All*, Oxford University Press, New York.

64   Grasses between spinifex – Bradstock et al. (2012) page 197. Fires sometimes
      kill spinifex.

65   Bird et al. (2004) *Australian Aboriginal Studies* 2004 (1): 90–96.

66   Federal webpage – environment.gov.au/cgi-bin/sprat/public/publicspecies.pl?taxon_
      id=59293.

67   Johnson (2006). See also van der Kaars (1991) *Palaeogeography, Palaeoclimatology,
      Palaeoecology*, 85: 239–302. Sakaguchi et al. (2011) provide a reason to think that Johnson
      overstated the level of treelessness.

68   More questions than answers – Bird et al. (2013); Global review – Johnson (2009)
      ; Assuming that most of the megafauna died about 46 000 years ago from human
      impacts, they vanished during a period that was colder & drier than today.

69   Woinarski et al. (2005; Dostine et al. (2001).

70   Invasive pasture grasses – Rossiter et al. (2003) *Diversity and Distributions* 9: 169–
      176, Low (2011), Woinarski & Legge (2013), Russell-Smith et al. (2003); Queensland
      report – Csurhes (2005). An Assessment of the Potential Impact of *Andropogon gayanus*
      (Gamba Grass) on the Economy, Environment and People of Queensland. Queensland
      Department of Natural Resources, Brisbane. The sentence I have quoted was removed
      from a later version of the report, perhaps because I kept quoting it.

71   Killing red gums – Friedel et al. (2006) *Buffel grass: both friend and foe*, Desert
      Knowledge Cooperative Research Centre, Alice Springs; Bottle tree rainforest – Low
      (2011) page 171.

72   Low (2008).

73   Unsurpassed growth – Williams & Biswas (2010) *Commercial Potential of Giant Reed
      for Pulp, Paper and Biofuel Production*, Rural Industries Research and Development
      Corporation, Canberra; King et al. (2013) *BioScience*, 63(2):102–117; Trees killed
      – Coffman et al. (2010) Biol Invasions 12:2723–2734; One company – ENEnergy
      enenergy.businesscatalyst.com/energy-market/our-projects.html.

Life in a Liquid Landscape

1   Estimates of the numbers of seabirds on earth range from 0.7 billion to 1.18 billion
      Brooke (2004; Riddick et al. (2012), a large proportion of which feed far out at sea.

2   Decade at sea – Weimerskirch et al. (2014); Lindsey (2008).

3   Commute between continents – Raymond et al. (2010), Einoder et al. (2011) *Austral
      Ecology* 36: 461–75.

4   Except for Abbot's booby on Christmas Island (a species that was once more
      widespread), those seabirds that are endemic to 'Australia', however it is defined, are
      very similar to species found elsewhere. The short-tailed shearwater, for example, is
      difficult to distinguish in the field from the sooty shearwater.

5   Storm-petrels closer to albatrosses – *Christidis & Boles (2008)*.

6   Robertson (1993) *Emu* 93: 269–76.

7   Unihemispheric sleep – Rattenborg et al. (2000) *Neuroscience and Biobehavioral
      Reviews* 24: 817–42.

8   Seabird expert Janos Hennicke told me that seabird experts assume that seabirds sleep
      on the wing, but research has not been undertaken because of the enormous logistic

difficulties. Day & night – Weimerskirch et al. (2003) *Nature* 421: 333–34; Christmas Island – Janos Hennicke pers. comm.

9   Swifts – Tarburton & Kaiser (2001) *Ibis* 143: 255–263.; Airman – Lack (1959) *Swifts in a Tower.* Chapman and Hall, London.

10   HANZAB.

11   Sooty albatross – Weimerskirch & Guionnet (2002) *Ibis* 144: 40–50; Bennett (1860).

12   Gulf wide – Lindsay (2008). Albatrosses were split in a 1998 book chapter that provided no proper evidence to justify this, and which admitted that conservation was a motivation.; Proverb – www.worldbirdnames.org.

13   Petrel attributes – Warham (1990; Warham (1996).

14   Krill betray – Nevitt et al. (2004) *Journal of Experimental Biology* 207: 3537–44; Lured 7 kilometres – Verheyden & Jouventin (1994) *Auk* 111(2): 285–91.

15   Oil – Olsen & Joseph (2011); 33–day run – Warham (1990); Vast reaches – Raymond et al. (2010).

16   Albatross chick and egg – Warham (1990), Lindsay (2008), HANZAB; First breed – Weimerskirch et al. (2014), HANZAB; Shearwaters – Skira (1991) *Corella* 15(2): 45–52.

17   60 million years – Slack et al. (2006). This was a stem penguin, but recogniseable in form; Sister group – DNA trees usually (but not always) place penguins with petrels, implying that each lineage is of equal age. Penguins diverge more from typical bird form than petrels, making it very possible that crown petrels are older than crown penguins.

18   Exploring larger area – Weimerskirch et al. (2014); The biggest difference between hemispheres is penguins only in the south and alcids (auks, puffins) only in the north.

19   Tropical diversity – Mönkkönen et al. (2006) *Global Ecology and Biogeography* 15: 290–302, Evans et al. (2005) *Biological Reviews* 80:1–25.

20   Armchair science – Davies et al. (2009); 2001 study – Karpouzi (2005) Modelling and mapping trophic overlap between fisheries and the world's seabirds. MSc thesis, University of British Columbia.

21   Species-energy theory – Chown and Gaston (1999; Chown et al. (1998; Davies et al. (2010); World's windiest – Toggweiler and Russell (2008).

22   Westerlies – Toggweiler & Russell (2008), Shulmeister et al. (2004) *Quaternary International* 118–9: 23–53; Antarctic Circumpolar Current – Tynan (1998) *Nature* 392: 708–10, Toggweiler & Russell (2008).

23   Becalmed – Lindsay (2008), Warham (1996).

24   Albatross travel – Murray et al. (2002) *Emu* 102: 377–85, Murray et al. (2003) *Emu* 103: 111–20 & 105: 59–65; 1000 km a day – Murray et al. (2003) *Emu* 103: 111–20; Circled the globe – Croxall et al. (2005) *Science* 307: 249–50, also Weimerskirch et al. (2014); Inflatable air cells – Gould (1865).

25   Upwellings – for example Raymond et al. (2010) *PLoS ONE* 5(6): e10960.

26   Unproductive – Gaston (2004), Jaquemet et al. (2004) *Marine Ecology Progress Series* 268: 281–292; Desert – Steadman (2006) page 402.

27   Seabird wings – Gaston (2004).

28   Capital – Gaston (2004); Predators – Tennyson (2010); Billions of seabirds – Lee et al. (2010).

29 Zealandia – Gordon et al. (2010), Wallis & Trewick 2009; Australia loses islands –
   Whiteway (2009) *Australian Bathymetry and Topography Grid*, June. 2009. Geoscience
   Australia Record 2009/21, Voris (2000) *Journal of Biogeography* 27: 1153–67.

30 Shelf slopes steeply, islands lost – Tony Nicholas, Geoscience Australia, pers. comm.
   Islands survived in north-western Australia, and a few may have emerged in the Coral
   Sea. See also Heap & Harris (2008) *Australian Journal of Earth Sciences* 55(4): 555–85;
   Zealandia too deep – Gordon et al. (2010) Table 2. Zealandia's outer islands are only
   a few million years old, from the dates I have seen, but that still makes them far older
   than Australia's continental islands.

31 English biologists – Croxall et al. (2012); See also Wilson (2006).

32 Indonesia – Gaston (2004), Croxall et al. (2012) (absent from Table 1); Australia four
   – short-tailed shearwater, shy albatross (by some species definitions), black-faced
   cormorant, pacific gull.

33 Kaikoura – De Leo et al. (2010) *Proceedings of the Royal Society of London B* 277: 2783–
   92.

34 Bass & Flinders – Boland (2000); Genetically – Abbott and Double (2003).

35 Penguin DNA – Overeem (2008) *Conservation Genetics* 9: 893–905; Fairy prion DNA
   on Albatross island also points to recent colonisation, as does sooty tern DNA on
   the Great Barrier Reef – see Ovenden et al. (1991) *Auk* 108 (3): 688–694 & Peck &
   Congdon (2004) *Journal of Avian Biology* 35: 327–35.

36 Mammals on islands – Courchamp et al. (2003) *Biol. Rev.* (2003): 78: 347–83. Rats –
   Jones et al. (2008) *Conservation Biology* 22(1): 16–26; Tristan albatross – Wanless et al.
   (2009) *Biological Conservation* 142:1710–1718; Major shift – Steadman (2006) page 401
   notes that most seabirds now nest on small uninhabited islands difficult for humans to
   access and where rats may be absent.

37 Steadman (2006); Rail DNA – Kirchman (2012) *Auk* 129(1):56–69.

38 Steadman (2006).

39 Polynesian innovations – Diamond (2000) *Nature* 403: 709–10.

40 Bonyhady (2000), Medway (2002) *Notornis* 49: 246–58.

41 Ibid.

42 Entice petrels back – Norfolk Island Natural Heritage Restoration Consortium (2006?)
   Norfolk Island Restored.

43 Robinson – Boland (2000); Gannets – Warham (1958) *Emu* 58(5): 339–69; Macquarie –
   Cumpston (1968); Rat Island – Nic Dunlop pers. comm. & Dunlop & Rippey (20) *Rat
   Island Recovery Project: A feasibility study*, Report for WA Department of Fisheries; Lady
   Elliott – Daley and Griggs (2006); Mammals a problem – Garnett et al. (2011).

44 Three-quarters – Ross et al. (1995); DNA – Austin et al. (1994) *Auk* 111(1): 70–9;
   No people – Mulvaney and Kamminga (1999).

45 Drake – Sparks & Soper (1987) *Penguins*, Macmillan, Melbourne.

46 Philopatry – See for example Morris-Pocock et al. (2012) *Conservation Genetics* 13: 1469–81.

47 Vlietstra & Parga (2002) *Marine Pollution Bulletin* 44: 945–55.

48 Passerine cognition – Zelenitsky et al. (2011).

49 Pisonia – Burger (2005) *Journal of Tropical Ecology* 21: 263–71; Peppercress – Gillham
   (1961) *Journal of Ecology* 49(2): 289–300.

50   Gillham (1960) *Australian Journal of Botany* 8: 277–317.

51   Norton et al. (1997) *Biodiversity and Conservation* 6: 765–85.

52   Foliage nitrogen – Fukami et al. (2006) *Ecology Letters* 9: 1299–1307; Norfolk – Richard Holdaway pers. comm.; 300 Tonnes – Norfolk Island Natural Heritage Restoration Consortium (2006?) *Norfolk Island Restored*; Macquarie nitrogen – Erskine et al. (1998) *Oecologia* 117: 187–93; Bacterial Testing – staff conducting tests told me this.

53   Lichens, show petrels – Ovstedal & Lewis-Smith (2001) *Lichens of Antarctica and South Georgia*. Cambridge University Press, Cambridge; Mosses – ww.bbc.co.uk/nature/18704332.

54   Skua's breast – Greenslade (1990) *Papers and Proceedings of the Royal Society of Tasmania* 124(1): 35–50; *Acaena* & seabirds – Warham (1990).

55   Hundreds die – Walker (1991); Frigate victim – A ranger on Christmas Island told me this, but not whether it applied to *P. grandis* or *P. umbellifera* which both grow there.

56   Once found widely – Pratt et al. (2009) *Bulletin of the British Ornithologists' Club* 129(2): 87–91.

57   Reville et al. (1990) *Biological Conservation* 51: 23–38.

58   No petrels in trees – Gaston (2004) Christmas Island does have ground-nesting brown boobies and red-tailed tropicbirds, which are large enough to see off crabs. They choose rocky nest sites of low value to foraging crabs.

59   Opposable toes – Gaston (2004). The hind toes of tropic-birds are not used for perching, and it may be the position of their legs that allows them to perch where petrels can't.

A Continent Compared

1   Shared birds – worldbirdnames.org/ioc-lists/master-list/. Because of the trend in taxonomy towards splitting the number of shared species is falling, with for example the osprey split into a very similar eastern and western osprey, but whether all these splits survive the test of time remains to be seen.

2   Genus *Rattus* evolved in South East Asia & colonised Australia long before *Rattus rattus* reached Europe in Roman times, followed by *R. norvegicus*.

3   Roshier et al. (2008) *Ibis* 150(3): 474–84.

4   Ornithologist – Newton (2003); Roshier cited – Cumming et al. (2008) *Ecology and Society* 13(2): 26.

5   Robin et al. (2009).

6   Principal driver – Krauss et al. (2009). See also Woinarski et al. (2000); Spotted gums (*Corymbia maculata*) – Pook et al. (1997) *Australian Journal of Botany* 45: 737–55; Most swift parrots migrate from Tasmania to Victoria, but when southern flowering fails they go much further north; More erratic – The Pleistocene ice age cycles reduced climatic stability and increased aridity.

7   Park managers – Brockett et al. (2001) *International Journal of Wildland Fire* 10: 169–83.

8   New Zealand may have retained a later connection to Australia as noted in chapter 7 – see Gaina et al. (1998); Eucalypts & banksias – Lee et al. (2001) *Australian Journal of Botany* 49: 341–56.

9   The website endemia.nc shows these plants. In the macquis shrublands in New
    Caledonia I had difficulty finding birds that were not honeyeaters and estimate that
    well over 90% were grey-eared honeyeaters.

10 *Glycaspis* spread – Valente & Hodkinson (2009) *Journal of Applied Entomology* 133:
    315–17.

11 Pardalote nests – HANZAB; Attacks, death, dive below ground – Woinarski (1984).
    Pardalotes occasionally next in tree holes.

12 Eguchi et al. (2013) *Emu* 113: 77–83.

13 Apostlebirds – Elliott (1938) *Emu* 38: 30–49, HANZAB; New Zealand – Morgan et al.
    (2005).

14 Honeyeater diets – HANZAB; Lorikeets on seed – Crome & Shields (1992). They do
    very occasionally eat foliage insects, but these cannot be considered normal lorikeet
    dies in most places; Woodswallows – HANZAB; Treecreepers & nectar – HANZAB,
    Doer & Doer (2002) *Corella* 26(1): 22–23; Sap – Longmore (1991).

15 DNA studies – Barker et al. (2004), Aggerbeck et al. (2013), etc.; Foot design – Olsen
    & Joseph (2011); Hang to sleep – HANZAB; Ames (1987) *Emu* 87: 192–5, also Schodde
    (2006); Vibrate – Olsen & Joseph (2011).

16 American biologist – Ricklefs (2005) *American Naturalist* 165 (6): 51–659.

17 Close relatives – Longmore (1991).

18 A genus can be considered a design, a blue-print. The oldest honeyeater fossils are
    Miocene Boles (2005), compared with Cretaceous grevillea and waratahs (Hill et al.
    1991).

19 Eocene eucalypts & banksias – Gandolfo et al. (2011), McNamara and Scott (1983);
    Miocene Europe had stem songbirds coexisting crown songbirds – see Manegold et
    al. (2004) *Auk* 121(4): 1155–60 and this may have been true as well in Eocene Australia,
    but it is more parsimonious to assume that Walter's fossil predates the emergence of
    songbirds including treecreepers.

20 American – Bock (1963) *Condor* 65(2): 91–125; More feathers – Schodde (2006),
    Schodde and Mason (1999); slow lives – HANZAB; Doubts raised – Christidis & Boles
    (2008).

21 Regents like silky oak – HANZAB, Frith & Frith (2004). Bowerbirds are only reported
    using a few nectar sources; Dettmann and Jarzen (1991) matched fossil pollen to silky
    oak & to *G. exul* in New Caledonia.

22 Swift parrot relatives – Schweizer et al. (2013) *Zoologica Scripta* 42: 13–27.

23 Bird pollinated trees are common outside Australia, for example *Erythrina* and
    Embothrium species, but they are scattered through forests rather than dominating
    them. *Metrosideros* distribution – see Wright et al. (2000) *Proceedings of the National
    Academy of Sciences of the United States of America* 97(8): 4118–23; Zambezi teak
    woodlands are confined to very arid regions and produce far less nectar than Australian
    trees because they do not produce large inflorescences. Most proteas are large shrubs
    but *Protea nitida* is a tree that forms woodlands.

24 Birds on the deepest branches tend to be large, & the relationship between deep
    branches & leaf consumption may reflect nothing more than this; Geese shot –
    Eberhard & Pearse (1981) *Australian Wildlife Research* 8: 147–62.

25 Mihirungs usurped – Murray & Rich (2004).

26 Bowerbirds – Frith and Frith (2004), HANZAB.

27 Poverty of mammals – **Hortal** et al. (2008) *Journal of Biogeography* 35: 1202–14; Low predation – Croft *Evolutionary Ecology Research*, 2006, 8: 1193–214. But see Wroe (2002) *Australian Journal of Zoology* 50: 1–24.

28 Bones – Baird (1983) *Emu* 84: 119–23; *Australlus* – Worthy & Boles (2011). They are the oldest crown rails but older stem rails are known (Mayr 2009).

29 Creodonts, mesonychids – Van Valkenburgh (1999) *Annual Review of Earth and Planetary Sciences* 27:463–93.

30 Reptile species – Steve Wilson pers. comm. Numbers are expected to go much higher when more genetic analyses are done – Patrick Couper pers. comm.; New lizard species – *Menetia timlowi, M. sadlieri*; Pianka (1989) *American Naturalist* 134(3): 344–64.

31 Le Souef (1923) *Australian Zoologist* 3(3):108–11; South America – Croft (2001) *Diversity and Distributions* 7: 271–87; Africa – Immigration can be deduced from these mammals having much older fossils outside Africa.

32 Native-hens may have been eliminated by the dingoes brought by humans rather than by humans directly; Nightingales – Holt et al. (2010) *Ibis* 152: 335–46; Sherbrooke – http://www.smh.com.au/environment/deer-to-be-culled-by-hunters-in-parks-including-sherbrooke-forest-20140304-345b3.html.

33 Soon Spread – Bennett (1860); Bottlebrush stand – Paton DC (1995). *Overview of feral and managed honey bees in Australia,* Australian Nature Conservation Agency, Canberra. See also Paton (1986); Pollen mobility can be expected to drop from honeybees replacing birds because their movements from the hive are so regular compared to bird movements, resulting in a low incidence of pollen transfer between trees (Paton 1986); Asian honeybees – daff.gov.au/animal-plant-health/pests-diseases-weeds/the-asian-honey-bee-in-australia; Bumblebees – Hingston & McQuillan (1998) *Australian Journal of Ecology* 23: 539–49.

34 The 'finch' examples include the hawfinch, which can crack cherry stones.

35 Country folk – Leach (1911); Plantations – McInnes & Carne (1978) *Australian Wildlife Research* 5: 101–21.

36 Striped possums & lemurs – Beck (2009) *Biological Journal of the Linnean Society* 97: 1–17; The yellow-bellied glider is a sap-specialist, and sap is taken by sugar and squirrel gliders and by Leadbeater's possum – Van Dyck & Strahan (2008).

37 Extraordinary large brains – Cnotka et al. (2008) *Neuroscience Letters* 433: 241–45; Woodpecker brains – Sultan (2005) *Current Biology* 15(17): R649–50. As birds that are seldom kept in captivity they have not been tested for intelligence.

38 White-throated treecreepers were recorded in mangroves once (HANZAB).

39 Noske (1996), Ford (1982) *Emu* 82: 12–23.

40 Ibid.

41 Fish – Larson et al. (2003) *Zootaxa* 3616 (2): 135–50.

42 Owls are missing from Riversleigh's cave deposits (Walter Boles pers. comm.), showing up in Mt Etna cave deposits in the Pliocene. Some *Tyto* and *Ninox* species roost in caves (HANZAB). They have an extremely old fossil record (Mayr 2014). Australia's only stock, the black-necked, occurs as well in Asia.

43   Lauded – Longmore (1991); 66000 seeds – Ward & Paton (2007) *Austral Ecology* 32: 113–21.

44   A newcomer, entry date – Reid (1987) *Emu* 87: 130–31; Spiny-cheeked honeyeaters will take mistletoe in large amounts but they are also inland birds.

45   Quick digestion – Olsen & Joseph (2011), Murphy et al. (1993) *Oecologia* 93:171–76. No Asian flowerpecker is as specialised as the mistletoebird, but having said that, it does take other foods.

46   Moyle et al. (2009).

47   Sarus cranes also occur in north Queensland but were long overlooked because they are so similar to brolgas. There have evidently been two invasion of Australia by *Grus* cranes.

48   Egret – HANZAB; Chat – Coates (1985); Woodpecker – Trainor (2005) *Emu* 105: 127–35.

49   Northern Territory – Don Franklin (pers. comm.) has seen birds taking nectar from all but a few rare eucalypts.

50   DNA test – Cracraft & Feinstein (2000); Other bird families that stand out for variability are tanagers and vangas.

51   Keast (1984).

52   Third branch – See Agger. They constitute the fourth branch if treecreepers and bowerbirds belong on separate branches, as seems very likely; 280 species – Gardner et al. (2010).

53   First songbirds in Europe – Manegold (2008); Gerygone DNA – Nya´ri & Joseph (2012) *PLoS ONE* 7(2): e31840.

54   DNA – Joseph et al. (2014). The blue-faced honeyeater has, as its closest relatives, seven small *Melithreptus* species with colour patterns that match it to greater or less degree, but to keep the story simple I have only mentioned one.

55   Honeyeater tree – Josepg (2014); Honeyeaters that don't take nectar include the insectivorous green-backed honeyeater, some chats, and the fruit-eating smoky honeyeaters.

56   Black butcherbird, DNA – Kearns et al. (2013).

57   HANZAB, Hughes et al. (2003), Hughes et al. (1996) *Emu* 96:65–70.

58   Extreme winters are not a feature of the southernmost parts of the Palaearctic and Nearctic.

59   Some birds do survive the hard winters without migrating, including seed-caching jays.; 80 per cent of birds – Newton (2003); Newton (2003).

60   Currawong increases – Major and Parsons (2010), Remes et al. (2012); Predation, currawongs, fruit – Remes et al. (2012), Parsons et al. (2006); Bowerbirds – Laing (2013).

61   One biologist – Bass (1995) *Corella* 19: 127–32; Another – Debus (2006) *Pacific Conservation Biology* 12: 279–87.

62   Neotropical richest – Hawkins et al. (2007) *American Naturalist* 170: S16–S27; Darwin – *Voyage of the Beagle*.

63   Biologists – Pennington et al. (2004) *Philosophical Transactions of the Royal Society Of London B* 359: 1455–64.

64 More termites – Milewski et al. (1994) *Journal of Biogeography* 21: 529–43. Africa has much larger termite-feeding mammals (aardvark, aardwolf). Flying termites are presumably taken on occasion by many Australian birds, but next to nothing is recorded about this because birds seldom if ever aggregate in enough numbers to draw attention to their feeding.

65 African national parks – Bird lists can be found at sanparks.org.

66 Limited rainforest birds – Symes & Woodborne (2009) *African Journal of Ecology* 48: 984– 93.

67 Early filling – Jetz et al. (2012) *Nature* 491: 444–48; Schodde (2006).

People and Birds

1 Bronzed surfer – Saunders (1998) *Journal of Australian Studies* 22(56): 96–105.

2 McRobbie (2000) *20th Century Gold Coast People*, Gold Coast Arts Centre Press, Surfers Paradise.

3 Marshall (1966).

4 Report on trip – Heuman (1925) *Emu* 26: 134–36.

5 Maatsuyker – Vanderwal & Horton (1984) Coastal Southwest Tasmania, *Terra Australis* 9 Department of Prehistory, Research School *of* Pacific Studies, Australian National University, Canberra; Cape Grim – Horton (ed.) (1994) *The Encyclopaedia of Aboriginal Australia*, Aboriginal Studies Press, Canberra.

6 Two-thirds – Seabird numbers are estimated in Baker et al. (2002) *Emu* 102: 71–97; Sydney Cove – Gilmore (1969), Skira (1990).

7 Bass Strait – Gilmore (1969), Skira (1990).

8 Skira (1995) *Natural History* 104(8): 24–35. Slither seal-like – Begg & Begg (1979) *The world of John Boultbee*, Whitcoulls, Christchurch.

9 *Skira (1992); Skira (1990); Serventy (1969); Serventy et al. (1971), Gould (1865);* Campbell (1901).

10 Ibid; Omega-3 – Woodward et al. (1995) *Australian Journal of Nutrition and Dietetics* 52(2): 87–91.

11 Skira (1990), Skira (1992), Serventy (1969), Warham (1996).

12 Flinders – Gould (1865); Less than 20 million – Baker et al. (2002) *Emu* 102:71–97 estimated 13–16.5 million. Banding studies show that more than half the chicks die before they breed, mainly on their first migration, so the harvest rate has less impact on the population than it might.; Climatic anomalies – Napp & Hunt (2001) *Fisheries Oceanography* 10(1): 61–8, Peter & Dooley (2014) *Birdlife Australia* 3(1): 24–7.

13 [Kissock] Anonymous (1903) *Emu* 2(3): 125–39.

14 Campbell (1901). (He quotes from his *Australasian* articles in this book.)

15 Campbell (1904) *Victorian Naturalist* 20(12): 166–73; Gibson-Carmichael – Campbell (1909) *Emu* 8: 207–10.

16 By 1913 Campbell was concerned about excessive egging and calling for controls, see *Emu* 12: 271–4.

17 Leach (1929).

18 Robin (2001).

19   Pelicans migrate – Trainor et al. (2007) *The Birds of Timor-Leste*, BirdLife International
      & Dove; Banks – in Pizzey (1983) *Stories of Australian Birds*, Currey O'Neil, Melbourne;
      Ducks – Dickison (1932) *Emu* 31: 175–96.

20   Cassowary – Miller (1962). The cassowary is only known from one undated bone so
      its cause of extinction cannot be determined; Mallee-fowl, coucal – Boles (2006);
      Torres Strait – Walker (1991).

21   Dawson (1881).

22   Kites – Olsen & Joseph (2011).

23   Sturt (1849) *Narrative of an expedition into Central Australia*, T. and W. Boone, London;
      Leichhardt (1847) *Journal of an Overland Expedition in Australia*, T. & W. Boone,
      London; Pools laced, bowerbirds – Thomson (1935).

24   Delaney et al. (2009).

25   Garnett et al. (2011).

26   Cumpston (1968).

27   Guano in Queensland – Daley & Griggs (2006).

28   Breeden & Breeden (1973) *Wildlife of Eastern Australia*, Collins, Sydney; Cardwell –
      Thorsborne et al. (1988) *Emu* 88: 1–8.

29   North (1901–4), Campbell (1901).

30   Beeton – Olsen (2006) *National Library of Australia News* 16(9): 7–10; Lawes (1882)
      Popular Science Monthly 20: 324–32.

31   See *aea-emu.org*. An internet search on 'emu farms' and 'India' produces many articles.

32   Leigh (1840) *Reconnoitering Voyages and Travels, with Adventures in the new colonies of
      South Australia*, Smith, Elder & Co, London.

33   Closed Season – Campbell (1903) *Emu* 2: 187–94; Littler (1901).

34   Geese – Delaney & Saalfeld (2009); Muttonbirding – Skira (1990); French scientists –
      Weimerskirch & Cherel (1998) *Marine Ecology Progress Series* 167: 261-74.

35   duck.org.au/articles/2011-article.

36   Bauer & English (2011) *Conservation through Hunting: An environmental paradigm
      change in NSW*, Game Council, Orange.

37   Bowerbirds, apostlebirds, magpies – North (1901–4).

38   Magpie nest – Campbell (1901); Butcherbirds – Legge (1902) *Emu* 1(3): 82–86.

39   Crystal Palace – Anonymous (1902) *Emu* 1(3): 148; London Zoo – Matthews (1907)
      Emu 6: 195–96.

40   North (1901–04).

41   Lab rat – See Griffith & Buchanan (2010) *Emu* 110: v–xii; A Web of Science search on
      the zebra finch brings up topics mentioned; Genome sequenced – *Balakrishnan* (2010)
      Emu 110: 233–41.

42   Morris (1954) *Behaviour* 6: 271–322.

43   Olsen & Joseph (2011), Perfito (2010) *Emu* 110: 199–208.

44   Remarked in 1932 – Macgilvray (1932) *Emu* 31(3): 169–74.

45   Australian Companion Animal Council (200?) *Australian and their Pets: The Facts*,
      ACAC, St Leonards, NSW; Chisholm (1922).

46   Reach for the gun – Marshall (1966); Bounties – HANZAB; North (1901–4).

47   Carcases – Lunney & Leary (1988) *Australian Journal of Ecology* 13: 67–92; Naturalist –

Darke (1977) *Wildlife in Australia* 14(1): 2–5; Locusts – Szabo et al. (2003) *Wingspan* 13: 10–15.

48 Mooney (2013) *Wildlife Australia* 50(4): 9–13.

49 Bounty – from Queensland Prickly Pear Land Commission annual reports; In Queensland – Details taken from the list of Damage Mitigation Permits issued from 2007–10.

50 Griffiths (1996) *Hunters and Collectors*. Cambridge University Press, Cambridge.

51 Campbell boasted – Marshall (1966); His book – Campbell (1901); Heuman (1925) *Emu* 26: 134–36.

52 Fancy terms – Campbell (1904) *Emu* 3: 168–71; Cameras captured men – see Robin (2001); Oologist's paradise – White (1909) *Emu* 9: 39–41; A participant recalls – Serventy (1972).

53 Mattingley (1907) *Emu* 7(2) 65–71.

54 Campbell (1901); North (1901–4).

55 Ashby (1923) *Emu* 22: 210–6, Ashby (1928) *Emu* 27: 169–72; Chisholm (1928) *Emu* 27: 172.

56 Chisholm (1923) *Emu* 22: 311–15.

57 Serventy (1972).

58 Robin (2001).

59 Harrison (1908) *Australian Naturalist* 1(2): 155–56; Lucas – Anon (1908) *Emu* 8: 86–91; McMichael, Deakin – Robin (2001); Campbell (1905) *Emu* 5: 7–12.

60 Hall (1907); Chisholm (1918) 17: 239–43.

61 *Wildlife in Australia* 3(2) various articles (1966).

62 Tails sold – North (1901–4), also HANZAB, Chisholm (1960) *The Romance of the Lyrebird*, Angus & Robertson, Sydney; Leach (1929).

63 Campbell – Marshall (1966); Slow breeding – HANZAB.

64 Robin (2001), Leach (1929).

65 Barrett (193?) *Australia's Wonder Animals*. Sun News-Pictorial, Melbourne.

66 Sharland (1944) *Emu* 44: 64–71; 200 tonnes – Ashton & Bassett (1997) *Australian Journal of Ecology* 22: 383–94; World Heritage – Mallick and Driessen (2010); Orchid – Threatened Species Section (2006). *Flora Recovery Plan: Tasmanian Threatened Orchids 2006–2010*, Department of Primary Industries, Water and Environment, Hobart.

67 Leach (1929).

68 Koala intelligence, ratio – Martin & Handasyde (1999) *The Koala*. University of New South Wales Press, Sydney, & Low (2002).

69 sea-eaglecam.org/news-2012.html.

70 Birds killed by clearing – Cogger et al. (2003) *Impacts of Land Clearing on Australian Wildlife in Queensland*, WWF; Philip Island resident – Jan Fleming.

71 Social scientists – Manfredo et al. (2003) *Human Dimensions of Wildlife* 8(4): 287–306; Littler (1901) (1902) *Emu*. He did agree with killing birds on farms.

72 Wallace (1869) *The Malay Archipelago*, Macmillan and Co, London.

73 Sheard (1999).

74 One birder – Dooley (2013) *Australian Birdlife* 2(2): 48–49; Dooley (2005).

75 Connell (2009).

76   80 Mile Beach – Low (2005) *Wingspan* 15(3): 8–11.

77   Research by Janos – Hennicke & Weimerskirch (2014) *Marine Ecology Progress Series* 499: 259–73.

78   Carter – Robinson & Dooley (2009) *Wingspan* 19(4): 15.

79   Wood – Ley (2009) *Wingspan* 19(3):12; jboyd.net.

80   Robinson (1993) *Wingspan* 9 (March): 1–3, 20–21. Many articles highlighting woodland bird declines appeared in *Wingspan*, the Birds Australia magazine.

81   Jones & Buckley (2001) *Birdwatching Tourism in Australia*, Wildlife Tourism Research Report Series: No. 10. Cooperative Research Centre for Sustainable Tourism; RSPB, American birdwatchers – Connell (2009).

82   Main gesture – see Jones (2011); Surveys – Jones (2011). Because people who feed birds are more likely to return survey forms than those who don't I am skeptical about surveys showing that more than a third of households feed birds, and even that figure seems too high.

83   Parker (1901) *Emu* 1(3):112–18.

84   Cannon (1999) *Bird Conservation International* 9:287–97, see also Jones (2011).

85   shopping.rspb.org.uk accessed December 2011; In the US – Jones (2011).

86   birdsinbackyards.net/feed-or-not-feed-0.

87   environment.nsw.gov.au/animals/KeepingWildlifeWild.htm; Not addicted – see Jones (2011).

88   Feed meat - Jones (2011), also Rollinson et al. (2003).

89   Cholesterol - Ishigame et al. (2006); Sydney study – Parsons et al. (2006); In Brisbane – Rollinson et al. (2003).

90   Cockatoos – Neill (2008) *Ranges Trader Mail* 17 June 2008.

91   Winter aid – Jones (2011),Jones and Reynolds (2008); Skewed – Catterall et al. (2010), Major & Parsons (2010), Parsons et al. (2006); Large urban parrots & songbirds – Loyn & Menkhorst (2011), Major & Parsons (2010).

92   Howard & Jones (2004) In *Urban Wildlife: More than Meets the Eye*, (Eds S. K. Burger and D. Lunney.) Royal Zoological Society of NSW: Sydney.

93   Low (2002).

94   Pizzey (1988).

95   birdsinbackyards.net.

96   Spinebills – Loyn & Menkhorst (2011).

97   Dixon (2002), chicken.org.au.

98   Senate Select Committee on Animal Welfare (1990), voiceless.org.au, Baxter (1994) *Veterinary Record* 134:614–19, Dixon (2002).

99   Dixon (2002). European ban – Pickett (2006) *The Way Forward for Europe's Egg Industry*, Compassion in World Farming Trust, Petersfield, Hampshire.

100   Debate – 2008 Janine Haines Memorial Lecture; Rose (2013) *Wildlife Australia* 50(2): 33–35.

101   National Parks – Booth (2013) *Wildlife Australia* 50(1): 22–28.

102   Birdlife Australia (2012) Submission to the Environment and Communications References Committee for inquiry on 'The effectiveness of threatened species and ecological communities' protection in Australia'.

103 Woodpecker – Possingham et al. (2010) *Decision Point* 37: 2–3; Get nothing – Garnett et al. (2003) *BioScience* 53(7): 658–65.

104 Canberra birds – Doug Laing pers. comm.; Laing (2013).

105 The hole is 25 cm deep, large enough to accommodate a football; Jack (1963) *A List of the Birds of Brisbane*, N. Jack, Brisbane.

106 Around Sydney – Hindwood and McGill (1958),Hindwood and McGill (1991); Eating Sydney – 2 November 2009 article by John Huxley.

107 Biologists – Major & Parsons (2010) *Emu* 110: 92–103; Missing birds – Hindwood & McGill (1991).

108 Melbourne – Loyn and Menkhorst (2011); Brisbane – Catterall et al. (2010).

109 IUCN – http://cmsdocs.s3.amazonaws.com/summarystats/2013_2_RL_Stats_Table5. pdf (2013 figures); Evidence everywhere – Szabo et al. (2011) *Emu* 111: 59–70 document increases of some large birds, including parrots, in South Australian woodlands.

Bibliography

Abbott CL and Double MC (2003) Genetic structure, conservation genetics and evidence of speciation by range expansion in shy and white-capped albatrosses, *Molecular Ecology* 12: 2953–62.

Aggerbeck MJ, Fjeldså J, Christidis L, et al. (2014) Resolving deep lineage divergences in core corvoid passerine birds supports a proto-Papuan island origin, *Molecular Phylogenetics and Evolution* 70: 272–85.

Anderson T (1849) On a new species of manna from New South Wales, *Edinburgh New Philosophical Journal* 47: 132–9.

Angas GF (1847) *Savage Life and Scenes in Australia and New Zealand.* Smith, Elder, London.

Archer M, Hand S and Godthelp H (1996) *Riversleigh,* Reed, Melbourne.

Banfield EJ (1908) *Confessions of a Beachcomber,* T. Fisher, Unwin, London.

Barbour RC, Crawford AC, Henson M, et al. (2008) The risk of pollen-mediated gene flow from exotic *Corymbia* plantations into native *Corymbia* populations in Australia, *Forest Ecology and Management* 256: 1–19.

Barker FK, Barrowclough GF and Groth JG (2002) A phylogenetic hypothesis for passerine birds: Taxonomic and biogeographic implications of an analysis of nuclear DNA sequence data, *Proceedings of the Royal Society B* 269: 295–308.

Barker FK, Cibois A, Schikler P, et al. (2004) Phylogeny and diversification of the largest avian radiation, *Proceedings of the National Academy of Sciences of the United States of America* 101: 11040–45.

Basden R (1966) The composition, occurrence and origin of lerp, the sugary secretion of Eurymela distincta (Signoret), *Proceedings of the Linnean Society of New South Wales* 91: 44.

Beadle NCW, Evans OD, Carolin RC, et al. (1982) *Flora of the Sydney Region,* Reed, French's Forest, NSW.

Beehler BM, Pratt TK and Zimmerman DA (1986) *Birds of New Guinea,* Princeton University Press, Princeton, New Jersey.

Bell HL (1982) A bird community of lowland rainforest in New Guinea. 1. Composition and density of the avifauna, *Emu* 82: 24–41.

Bell Miner Associated Dieback Working Group (2004) *Bell miner Associated Dieback (BMAD) Strategy,* Department of Environment and Conservation (NSW), Coffs Harbour.

Bennett G (1860) *Gatherings of a Naturalist in Australasia,* Van Voorst, London.

Bird MI, Hutley LB, Lawes MJ, et al. (2013) Humans, megafauna and environmental change in tropical Australia, *Journal of Quaternary Science* 28: 439–52.

Boland C (2000) Shy albatross, *Nature Australia* 26: 22.

Boles WE (1995) Preliminary analysis of the Passeriformes from Riversleigh, northwestern Queensland, Australia, with the description of a new species of lyrebird, *Courier Forschungsinstitut Senckenberg* 181: 163–70.

—— (1999) A new songbird (Aves: Passeriformes: Oriolidae) from the Miocene of Riversleigh, northwestern Queensland, Australia, *Alcheringa* 23: 51–56.

—— (2005) Fossil honeyeaters (Meliphagidae) from the Late Tertiary of Riversleigh, north-western Queensland, *Emu* 105: 21–26.

—— (2006) The avian fossil record of Australia: An overview. In: Merrick JR, Archer M, Hickey GM, et al. (eds) *Evolution and Biogeography of Australasian Vertebrates.* Australian Scientific Publishing, Sydney.

Bomford M and Sinclair R (2002) Australian research on bird pests: impact, management and future directions, *Emu* 102: 29–45.

Bond WJ and Midgley JJ (1995) Kill thy neighbor – an individualistic argument for the evolution of flammability, *Oikos* 73: 79–85.

Bond WJ, Woodward FI and Midgley GF (2005) The global distribution of ecosystems in a world without fire, *New Phytologist* 165: 525–37.

Bonyhady T (2000) *The Colonial Earth,* Miegunyah Press, Carlton, Vic.

Bowman D, Brown GK, Braby MF, et al. (2010) Biogeography of the Australian monsoon tropics, *Journal of Biogeography* 37: 201–16.

Bowman DMJS (2000) *Australian Rainforests: Islands of Green in a Land of Fire,* Cambridge University Press, Cambridge, UK.

Bradstock RA, Williams RJ and Gill AM (2012) *Flammable Australia: Fire Regimes, Biodiversity and Ecosystems in a Changing World,* CSIRO Publishing, Melbourne.

Brooke MD (2004) The food consumption of the world's seabirds, *Proceedings of the Royal Society B* 271: S246–S48.

Bryant N (1985) Outdoors; bird smuggling is on the rise. *New York Times.* New York.

Burbidge NT (1960) The phytogeography of the Australian region, *Australian Journal of Botany* 8: 75–211.

Byrne M, Steane DA, Joseph L, et al. (2011) Decline of a biome: evolution, contraction, fragmentation, extinction and invasion of the Australian mesic zone biota, *Journal of Biogeography* 38: 1635–56.

Caley G (1966) *Reflections on the Colony of New South Wales,* Lansdowne, Melbourne.

Campbell AJ (1901) *Nests and Eggs of Australian Birds.* Sheffield : Printed for the author by Pawson & Brailsford, Sheffield.

Carpenter RJ and Jordan GJ (1997) Early Tertiary macrofossils of Proteaceae from Tasmania, *Australian systematic Botany* 10: 533–63.

Catterall CP, Cousin JA, Piper S, et al. (2010) Long-term dynamics of bird diversity in forest and suburb: decay, turnover or homogenization?, *Diversity and Distributions* 16: 559–70.

Chansigaud V (2010) *All About Birds,* Princeton University Press, Princeton.

Chapman GV (1969) *Common City Birds of Australia,* Periwinkle Books, Melbourne.

Chiari Y, Cahais V, Galtier N, et al. (2012) Phylogenomic analyses support the position of turtles as the sister group of birds and crocodiles (Archosauria), *BMC Biology* 10: 65.

Chisholm AH (1922) *Mateship with Birds,* Whitcombe & Tombs, Melbourne.

Chown SL and Gaston KJ (1999) Patterns in procellariiform diversity as a test of species-energy theory in marine systems, *Evolutionary Ecology Research* 1: 365–73.

Chown SL, Gaston KJ and Williams PH (1998) Global patterns in species richness of pelagic seabirds: the Procellariiformes, *Ecography* 21: 342–50.

Christidis L and Boles W (2008) *Systematics and Taxonomy of Australian Birds,* CSIRO Publishing, Melbourne.

Christidis L and Schodde R (1991) Relationships of Australo-Papuan songbirds – protein evidence, *Ibis* 133: 277–85.

Clarke MF and Schedvin N (1999) Removal of bell miners *Manorina melanophrys* from *Eucalyptus radiata* forest and its effect on avian diversity, psyllids and tree health, *Biological Conservation* 88: 111–20.

Coates BJ (1985) *The Birds of Papua New Guinea: Including the Bismarck Archipelago and Bougainville,* Dove, Brisbane.

Cockburn A. (1998) Evolution of helping behavior in cooperatively breeding birds. *Annual Review of Ecology and Systematics.* 141–77.

—— (2003) Cooperative breeding in oscine passerines: does sociality inhibit speciation?, *Proceedings of the Royal Society B* 270: 2207–14.

Collias NE (1997) On the origin and evolution of nest building by passerine birds, *Condor* 99: 253–70.

Connell J (2009) Birdwatching, twitching and tourism: Towards an Australian perspective, *Australian Geographer* 40: 203–17.

Corlett RT (1998) Frugivory and seed dispersal by vertebrates in the Oriental (Indomalayan) Region, *Biological Reviews of the Cambridge Philosophical Society* 73: 413–48.

Costin AB (1979) *Kosciusko Alpine Flora,* CSIRO/Collins Australia, Melbourne.

Cracraft J (2001) Avian evolution, Gondwana biogeography and the Cretaceous–Tertiary mass extinction event, *Proceedings of the Royal Society B* 268: 459–69.

Cracraft J and Feinstein J (2000) What is not a bird of paradise? Molecular and morphological evidence places *Macgregoria* in the Meliphagidae and the Cnemophilinae near the base of the corvoid tree, *Proceedings of the Royal Society B* 267: 233–41.

Crome F and Shields JS (1992) *Parrots and Pigeons of Australia,* Angus and Robertson, Sydney.

Croxall JP, Butchart SHM, Lascelles B, et al. (2012) Seabird conservation status, threats and priority actions: a global assessment, *Bird Conservation International* 22: 1–34.

Cumpston JS (1968) *Macquarie Island,* Antarctic Division, Department of External Affairs, Canberra.

Daley P and Griggs P (2006) Mining the reefs and cays: coral, guano and rock phosphate extraction in the Great Barrier Reef, Australia, 1844–1940, *Environment and History* 12: 395–433.

Davies RG, Irlich UM, Chown SL, et al. (2010) Ambient, productive and wind energy, and ocean extent predict global species richness of procellariiform seabirds, *Global Ecology and Biogeography* 19: 98–110.

Dawkins R (2004) *The Ancestor's Tale,* Houghton Mifflin, Boston.

Dawson J (1881) *Australian Aborigines: The Languages and Customs of Several Tribes of Aborigines in the Western District of Victoria, Australia,* G. Robertson, Melbourne.

Dawson J and Lucas R (2000) *Nature Guide to the New Zealand Forest,* Random House, Auckland.

del Hoyo J, Elliott A and Christie DA (2008) *Handbook of the Birds of the World. Vol 13. Penduline Tits to Shrikes,* Lynx Edicions, Barcelona.

del Hoyo J, Elliott A and Sargatal J (1997) *Handbook of the Birds of the World. Vol 4. Sandgrouse to Cuckoos,* Lynx Edicions, Barcelona.

Delaney R, Y. F and Saalfeld K (2009) *Management Program for the Magpie Goose (Anseranas semipalmata) in the Northern Territory of Australia, 2009–2014,* Northern Territory Department of Natural Resources, Environment, the Arts and Sport, Darwin.

Dettmann ME and Jarzen DM (1991) Pollen evidence for Late Cretaceous differentiation of Proteaceae in southern polar forests, *Canadian Journal of Botany* 69: 901–06.

—— (1998) The early history of the Proteaceae in Australia: The pollen record, *Australian Systematic Botany* 11: 401–38.

Diamond J and Bond AB (2003) A comparative analysis of social play in birds, *Behaviour* 140: 1091–115.

Dixon J (2002) *The Changing Chicken: Chooks, Cooks and Culinary Culture,* University of New South Wales Press, Sydney.

Donkin RA (1980) *Manna: An Historical Geography,* Junk, The Hague.

Dooley S (2005) *The Big Twitch,* Allen & Unwin, Sydney.

Dostine PL, Johnson GC, Franklin DC, et al. (2001) Seasonal use of savanna landscapes by the Gouldian finch, *Erythrura gouldiae,* in the Yinberrie Hills area, Northern Territory, *Wildlife Research* 28: 445–58.

Dow DD (1977) Indiscriminate interspecific aggression leading to almost sole occupancy of space by a single species of bird, *Emu* 77: 115–21.

Driskell AC and Christidis L (2004) Phylogeny and evolution of the Australo-Papuan honeyeaters (Passeriformes, Meliphagidae), *Molecular Phylogenetics and Evolution* 31: 943–60.

Dunlop M and Brown PR (2008) *Implications of Climate Change for Australia's National Reserve System: A Preliminary Assessment,* Department of Climate Change, Canberra.

Dyke GJ and Van Tuinen M (2004) The evolutionary radiation of modern birds (Neornithes): reconciling molecules, morphology and the fossil record, *Zoological Journal of the Linnean Society* 141: 153–77.

Edwards SV and Boles WE (2002) Out of Gondwana: The origin of passerine birds, *Trends in Ecology and Evolution* 17: 347–49.

Elzanowski A and Boles WE (2012) Australia's oldest anseriform fossil: A quadrate from the early Eocene Tingamarra fauna, *Palaeontology* 55: 903–11.

Elzanowski A and Stidham TA (2011) A Galloanserine quadrate from the Late Cretaceous Lance Formation of Wyoming, *Auk* 128: 138–45.

Emery NJ (2006) Cognitive ornithology: the evolution of avian intelligence, *Philosophical Transactions of the Royal Society of London B Biological Sciences* 361: 23–43.

Environment and Natural Resources Committee M (1995) *Problems in Victoria caused by*

Long-billed Corellas, Sulphur-crested Cockatoos and Galahs, Melbourne.

Ericson PGP (2008) Current perspectives on the evolution of birds, *Contributions to Zoology* 77: 109–16.

—— (2012) Evolution of terrestrial birds in three continents: Biogeography and parallel radiations, *Journal of Biogeography* 39: 813–24.

Ericson PGP, Anderson CL, Britton T, et al. (2006) Diversification of Neoaves: Integration of molecular sequence data and fossils, *Biology Letters* 2: 543–47.

Ericson PGP, Christidis L, Cooper A, et al. (2002) A Gondwanan origin of passerine birds supported by DNA sequences of the endemic New Zealand wrens, *Proceedings of the Royal Society B* 269: 235–41.

Ericson PGP, Irestedt M and Johanson US (2003) Evolution, biogeography, and patterns of diversification in passerine birds, *Journal of Avian Biology* 34: 3–15.

Feare C and Craig A (1999) *Starlings and Mynas,* Princeton University Press, Princeton, New Jersey.

Fitzsimons J, Legge S, Traill B, et al. (2010) *Into Oblivion: The Disappearing Mammals of Northern Australia,* The Nature Conservancy, Melbourne.

Flannery TF (1994) *The Future Eaters: An Ecological History of the Australasian Lands and People,* Reed Books, Sydney.

—— (1995) *Mammals of New Guinea,* Reed Books, Sydney.

Fleischer RC, James HF and Olson SL (2008) Convergent Evolution of Hawaiian and Australo-Pacific Honeyeaters from Distant Songbird Ancestors, *Current Biology* 18: 1927–31.

Flinders M (1814) *A Voyage to Terra Australis, Undertaken for the Purpose of completing the Discovery of that Vast Country and prosecuted in the years 1801, 1802, and 1803,* G. and W. Nicol, London.

Ford HA (1989) *Ecology of Birds: An Australian Perspective,* Surrey, Beatty & Sons, Sydney.

Ford HA, Barrett GW, Saunders DA, et al. (2001) Why have birds in the woodlands of Southern Australia declined?, *Biological Conservation* 97: 71–88.

Ford HA, Paton DC and Forde N (1979) Birds as pollinators of Australian plants, *New Zealand Journal of Botany* 17: 509–19.

Franklin DC (1999) Evidence of disarray amongst granivorous bird assemblages in the savannas of northern Australia, a region of sparse human settlement, *Biological Conservation* 90: 53–68.

Franklin DC, Garnett ST, Luck GW, et al. (2014) Relative brain size in Australian birds, *Emu* In press.

Franklin DC and Noske RA (1999) Birds and nectar in a monsoonal woodland: Correlations at three spatio-temporal scales, *Emu* 99: 15–28.

—— (2000) Nectar sources used by birds in monsoonal north-western Australia: A regional survey, *Australian Journal of Botany* 48: 461–74.

Franklin DC, Woinarski JCZ and Noske RA (2000) Geographical patterning of species richness among granivorous birds in Australia, *Journal of Biogeography* 27: 829–42.

Frith CB, Beehler BM and Cooper WT (1998) *The Birds of Paradise: Paradisaeidae,* Oxford University Press, Oxford.

Frith CB and Frith DW (2004) *The Bowerbirds: Ptilonorhynchidae,* Oxford University Press,

Oxford.

—— (2010) *Birds of paradise: nature, art & history,* Frith & Frith, Malanda, Qld.

Frith HJ (1982) *Pigeons and Doves of Australia,* Rigby, Adelaide.

Frith HJ and Calaby JH (1976) *Proceedings of the 16th International Ornithological Congress,* Australian Academy of Science, Canberra.

Frith HJ, Crome FHJ and Brown BK (1977) Aspects of biology of Japanese snipe *Gallinago hardwickii, Australian Journal of Ecology* 2: 341–68.

Gammage B (2011) *The Biggest Estate on Earth: How Aborigines Made Australia,* Allen & Unwin, Sydney.

Gandolfo MA, Hermsen EJ, Zamaloa MC, et al. (2011) Oldest Known *Eucalyptus* Macrofossils Are from South America, *Plos One* 6.

Gardner JL, Trueman JWH, Ebert D, et al. (2010) Phylogeny and evolution of the Meliphagoidea, the largest radiation of Australasian songbirds, *Molecular Phylogenetics and Evolution* 55: 1087–102.

Garnett S, Szabo J and Duston G (2011) *The Action Plan for Australian Birds 2010.* CSIRO Publishing, Melbourne.

Gaston A (2004) *Seabirds: A Natural History* T & A D Poyser, London.

Gaze PD and Clout MN (1983) Honeydew and its importance to birds in beech forests of South Island, New Zealand, *New Zealand Journal of Ecology* 6: 33–37.

Gentilli J (1949) Foundations of Australian bird geography, *Emu* 49: 85–129.

Gilliard ET (1969) *Birds of Paradise and Bower Birds,* Weidenfeld & Nicolson, London.

Gilmore A. (1969) Fishing in Bass Strait, in *Bass Strait, Australia's last frontier.* Sydney, Australian Broadcasting Commission.

Gonzalez J, Duttmann H and Wink M (2009) Phylogenetic relationships based on two mitochondrial genes and hybridization patterns in Anatidae, *Journal of Zoology* 279: 310–18.

Goodfellow D and Stott M (2001) *Birds of Australia's Top End,* Scrubfowl Press, Parap, N.T.

Gordon DP, Beaumont J, MacDiarmid A, et al. (2010) Marine biodiversity of Aotearoa New Zealand, *Plos One* 5.

Gould J (1865) *Handbook to The Birds of Australia.* The author, London.

Greenberg R, Caballero CM and Bichier P (1993) Defense of homopteran honeydew by birds in the Mexican highlands and other warm temperate forests, *Oikos* 68: 519–24.

Hackett SJ, Kimball RT, Reddy S, et al. (2008) A phylogenomic study of birds reveals their evolutionary history, *Science* 320: 1763–68.

Hall L and Richards G (2000) *Flying Foxes: Fruit and Blossom Bats of Australia,* University of New South Wales Press, Sydney.

Hartshorne C (1973) *Born to Sing: An Interpretation and World Survey of Bird Song,* Indiana University Press, Bloomington.

Heinsohn R (2008) *Life in the Cape York Rainforest.* CSIRO Publishing, Melbourne.

Heinsohn T and Hope G. (2006) The Torresian connections: Zoogeography of New Guinea. In: Merrick J, Archer M, Hickey G, et al. (eds) *Evolution and Biogeography of Australasian Vertebrates.* Auscipub, Sydney.

Henderson J (1851) *Excursions and Adventures in New South Wales,* Shoberl, London.

Higgins PJ (1999) *Handbook of Australian, New Zealand & Antarctic birds. Volume 4: Parrots to Dollarbird,* Oxford University Press, Melbourne.

Higgins PJ, Christidis L and Ford HA. (2008) Family Meliphagidae. In: del Hoyo J, Elliott A and Christie DA (eds) *Handbook of the Birds of the World. Vol 13. Penduline Tits to Shrikes.* Barcelona, Lynx Edicions.

Higgins PJ and Davies SJJF (1996) *Handbook of Australian, New Zealand & Antarctic Birds. Volume 3: Snipe to Pigeons,* Oxford University Press, Melbourne.

Higgins PJ and Peter JM (2002) *Handbook of Australian, New Zealand & Antarctic Birds. Volume 6: Pardalotes to Shrike-thrushes,* Oxford University Press, Melbourne.

Higgins PJ, Peter JM and Cowling SJ (2006) *Handbook of Australian, New Zealand & Antarctic Birds. Volume 7: Boatbills to Starlings,* Oxford University Press, Melbourne.

Higgins PJ, Peter JM and Steele WK (2001) *Handbook of Australian, New Zealand & Antarctic Birds. Volume 6: Tyrant-flycatchers to Chats,* Oxford University Press, Melbourne.

Hill RS (1998) Fossil evidence for the onset of xeromorphy and scleromorphy in Australian Proteaceae, *Australian Systematic Botany* 11: 391–400.

Hill RS, Truswell EM, McLoughlin S, et al. (1999) Evolution of the Australian flora: Fossil evidence, In *Flora of Australia Volume 1, 2nd edition.* ABRS/CSIRO Australia, Melbourne.

Hindwood KA and McGill AR (1958) *The Birds of Sydney, County of Northumberland, New South Wales,* Royal Zoological Society of N.S.W, Sydney.

—— (1991) *The Birds of Sydney, County of Northumberland, New South Wales 1770–1989,* Surrey Beatty & Sons, Sydney.

Hocknull SA, Zhao JX, Feng YX, et al. (2007) Responses of Quaternary rainforest vertebrates to climate change in Australia, *Earth and Planetary Science Letters* 264: 317–31.

Holmes RT and Recher HF (1986) Determinants of guild structure in forest bird communities – an intercontinental comparison, *Condor* 88: 427–39.

Hooker JD (1859) *On the Flora of Australia, Its Origin, Affinities, and Distribution: Being an Introductory Essay to the Flora of Tasmania.* Lovell Reeve, London.

Hopper SD (2009) OCBIL theory: towards an integrated understanding of the evolution, ecology and conservation of biodiversity on old, climatically buffered, infertile landscapes, *Plant and Soil* 322: 49–86.

Hughes JM, Mather PB, Toon A, et al. (2003) High levels of extra-group paternity in a population of Australian magpies *Gymnorhina tibicen:* evidence from microsatellite analysis, *Molecular Ecology* 12: 3441–50.

Hutton FW (1896) Theoretical Explanations of the Distribution of Southern Faunas, *Proceedings of the Linnean Society of New South Wales* 21: 36–47.

Hutton FW and Drummond J (1909) *Animals of New Zealand; An Account of the Dominion's Air-breathing Vertebrates,* Whitcombe and Tombs, Christchurch.

Ishigame G, Baxter GS and Lisle AT (2006) Effects of artificial foods on the blood chemistry of the Australian magpie, *Austral Ecology* 31: 199-207.

Jacobs BF, Kingston JD and Jacobs LL (1999) The Origin of Grass-Dominated Ecosystems, *Annals of the Missouri Botanical Gardens* 86: 590–643.

Johnson C (2006) *Australia's Mammal Extinctions: A 50,000 year history,* Cambridge University Press, Cambridge.

Johnson CN (2009) Ecological consequences of Late Quaternary extinctions of megafauna, *Proceedings of the Royal Society B* 276: 2509–19.

Johnson KR (1993) Extinctions at the Antipodes, *Nature* 366: 511–12.

Jones D (2002) *Magpie Alert: Learning to Live with a Wild Neighbour*, University of New South Wales, Sydney.

—— (2011) An appetite for connection: why we need to understand the effect and value of feeding wild birds, *Emu* 111: I–VII.

Jones DN and Reynolds SJ (2008) Feeding birds in our towns and cities: a global research opportunity, *Journal of Avian Biology* 39: 265–71.

Jones HB (1853) *Adventures in Australia, in 1852 and 1853,* Bentley, London.

Jønsson KA, Bowie RCK, Nylander JAA, et al. (2010) Biogeographical history of cuckoo-shrikes (Aves: Passeriformes): transoceanic colonization of Africa from Australo-Papua, *Journal of Biogeography* 37: 1767–81.

Jønsson KA, Fabre PH, Ricklefs RE, et al. (2011) Major global radiation of corvoid birds originated in the proto-Papuan archipelago, *Proceedings of the National Academy of Sciences of the United States of America* 108: 2328–33.

Jønsson KA and Fjeldså J (2006) Determining biogeographical patterns of dispersal and diversification in oscine passerine birds in Australia, Southeast Asia and Africa, *Journal of Biogeography* 33: 1155–65.

Jordan GJ (2001) An investigation of long-distance dispersal based on species native to both Tasmania and New Zealand, *Australian Journal of Botany* 49: 333–40.

Joseph L, Toon A, Nyari AS, et al. (2014) A new synthesis of the molecular systematics and biogeography of honeyeaters (Passeriformes: Meliphagidae) highlights biogeographical and ecological complexity of a spectacular avian radiation, *Zoologica Scripta* 43: 235–48.

Kaplan G (2004) *Australian Magpie: Biology and Behaviour of an Unusual Songbird,* CSIRO, Melbourne.

Kearns AM, Joseph L and Cook LG (2013) A multilocus coalescent analysis of the speciational history of the Australo-Papuan butcherbirds and their allies, *Molecular Phylogenetics and Evolution* 66: 941–52.

Keast A. (1984) Contemporary ornithogeography: the Australian avifauna, its relationships and evolution. In: Archer M and Clayton G (eds) *Vertebrate Zoogeography & Evolution in Australasia: Animals in Space & Time.* Hesperian Press, Perth.

Kemp AC, Herholdt JJ, Whyte I, et al. (2001) Birds of the two largest national parks in Africa, *South African Journal of Science* 97: 393–403.

Kemp NJ and Burnett JB (2003) *Biodiversity Risk Assessment and Recommendations for Risk Management of Long-tailed Macaques (Macaca fascicularis) in New Guinea,* Indo-Pacific Conservation Alliance, Washington DC.

Kofron CP (1999) Attacks to humans and domestic animals by the southern cassowary (*Casuarius casuarius johnsonii*) in Queensland, Australia, *Journal of Zoology* 249: 375–81.

Krauss SL, He T, Barrett LG, et al. (2009) Contrasting impacts of pollen and seed dispersal on spatial genetic structure in the bird-pollinated *Banksia hookeriana, Heredity* 102: 274–85.

Kristiansen J (1996) *Biogeography of Freshwater Algae,* Kluwer Academic Publishers, Dordrecht.

Ksepka DT, Clarke JA and Grande L (2011) Stem parrots (Aves, Halcyornithidae) from the
    Green River formation and a combined phylogeny of pan-Psittaciformes, *Journal of
    Paleontology* 85: 835–52.

Laing D (2013) Canberra birds: A century of change, *Wildlife Australia* 50: 14–7.

Lamont BB, Enright NJ, Witkowski ETF, et al. (2007) Conservation biology of banksias:
    insights from natural history to simulation modelling, *Australian Journal of Botany* 55:
    280–92.

Lamont BB, Le Maitre DC, Cowling RM, et al. (1991) Canopy seed storage in woody plants,
    Botanical Review 57: 277–317.

Leach JA (1911) *An Australian Bird Book: A Pocket Book for Field Use,* Whitcombe & Tombs,
    Melbourne.

—— (1929) *Australian Bird Book: with supplement: A Complete Guide to the Identification
    of Australian Birds.* 7th edition. Whitcombe & Tombs, Melbourne.

—— (1929) The lyrebird – Australia's wonder-bird, *Emu* 28: 199–214.

Lee WG, Wood JR and Rogers GM (2010) Legacy of avian-dominated plant-herbivore
    systems in New Zealand, *New Zealand Journal of Ecology* 34: 28–47.

Lindsey T (2008) *Albatrosses.* CSIRO Publishing, Melbourne.

Littler F (1901) Bird conservation, *Emu* 1: 10–12.

Livingstone D (1857) *Missionary Travels and Researches in South Africa.* Ward Lock, London.

Lockwood JL and Gilroy JJ (2004) The portability of foodweb dynamics: reassembling
    an Australian eucalypt-psyllid-bird association within California, *Global Ecology and
    Biogeography* 13: 445–50.

Longmore W (1991) *Honeyeaters & Their Allies of Australia,* Collins/Angus & Robertson,
    Sydney.

Low T (1999) *Feral Future,* Penguin, Melbourne.

—— (2002) *The New Nature,* Penguin, Melbourne.

—— (2008) *Climate Change and Invasive Species: A Review of Interactions,* Biological
    Diversity Advisory Committee, Canberra.

—— (2011) *Climate Change and Queensland Biodiversity,* Department of Environment
    and Resource Management, Queensland Government, Brisbane.

Loyn RH (2002) Patterns of ecological segregation among forest and woodland birds in
    south-eastern Australia, *Ornithological Science* 1: 7–27.

Loyn RH and Menkhorst PW (2011) The bird fauna of Melbourne: Changes over a century
    of urban growth and climate change, using a benchmark from Keartland (1900),
    Victorian Naturalist 128: 210–31.

Loyn RH, Runnalis RG, Forward GY, et al. (1983) Territorial bell miners and other birds
    affecting populations of insect prey, *Science* 221: 1411–13.

Luo ZX, Ji Q, Wible JR, et al. (2003) An early Cretaceous tribosphenic mammal and
    metatherian evolution, *Science* 302: 1934–40.

Mac Nally R, Bowen M, Howes A, et al. (2012) Despotic, high-impact species and the
    subcontinental scale control of avian assemblage structure, *Ecology* 93: 668–78.

Maiden JH (1889) *The Useful Native Plants of Australia (including Tasmania),* Trubner,
    London.

Major RE and Parsons H (2010) What do museum specimens tell us about the impact of urbanisation? A comparison of the recent and historical bird communities of Sydney, *Emu* 110: 92–103.

Malekian M, Cooper SJB, Norman JA, et al. (2010) Molecular systematics and evolutionary origins of the genus *Petaurus* (Marsupialia: Petauridae) in Australia and New Guinea, *Molecular Phylogenetics and Evolution* 54: 122–35.

Mallick SA and Driessen MM (2010) *Review, Risk Assessment and Management of Introduced Animals in the Tasmanian Wilderness World Heritage Area. Nature Conservation Report 10/01*, Resource Management and Conservation Division, Department of Primary Industries and Water, Hobart.

Manegold A (2008) Passerine diversity in the late Oligocene of Germany: Earliest evidence for the sympatric coexistence of Suboscines and Oscines, *Ibis* 150: 377–87.

Marchant S and Higgins PJ (1990) *Handbook of Australian, New Zealand & Antarctic birds. Volume 1: Ratites to Ducks.* , Oxford University Press, Melbourne.

—— (1993) *Handbook of Australian, New Zealand & Antarctic birds. Volume 2: Raptors to Lapwings,* Oxford University Press, Melbourne.

Marriner GR (1906) Notes on the Natural History of the Kea, with Special Reference to its Reputed Sheep-killing Propensities, *Transactions and Proceedings of the Royal Society of New Zealand,* 39: 271–305.

Marshall AJ (1966) *The Great Extermination: A Guide to Anglo-Australian Cupidity, Wickedness & Waste,* Heinemann, London.

Martin HA (1998) Tertiary climatic evolution and the development of aridity in Australia, *Proceedings of the Linnean Society of New South Wales* 119: 115–36.

Matthew WD (1915) Climate and evolution, *Annals of the New York Academy of Sciences* 24: 171–416.

Mayhew H (1864) *London Labour and the London Poor,* C. Griffin, London.

Mayr E (1944) Timor and the colonization of Australia by birds, *Emu* 44: 113–30.

Mayr E and Diamond JM (2001) *The Birds of Northern Melanesia: Speciation, Ecology, & Biogeography,* Oxford University Press, New York.

—— (2004) Old world fossil record of modern-type hummingbirds, *Science* 304: 861–64.

—— (2009) *Paleogene Fossil Birds,* Springer, Heidelberg.

—— (2013) The age of the crown group of passerine birds and its evolutionary significance – molecular calibrations versus the fossil record, *Systematics and Biodiversity* 11: 7–13.

McCormack JE, Harvey MG, Faircloth BC, et al. (2013) A phylogeny of birds based on over 1,500 loci collected by target enrichment and high-throughput sequencing, *Plos One* 8.

McGlone MS (2005) Goodbye Gondwana, *Journal of Biogeography* 32: 739–40.

McGoldrick JM and Mac Nally R (1998) Impact of flowering on bird community dynamics in some central Victorian eucalypt forests, *Ecological Research* 13: 125–39.

McGowan H, Marx S, Moss P, et al. (2012) Evidence of ENSO mega-drought triggered collapse of prehistory Aboriginal society in northwest Australia, *Geophysical Research Letters* 39.

McKinnon GE, Jordan GJ, Vaillancourt RE, et al. (2004) Glacial refugia and reticulate evolution: The case of the Tasmanian eucalypts, *Philosophical Transactions of the Royal Society of London B* 359: 275–84.

McKinnon GE, Vaillancourt RE, Jackson HD, et al. (2001) Chloroplast sharing in the

Tasmanian eucalypts, *Evolution* 55: 703–11.

McNamara KJ and Scott JK (1983) 1983. A new species of *Banksia* (Proteaceae) from the Eocene Merlinleigh Sandstone of the Kennedy Range, Western Australia, *Alcheringa* 7: 185–93.

Miller AH (1962) The history and significance of the fossil *Casuarius lydekkeri, Records of the Australian Museum* 25: 235–38.

Mooney SD, Harrison SP, Bartlein PJ, et al. (2011) Late Quaternary fire regimes of Australasia, *Quaternary Science Reviews* 30: 28–46.

Moore GF (1884) *Diary of Ten Years Eventful Life of an Early Settler in Western Australia,* M. Walbrook, London.

Morgan D, Waas JR and Innes J (2005) Magpie interactions with other birds in New Zealand: Results from a literature review and public survey, *Notornis* 52: 61–74.

Morris EE (1898) *Austral English,* Macmillan, London.

Morton SR and James CD (1988) The diversity and abundance of lizards in arid Australia: A new hypothesis, *American Naturalist* 132: 237–56.

Morton SR, Smith DMS, Dickman CR, et al. (2011) A fresh framework for the ecology of arid Australia, *Journal of Arid Environments* 75: 313–29.

Moyle RG, Filardi CE, Smith CE, et al. (2009) Explosive Pleistocene diversification and hemispheric expansion of a "great speciator", *Proceedings of the National Academy of Sciences of the United States of America* 106: 1863–68.

Mulvaney DJ and Kamminga J (1999) *Prehistory of Australia,* Allen & Unwin, Sydney.

Mundy GC (1852) *Our Antipodes,* Richard Bentley, London.

Murphy BP, Bradstock RA, Boer MM, et al. (2013) Fire regimes of Australia: a pyrogeographic model system, *Journal of Biogeography* 40: 1048–58.

Murray P and Rich PV (2004) *Magnificent Mihirungs,* Indiana University Press, Bloomington.

Mutch RW (1970) Wildland fires and ecosystems – a hypothesis, *Ecology* 51: 1046–51.

Nelson G and Ladiges PY (2001) Gondwana, vicariance biogeography and the New York School revisited, *Australian Journal of Botany* 49: 389–409.

Newton A (1893) *A Dictionary of Birds,* A. and C. Black, London.

Newton I (2003) *The Speciation and Biogeography of Birds,* Academic Press, Amsterdam.

Nguyen JMT, Worthy TH, Boles WE, et al. (2013) A new cracticid (Passeriformes: Cracticidae) from the Early Miocene of Australia, *Emu* 113: 374–82.

Nielsen L (1996) *Birds of Queensland's Wet Tropics and Great Barrier Reef,* Gerard Industries, Bowden, South Australia.

Norman JA, Ericson PGP, Jonsson KA, et al. (2009) A multi-gene phylogeny reveals novel relationships for aberrant genera of Australo-Papuan core Corvoidea and polyphyly of the Pachycephalidae and Psophodidae (Aves: Passeriformes), *Molecular Phylogenetics and Evolution* 52: 488–97.

North AJ (1901–4) *Nests and Eggs of Birds found Breeding in Australia and Tasmania. 4 volumes,* F.W.White, Sydney.

Noske RA (1996) Abundance, zonation and foraging ecology of birds in mangroves of Darwin Harbour, Northern Territory, *Wildlife Research* 23: 443–74.

Olsen P (2007) *Glimpses of Paradise: The Quest for the Beautiful Parrakeet,* National Library of Australia, Canberra.

Olsen P and Joseph L (2011) *Stray Feathers: Reflections on the Structure, Behaviour and Evolution of Birds,* CSIRO Publishing, Melbourne.

Orians GH and Milewski AV (2007) Ecology of Australia: the effects of nutrient-poor soils and intense fires, *Biological Reviews* 82: 393–423.

Parsons H, Major RE and French K (2006) Species interactions and habitat associations of birds inhabiting urban areas of Sydney, Australia, *Austral Ecology* 31: 217–27.

Paton DC (1980) The importance of manna, honeydew and lerp in the diets of honey-eaters, *Emu* 80: 213–26.

—— (1986) Honeyeaters and their plants in south-eastern Australia. In: Ford HA and Parton DC (eds) *The Dynamic Partnership: Birds and Plants in Southern Australia.* Government Printer, Adelaide.

Pearson DL (1977) Pan-tropical comparison of bird community structure on 6 lowland forest sites, *Condor* 79: 232–44.

Pepperberg IM (2006a) Cognitive and communicative abilities of Grey parrots, *Applied Animal Behaviour Science* 100: 77–86.

Pereira SL, Johnson KP, Clayton DH, et al. (2007) Mitochondrial and nuclear DNA sequences support a Cretaceous origin of Columbiformes and a dispersal-driven radiation in the Paleogene, *Systematic Biology* 56: 656–72.

Perkins R (1903) Vertebrata, *Fauna Hawaiiensis* 1: 365–466.

Piper SD and Catterall CP (2003) A particular case and a general pattern: Hyperaggressive behaviour by one species may mediate avifaunal decreases in fragmented Australian forests, *Oikos* 101: 602–14.

Pizzey G, Knight F and Menkhorst P (2007) *The Field Guide to the Birds of Australia,* HarperCollins, Sydney.

Plumwood V (2006) The concept of a cultural landscape; Nature, culture and agency in the land, *Ethics & the Environment* 11: 115–50.

Potts BM and Wiltshire RJE (1997) *Eucalypt genetics and genecology.* In: Williams JE and Woinarski JCZ (eds) *Eucalypt Ecology.* Cambridge, Cambridge University Press.

Primack RB and Corlett RT (2005) *Tropical Rain Forests: An Ecological and Biogeographical Comparison,* Blackwell Publishers, Madden.

Proctor M, Yeo P and Lack A (1996) *The Natural History of Pollination,* Harper Collins, London.

Rainforest Conservation Society of Queensland. (1986) *Tropical Rainforests of North Queensland: Their Conservation Significance,* Australian Government Publishing Service, Canberra.

Raymond B, Shaffer SA, Sokolov S, et al. (2010) Shearwater foraging in the Southern Ocean: The roles of prey availability and winds, *Plos One* 5(6): e10960.

Reilly PN and Chambers P (1988) *The Lyrebird: A Natural History,* New South Wales University Press, Sydney.

Remes V, Matysiokova B and Cockburn A (2012) Long-term and large-scale analyses of nest predation patterns in Australian songbirds and a global comparison of nest predation rates, *Journal of Avian Biology* 43: 435–44.

Rich PV and Thompson EM (1982) *The Fossil Vertebrate Record of Australasia,* Monash University, Melbourne.

Riddick SN, Dragosits U, Blackall TD, et al. (2012) The global distribution of ammonia
    emissions from seabird colonies, *Atmospheric Environment* 55: 319–27.
Ridley HN (1930) *The Dispersal of Plants Throughout the World,* L. Reeve, Ashford, Kent.
Robin L (2001) *Flight of the Emu: One Hundred Years of Australian Ornithology 1901–2001,*
    Melbourne University Press, Melbourne.
Robin L, Heinsohn R and Joseph L (2009) *Boom & Bust: Bird Stories for a Dry Country,*
    CSIRO Publishing, Melbourne.
Rollinson DJ, O'Leary R and Jones DN (2003) The practise of wildlife feeding in suburban
    Brisbane, *Corella* 27: 52–58.
Ross G, A. B, Brothers N, et al. (1995) *The Status of Australia's Seabirds,* Great Barrier Reef
    Marine Park Authority, Townsville.
Roth HL (1899) *The Aborigines of Tasmania,* F. King, Halifax, England.
Rowley I (1975) *Bird Life,* Collins, Sydney.
—— (1990) *Behavioural Ecology of the Galah,* Surrey Beatty & Sons, Sydney.
Russell-Smith J, Yates C, Edwards A, et al. (2003) Contemporary fire regimes of northern
    Australia, 1997–2001: Change since Aboriginal occupancy, challenges for sustainable
    management, *International Journal of Wildland Fire* 12: 283–97.
Sakaguchi S, Bowman DMJS, Prior LD, et al. (2013) Climate, not Aboriginal landscape
    burning, controlled the historical demography and distribution of fire-sensitive conifer
    populations across Australia, *Proceedings of the Royal Society B* 280: 20132182.
Saranathan V, Hamilton D, Powell GVN, et al. (2007) Genetic evidence supports song
    learning in the three-wattled bellbird *Procnias tricarunculata* (Cotingidae), *Molecular
    Ecology* 16: 3689–702.
Schodde R (1989) Origins, radiations and sifting in the Australasian Biota – Changing
    concepts from new data and old, *Australasian Systematic Botany Society Newsletter* 60
    (available online).
—— (2000) Charles G. Sibley – 1911–1998 – Obituary, *Emu* 100: 75–76.
—— (2006) Australasia's bird fauna today – origins and evolutionary development.
    In: Merrick JR, Archer M, Hickey GM, et al. (eds) *Evolution and biogeography of
    Australasian Vertebrates.* Sydney, Australian Scientific Publishing.
Schodde R and Christidis L (2014) Relics from Tertiary Australasia: undescribed families
    and subfamilies of songbirds (Passeriformes) and their zoogeographic signal, *Zootaxa*
    3786: 501–22.
Schodde R and Mason IJ (1999) *The Directory of Australian Birds,* CSIRO Publishing,
    Melbourne.
Schweizer M, Seehausen O and Hertwig ST (2011) Macroevolutionary patterns in
    the diversification of parrots: effects of climate change, geological events and key
    innovations, *Journal of Biogeography* 38: 2176–94.
Senate Select Committee on Animal Welfare (1990) *Intensive Livestock Production,*
    Australian Government Publishing Service, Canberra.
Serventy DL (1969) Mutton-birding. *Bass Strait, Australia's last frontier.* Australian
    Broadcasting Commission, Sydney.
—— (1972) A historical background of ornithology with special reference to Australia,
    Emu 72: 41–50.

Serventy DL, Serventy DL and Warham J (1971) *The Handbook of Australian Sea-Birds*, Reed, Sydney.

Sheard K (1999) A twitch in time saves nine: Birdwatching, sport, and civilizing processes, *Sociology of Sport Journal* 16: 181–205.

Sibley CG (1974) The relationships of the lyrebirds, *Emu* 74: 65–79.

Sibley CG and Ahlquist JE (1985) The phylogeny and classification of the Australo-Papuan passerine birds, *Emu* 85: 1–14.

—— (1986) Reconstructing bird phylogeny by comparing DNAs, *Scientific American* 254: 82–92.

—— (1990) *Phylogeny and Classification of Birds: A Study in Molecular Evolution*, Yale University Press, New Haven.

Sibley CG, Ahlquist JE and Monroe BL (1988) A classification of the living birds of the world based on DNA-DNA hybridization studies, *Auk* 105: 409–23.

Skira I (1992) *Commercial Harvesting of Short-tailed Shearwaters (Tasmanian Mutton-birds)*, in *Wildlife Use and Management*. Report of a Workshop for Aboriginal and Torres Strait Islander people. Australian Government Publishing Service, Canberra.

Skira IJ (1990) Human exploitation of the short-tailed shearwater (*Puffinus tenuirostris*), *Papers and Proceedings of the Royal Society of Tasmania* 124: 77–90.

Skutch, A. F. (1935). Helpers at the nest. *Auk* 52 (3): 257–273.

Slack KE, Jones CM, Ando T, et al. (2006) Early penguin fossils, plus mitochondrial genomes, calibrate avian evolution, *Molecular Biology and Evolution* 23: 1144–55.

Slater PJB and Mann NI (2004) Why do the females of many bird species sing in the tropics?, *Journal of Avian Biology* 35: 289–94.

Smith M. (2009) *Genyornis*: last of the dromornithids. In: Robin L, Heinsohn R and Joseph L (eds) *Boom and Bust: Bird Stories for a Dry Country*, CSIRO Publishing, Melbourne.

Sniderman JMK and Jordan GJ (2011) Extent and timing of floristic exchange between Australian and Asian rain forests, *Journal of Biogeography* 38: 1445–55.

Southerton SG, Birt P, Porter J, et al. (2004) Review of gene movement by bats and birds and its potential significance for eucalypt plantation forestry, *Australian Forestry* 67: 44–53.

Steadman DW (2006) *Extinction & Biogeography of Tropical Pacific Birds*, University of Chicago Press, Chicago.

Steffen WL (2009) *Australia's Biodiversity and Climate Change*, CSIRO Publishing, Melbourne.

Strömberg CAE (2011) Evolution of grasses and grassland ecosystems, *Annual Review of Earth and Planetary Sciences* 39: 517–44.

Stutchbury BJM and Morton ES (2001) *Behavioral Ecology of Tropical Birds*, Academic Press, San Diego.

Svensson L, Mullarney K and Zetterstrom D (2009) *Collins Bird Guide. 2nd edition*, HarperCollins, London.

Swadling P, Wagner R and Laba B (1996) *Plumes from Paradise*, Papua New Guinea National Museum, Boroko, Papua New Guinea.

Talent JA, Duncan PM and Handby PL (1966) Early Cretaceous feathers from Victoria, *Emu* 66: 81–6.

Tasmania (1855) *Tasmanian Contributions to the Universal Exhibition of Industry at Paris, 1855,* H. & C. Best, Hobart Town.

Tennyson AJD (2010) The origin and history of New Zealand's terrestrial vertebrates, *New Zealand Journal of Ecology* 34: 6–27.

Thomson DF (1935) *Birds of Cape York Peninsula,* Government Printer, Melbourne.

Tiffin H (2007) *Five Emus to the King of Siam: Environment and Empire,* Rodopi, Amsterdam.

Tiffney BH (2004) Vertebrate dispersal of seed plants through time, *Annual Review of Ecology Evolution and Systematics* 35: 1–29.

Toggweiler JR and Russell J (2008) Ocean circulation in a warming climate, *Nature* 451: 286–88.

Trease GE and Evans WC (1978) *Pharmacognosy,* Baillière Tindall, London.

Trewick SA and Gibb GC (2010) Vicars, tramps and assembly of the New Zealand avifauna: a review of molecular phylogenetic evidence, *Ibis* 152: 226–53.

Tzaros CL, Shimba T and Robertson P (2005) *Wildlife of the Box-Ironbark Country.* CSIRO Publishing, Melbourne.

Van Dyck S and Strahan R (2008) *The Mammals of Australia,* Reed New Holland, Sydney.

van Langevelde F, van de Vijver C, Kumar L, et al. (2003) Effects of fire and herbivory on the stability of savanna ecosystems, *Ecology* 84: 337–50.

VanDerWal J, Shoo LP and Williams SE (2009) New approaches to understanding late Quaternary climate fluctuations and refugial dynamics in Australian wet tropical rain forests, *Journal of Biogeography* 36: 291–301.

Vanstone VA and Paton DC (1988) Extrafloral nectaries and pollination of *Acacia pycnantha* Benth by birds, *Australian Journal of Botany* 36: 519–31.

Walker TA (1991) *Pisonia Island of the Great Barrier Reef,* National Museum of Natural History, Smithsonian Institute, Washington, DC.

Wallace A (1862) Narrative of search after birds of paradise, *Proceedings of the Zoological Society of London.* 30: 153–61.

Wallace AR (1865) On the Pigeons of the Malay Archipelago, *Ibis* 7: 365–400.

Wallis GP and Trewick SA (2009) New Zealand phylogeography: evolution on a small continent, *Molecular Ecology* 18: 3548–80.

Wallis IR, Keszei A, Henery ML, et al. (2011) A chemical perspective on the evolution of variation in *Eucalyptus globulus, Perspectives in Plant Ecology Evolution and Systematics* 13: 305–18.

Warham J (1990) *The Petrels: Their Ecology and Breeding Systems,* Academic Press, London.

—— (1996) *The Behaviour, Population Biology and Physiology of the Petrels,* Academic Press, London.

Weimerskirch H, Cherel Y, Delord K, et al. (2014) Lifetime foraging patterns of the wandering albatross: Life on the move!, *Journal of Experimental Marine Biology and Ecology* 450: 68–78.

Wilson K (2006) *The State of New Zealand's Birds 2006: Special Report: Seabirds,* Ornithological Society of New Zealand.

Woinarski JCZ (1984) Small birds lerp-feeding and the problem of honey-eaters, *Emu* 84: 137–41.

—— (1985a) Breeding biology and life history of small insectivorous birds in Australian forests: response to a stable environment?, *Proceedings of the Ecological Society of Australia* 14: 159–68.

—— (1985b) Foliage-gleaners of the treetops, the pardalotes. In: Keast A, Recher HF, Ford H, et al. (eds) *Birds of Eucalypt Forests and Woodlands: Ecology, Conservation, Management.* Sydney, Surrey Beatty.

Woinarski JCZ, Connors G and Franklin DC (2000) Thinking honeyeater: nectar maps for the Northern Territory, Australia, *Pacific Conservation Biology* 6: 61–80.

Woinarski JCZ, Cullen JM, Hull C, et al. (1989) Lerp-feeding in birds a smorgasbord experiment, *Australian Journal of Ecology* 14: 227–34.

Woinarski JCZ and Legge S (2013) The impacts of fire on birds in Australia's tropical savannas, *Emu – Austral Ornithology* 113: 319–52.

Woinarski JCZ, Recher HF and Majer JD. (1997) Vertebrates of eucalypt formations. In: Williams JE and Woinarski JCZ (eds) *Eucalypt Ecology.* Cambridge, Cambridge University Press.

Woinarski JCZ, Williams RJ, Price O, et al. (2005) Landscapes without boundaries: wildlife and their environments in northern Australia, *Wildlife Research* 32: 377–88.

Worthy TH (2012) A phabine pigeon (Aves: Columbidae) from Oligo-Miocene Australia, *Emu* 112: 23–31.

Worthy TH and Boles WE (2011) *Australlus,* a New Genus for *Gallinula disneyi* (Aves: Rallidae) and a Description of a New Species from Oligo-Miocene Deposits at Riversleigh, Northwestern Queensland, Australia, *Records of the Australian Museum* 63: 61–77.

Wright TF, Schirtzinger EE, Matsumoto T, et al. (2008) A multilocus molecular phylogeny of the parrots (Psittaciformes): Support for a Gondwanan origin during the Cretaceous, *Molecular Biology and Evolution* 25: 2141–56.

Yalden DW and Albarella U (2009) *The History of British Birds,* Oxford University Press, Oxford.

Yibarbuk D, Whitehead PJ, Russell-Smith J, et al. (2001) Fire ecology and Aboriginal land management in central Arnhem Land, northern Australia: a tradition of ecosystem management, *Journal of Biogeography* 28: 325–43.

Zacharin RF (1978) *Emigrant Eucalypts: Gum Trees as Exotics,* Melbourne University Press, Melbourne.

Zelenitsky DK, Therrien FO, Ridgely RC, et al. (2011) Evolution of olfaction in non-avian theropod dinosaurs and birds, *Proceedings of the Royal Society B* 278: 3625–34.

Acknowledgements

Because the information in this book was accumulated over more than two decades I cannot hope to acknowledge everyone who helped with information or company in the field. In the text I mention many biologists and naturalists who showed me birds or provided interviews or information for which I am very grateful. I won't repeat their names here, with the exception of Les Christidis and Richard Schodde, who each submitted to several hours of questioning, and Sara Legge, John Woinarski and Stephen Garnett, who responded patiently to many questions about grassland birds. I am not implying that any experts I consulted agree with my conclusions.

The Australian Antarctic Division awarded me an Antarctic Arts Fellowship to visit Antarctica and Macquarie Island, and I thank Rob Easther and Cathy Bruce in particular. In 2012 I benefited from a Churchill Fellowship to investigate climate change biology in the Northern Hemisphere, since birds were part of the focus. Roger Kitching, Griffith University, invited me to Sabah. The Hawaii Conservation Alliance was generous with accommodation in Honolulu.

Carol Booth helped in every way possible – from pointing out birds in the field to locating journal articles and editing several chapters. Doug Lang provided accommodation and excellent company in Zimbabwe and elsewhere and found me passage on a barge serving Torres Strait. Jayne Balmer and Cascade Cohousing provided accommodation and generous hospitality during multiple visits to Hobart. Simon and Rose Cavendish loaned their car for a Kimberley trip. Jenny Horwood proofread the text with great care. Terry Reis helped in many ways. Others I am grateful to include Steve Anyon-Smith, Patrick Couper, Neil Doran, Nic Dunlop, Owen Foley, Mark Holdsworth, Conrad Hoskin, Gerald Mayr, Geoff Monteith, Tony Nicholas, Damon Oliver, Lisa Preston, Mark Read, Robbie Ward.

People outside Australia who helped include Malcolm Ausden (UK), Maryanne Bache (US, Mexico), Jack Craw (New Zealand), Peter Gaze (New Zealand), Jack Jeffrey (Hawaii), Linda Kelly (Zimbabwe, US), Wilberforce Okeka (Kenya), Sonja Wipf and Christian Rixen (Switzerland). To those I have forgotten to mention I apologise. To minimise the carbon footprint most of my overseas travel was undertaken following invited attendance at conferences and workshops.

For the support that Penguin provided I am very grateful. My editor, Meredith Rose, showed a level of commitment that I suspect is rare in publishing these days, and which shines through in every aspect of this project and led to many improvements. Ben Ball as publisher was also very committed to producing the best possible book. Alex Ross provided an excellent cover and Adam Laszczuk a wonderful text design.

Photo Credits

A number after a photo indicates the text page on which it appears. Photos in the colour insert have no page numbers.

Eleanor Ager – Superb lyrebird

Anonymous – John Paterson (282) (Image reproduction courtesy of the Gold Coast City Council Local Studies Library, Image number LS-LSP-CD003-1MG0050)

John Anderson – Red wattlebird, rock parrot (127), splendid fairy-wren, striated pardalotes (36), western bowerbird

Archibald Campbell – Muttonbird egging (286)

Bruce Doran – Apostlebirds, chestnut-breasted mannikins (210)

Elizabeth Gould – Norfolk Island kaka

Jon Irvine – Belford's melidectes

Cyril Jerrard – Paradise parrot (213)

Stephen E. Kacir – Metallic starling

John Gerrard Keulemans – Kioea (23) (from Walter Rothschild's [1893-1900] *The Avifauna of Laysan and the Neighbouring Islands ...* R.H. Porter, London), New Zealand wrens (69) (from Walter Buller's *A History of the Birds of New Zealand* [1888] John Van Voorst, London)

Hessisches Landesmuseum, Darmstadt – *Parargornis* (163)

Nevil Lazarus – Cape sugarbird, spotted pardalote, wandering albatross, white-winged chough

Alan Liefting – Kea (117)

Tim Low – Brush-turkey, cassowary signs (138), cassowary, lerp (40)

Myopixia – Noisy miner

John Norling – Greater frigatebird (235)

Julian Robinson – Bell miner, blue-faced honeyeater, owlet-nightjar, Tasmanian native-hen

Terry Reis – Magpie-goose, pied imperial-pigeon, topknot pigeon (187)

Mark Sanders – Yellow-tailed black-cockatoo

Alwyn Simple – Black-chinned honeyeater, magnificent riflebird, radjah shelduck (159), rose-crowned fruit-dove.

Clement Tang – Magpie attacks cockatoo

Chris Tate – Baudin's black-cockatoo

Gary Tate – Musk duck

Bruce Thomson – Channel-billed cuckoo, noisy friarbird (27)

Chris Tzaros – Swift parrot, mallee emu-wren

Peter Wanders – Gibberbird, partridge pigeon

Lisbeth Wastra – Masked owl (271)

Index